Radio Wave Propagation and Channel Modeling for Earth–Space Systems

Radio Wave Propagation and Channel Modeling for Earth–Space Systems

Athanasios G. Kanatas

University of Piraeus, Greece

Athanasios D. Panagopoulos

National Technical University of Athens, Greece

CRC Press
Taylor & Francis Group
Boca Raton London New York

CRC Press is an imprint of the
Taylor & Francis Group, an **informa** business

CRC Press
Taylor & Francis Group
6000 Broken Sound Parkway NW, Suite 300
Boca Raton, FL 33487-2742

First issued in paperback 2020

ISBN 13: 978-0-367-57480-2 (pbk)
ISBN 13: 978-1-4822-4970-5 (hbk)

Library of Congress Cataloging-in-Publication Data

Names: Kanatas, Athanasios G., editor. | Panagopoulos, Athanasios D., 1975- editor.
Title: Radio wave propagation and channel modeling for Earth-space systems / editors, Athanasios G. Kanatas and Athanasios D. Panagopoulos.
Description: Boca Raton : Taylor & Francis, CRC press, 2015. | Includes bibliographical references and index.
Identifiers: LCCN 2015048511 | ISBN 9781482249705 (alk. paper)
Subjects: LCSH: Astronautics--Communication systems. | Radio wave propagation--Mathematical models. | Earth stations (Satellite telecommunication)
Classification: LCC TL3025 .R34 2015 | DDC 621.3841/56--dc23
LC record available at http://lccn.loc.gov/2015048511

Visit the Taylor & Francis Web site at
http://www.taylorandfrancis.com

and the CRC Press Web site at
http://www.crcpress.com

Contents

Preface

The evolution of Earth–space systems and networks in recent years has been accelerating at an extraordinary pace. The satellite systems have merged into a global integrated network providing telecommunication, navigation, broadcasting, and Earth observation services to the users. The reliable communication is constrained by the radio propagation effects of the Earth–space links. The accurate design of Earth–space systems requires a comprehensive knowledge of the various propagation media and phenomena that differ on the frequency and the type of application. The choice of the relevant channel models is crucial in the design process and constitutes an important step in performance evaluation and testing of Earth–space systems.

The subject of this book is built around the two characteristic cases of satellite systems: the fixed satellite and the mobile satellite systems. The book presents the state of the art in satellite channel modeling. The frequencies of interest range from 100 MHz to 100 GHz (from VHF to W band), whereas the use of optical free-space communications is envisaged.

This book includes chapters that analytically present the state of the art in channel modeling and characterization of next-generation multiple-antennas fixed and mobile satellite systems, as well as propagation phenomena and fade mitigation techniques. Chapters presenting the research and the well-accepted satellite community results for land mobile satellite and tropospheric attenuation time-series synthesizers are also included. Moreover, research advances on space–time tropospheric propagation fields and optical satellite communication channel models are presented. Aeronautical communications channel characteristics and modeling and the relative radio wave propagation campaigns and stratospheric channel model for various applications and frequencies are also covered. Finally, propagation effects on satellite navigation systems and the corresponding models are included in this book.

Editors

Athanasios G. Kanatas is a professor at the Department of Digital Systems and dean of the School of Information and Communication Technologies at the University of Piraeus, Greece. He earned his diploma in electrical engineering from the National Technical University of Athens (NTUA), Greece, in 1991, MSc in satellite communication engineering from the University of Surrey, Surrey, UK, in 1992, and PhD in mobile satellite communications from NTUA, Greece, in February 1997. From 1993 to 1994, Dr. Kanatas was with the National Documentation Center of the National Research Institute. In 1995, Dr. Kanatas joined SPACETEC Ltd. as a technical project manager for the VISA/EMEA VSAT Project in Greece. In 1996, he joined the Mobile Radio-Communications Laboratory as a research associate. From 1999 to 2002, he was with the Institute of Communication & Computer Systems, responsible for the technical management of various research projects. In 2000, he became a member of the board of directors of OTESAT S.A. In 2002, he joined the University of Piraeus as an assistant professor. From 2007 to 2009, he served as a Greek delegate to the Mirror Group of the Integral SatCom Initiative. Dr. Kanatas has published more than 130 papers in international journals and international conference proceedings. His current research interests include the development of new digital techniques for wireless and satellite communications systems, channel characterization, simulation, and modeling for mobile, mobile satellite, and future wireless communication systems, antenna selection and RF preprocessing techniques, new transmission schemes for MIMO systems, V2V communications, and energy-efficient techniques for wireless sensor networks. Dr. Kanatas has been a senior member of the IEEE since 2002. In 1999, he was elected the chairman of the Communications Society of the Greek Section of IEEE.

Athanasios D. Panagopoulos earned his diploma in electrical and computer engineering (summa cum laude) and PhD in engineering from the National Technical University of Athens (NTUA) in July 1997 and April 2002, respectively. From May 2002 to July 2003, he served in the Technical Corps of the Hellenic Army. In September 2003, he joined the School of Pedagogical and Technological Education as a part-time assistant professor. From January 2005 to May 2008, he was head of the Satellite Division of Hellenic Authority for the Information and Communication Security and Privacy (ADAE). From May 2008 to May 2013, he was a lecturer in the School of Electrical and Computer Engineering of NTUA, and he is now an assistant professor. Dr. Panagopoulos has published more than 135 papers in international journals and transactions and more than 190 papers in conference proceedings. He has also published more than 25 book chapters in international books. He is the recipient of the URSI General Assembly Young Scientist Award in 2002 and 2005. He is the editor of two international books on mobile and satellite communications. He is the corecipient of best paper awards in the IEEE RAWCON 2006 and IEEE ISWCS 2015 conferences. His research interests include radio communication systems design, wireless and satellite communications networks, and the propagation effects on multiple access systems and communication protocols. He participates in ITU-R and ETSI study groups and is a member of the Technical Chamber of Greece and a senior member of the IEEE. Dr. Panagopoulos has been the chairman of the IEEE Greek Communication Chapter since 2013. He serves on the editorial boards of *International Journal of Antennas and Propagation, International Journal of Vehicular Technology,* and *Physical Communication* (Elsevier). He is also an associate editor of *IEEE Transactions on Antennas and Propagation* and *IEEE Communication Letters.*

Contributors

Ana Vazquez Alejos earned her MS and PhD from the University of Vigo, Spain, in 2000 and 2006, respectively. She received the Ericcson Award for her master thesis in 2002 from the Spanish Association of Electrical Engineers as the best multimedia wireless project. Dr. Alejos was granted the Marie Curie International Outgoing Fellowship, with the outgoing phase conducted at the Klipsch School of ECE, New Mexico State University, New Mexico, USA, with the research focused on the measurement and modeling of propagation through dispersive media, and radar waveform generation. Her research work includes radio propagation, communication electronics, wideband radio channel modeling, multimedia wireless systems, waveform and noise code design, and radar. Dr. Alejos is a member of the IEEE and a reviewer of several IEEE and IET journals.

Pantelis-Daniel Arapoglou earned his diploma in electrical and computer engineering and PhD in engineering from the National Technical University of Athens (NTUA), Athens, Greece, in 2003 and 2007, respectively. From September 2008 to October 2010, he was involved in postdoctoral research on MIMO over satellite, jointly supported by the NTUA and the European Space Agency Research and Technology Centre (ESA/ESTEC), the Netherlands. From October 2010 to September 2011, he was a research associate at the Interdisciplinary Centre for Security, Reliability and Trust (SnT), University of Luxembourg. Since September 2011, he has been a communications system engineer at ESA/ESTEC, where he is technically supporting R&D activities and developments in the areas of satellite telecommunications, digital and optical communications, and high data rate telemetry for Earth observation applications. Dr. Arapoglou was a recipient of the Ericsson Award of Excellence in Telecommunications for his diploma thesis in 2004 and the URSI General Assembly Young Scientist Award in 2005. As a researcher, he has participated in the work of Study Group 3 of the ITU-R in SatNEx III and in COST Action IC0802. He is involved in ESA-funded SatNEx and the Optical Working Group of the CCSDS (Consultative Committee for Space Data Systems).

Laurent Castanet earned his BS in microwave engineering from TELECOM Bretagne in 1991 and MS in space telecommunications from TELECOM Paris in 1992. He earned his PhD from SUPAERO on "Fade Mitigation Techniques for New SatCom Systems Operating at Ka and V Bands." From 1994, he worked as a research engineer in the radiowave propagation field at ONERA Toulouse. His main research interests are Earth–space propagation and mitigation techniques to improve system performances. Dr. Castanet is now the head of the radio-communication and propagation research unit of the Electromagnetics and Radar Department of ONERA. Dr. Castanet teaches radiowave propagation and link analysis at SUPAERO and TELECOM Paris Engineering schools, at the University of Toulouse, and at the EuroSAE Training institution.

Edgar Lemos Cid earned his telecommunication engineering degree from the Universidade de Vigo, Vigo, Spain, in 2012. He has been working with the Radio Systems Group, Universidade de Vigo, and with the Experimental High Energy Physics Group, Universidade de Santiago de Compostela, as a researcher since 2012. His research interests include wideband and narrowband radio channel measurement, characterization, and modeling; and electronic VELO upgrade design of the LHCb CERN detector.

Nicolas Jeannin has worked as research engineer in the radiocommunication and propagation group of ONERA Toulouse since 2009. He earned his PhD in the field of tropospheric channel modeling from the University of Toulouse in 2008 and his engineering degree and MSc in image processing from ISAE/SUPAERO in 2005. Dr. Jeannin's current research interest concerns the impact and the modeling of the troposphere on satellite telecommunication systems and remote sensing instruments.

Charilaos Kourogiorgas earned his diploma in electrical and computer engineering from the National Technical University of Athens (NTUA), Athens, Greece, in 2009. In May 2015, he earned his PhD on channel modeling and performance evaluation of next-generation high data rate wireless terrestrial and satellite communication systems from the School of Electrical and Computer Engineering at the National Technical University of Athens. From October 2009 to June 2011, he was with the Department of Electromagnetism and Radar at Office National d'Études et Recherches Aérospatiales, Toulouse, France. Dr. Kourogiorgas has published more than 50 papers in international refereed journals and conferences. He was a participant at COST Actions IC0802 and IC1004 and has worked on three ESA projects and several national projects on satellite communications. Dr. Kourogiorgas was awarded the Chorafas scholarship for his PhD thesis in 2015 and the Young Scientist Award from URSI. Dr. Kourogiorgas is a member of the Technical Chamber of Greece. His research interests are channel modeling for satellite and terrestrial communication systems and system evaluation of next-generation wireless systems.

Konstantinos P. Liolis earned his diploma in electrical and computer engineering from the National Technical University of Athens (NTUA), Greece, in 2004, an MSc in electrical engineering from the University of California at San Diego (UCSD), USA, in 2006, and PhD in electrical engineering from NTUA in 2011. From 2004 to 2006, he was a research assistant at the California Institute for Telecommunications and Information Technology (Cal-IT2), San Diego, USA. From 2006 to 2008, he was a communication systems engineer at the European Space Agency, Research and Technology Centre (ESA/ESTEC), the Netherlands. From 2008 to 2012, he was a R&D project manager at Space Hellas S.A., Greece. Since 2012, he has been a space systems engineer at SES Techcom S.A., Luxembourg. Dr. Liolis has published more than 45 scientific papers in international peer-reviewed journals, conference proceedings, and book chapters, in areas mainly related to satellite communications systems and technologies. He also has numerous contributions to the ETSI, DVB, and ITU-R international standardization bodies, and has actively participated in more than 40 R&D and innovation projects on ICT, space and security technologies funded as part of EU, ESA, and national programs. He is a member of the Technical Chamber of Greece. He received the Best Student Paper Award in IEEE RAWCON 2006 and is listed in the *Who's Who in the World* 2010 Edition.

Lorenzo Luini earned his Laurea degree in telecommunications engineering in 2004 and PhD in information technology in 2009, both from Politecnico di Milano, Italy. He is currently an assistant professor at DEIB (Dipartimento di Elettronica, Informazione e Bioingegneria). Since 2004, his research activities have been focused on EM wave propagation through the atmosphere, both at radio and optical frequencies; physical modeling and synthesis of the meteorological environment; development and implementation of models for the remote sensing of atmospheric constituents using radiometric data; physical and statistical modeling for EM propagation applications; analysis and dimensioning of wireless terrestrial and SatCom (GEO, MEO, LEO) systems operating in the 1–100 GHz range; design and simulation of systems implementing fade mitigation techniques; assessment of the impact of the atmosphere on free space optical Earth–space (to satellite or deep space probes) systems; assessment of the impact of atmospheric constituents on Ka-band synthetic aperture radars; and analysis of the performance of space-borne GNSS receivers. Dr. Luini has been involved in European COST projects, in the European Satellite Network of Excellence (SatNEx), as well as in projects commissioned by the European Space Agency (ESA) and the U.S. Air Force Laboratory. Dr. Luini is the author of several contributions to international journals and conferences; he is an associate editor for *International Journal of Antennas and Propagation* and serves as a reviewer for several scientific journals (e.g., *IEEE Transactions on Antennas and Propagation* and *Radio Science*). He also worked as a system engineer in the Industrial Unit—Global Navigation Satellite System (GNSS) Department—at Thales Alenia Space Italia S.p.A.

Emmanouel T. Michailidis earned a BSc in electronics engineering in 2004 from the Piraeus University of Applied Sciences (TEI of Piraeus), Egaleo, Greece, an MSc in digital communications and networks in 2006 from the University of Piraeus, Piraeus, Greece, and a PhD in broadband wireless communications in 2011 from the University of Piraeus. Since 2012, he has been a postdoctoral researcher at the Telecommunication Systems Laboratory (TSL), Department of Digital Systems, School of Information and Communication Technologies, University of Piraeus. Since 2007, he has been a lab instructor at the Department of Electronics Engineering, School of Technological Applications, Piraeus University of Applied Sciences and an electronics and informatics instructor in vocational schools and vocational training institutes. Dr. Michailidis has published more than 30 papers in international journals and international conference proceedings. His current research interests include channel characterization, modeling, and simulation for future wireless communication systems and new transmission schemes for cooperative and MIMO systems. He serves on the editorial board of the *International Journal on Advances in Telecommunications*. Dr. Michailidis received the Best Paper Award at the International Conference on Advances in Satellite and Space Communications (SPACOMM) 2010 and was awarded by the Hellenic Ministry of Education, Research & Religious Affairs for contributing to Academic and Scientific Excellence.

Roberto Nebuloni earned his Laurea degree in electronic engineering and PhD in information engineering from the Politecnico di Milano, Milan, Italy in 1997 and 2004, respectively. In 2005, Dr. Nebuloni's joined the Italian National Research Council (CNR), where he is currently a researcher at the Institute of Electronics, Computer and Telecommunication Engineering (IEIIT), in Milan. Dr. Nebuloni's research interests include theoretical and experimental aspects of radio and optical wave propagation through the atmosphere and radar applications in the areas of telecommunications, meteorology, and environmental monitoring.

Fernando Pérez-Fontán earned his diploma in telecommunications engineering and PhD from the Technical University of Madrid, Madrid, Spain, in 1982 and 1992, respectively. He is a full professor at the Telecommunications Engineering School, University of Vigo, Vigo, Spain. Dr. Pérez-Fontán is the author of a number of international magazine and conference papers. His main research interest is in the field of mobile and fixed radio communications propagation channel modeling.

Roberto Prieto-Cerdeira is the GNSS principal R&D engineer in the GNSS Evolutions Programme and Strategy Division of the European Space Agency, where he is responsible for technology R&D and science activities for the evolution of European GNSS systems, in particular for the Galileo Second Generation. From 2005 to 2014, he was with the Electromagnetics and Space Environment Division of the European Space Agency, where he was responsible for activities related to radiowave propagation for GNSS and Mobile SatCom, and where he served as the cochair of the international SBAS-Ionospheric Working Group, ESA delegate in ITU-R Study Group 3, and the chair of the Commission G of the URSI Netherlands Committee. Dr. Prieto-Cerdeira has published over 100 papers in journals and conferences. He is a member of the IEEE and the Institute of Navigation (ION).

Manuel Garcia Sanchez (IEEE Student Member 1988 and IEEE Member 1993) joined the Department of Signal Theory and Communications, University of Vigo, in 1990. He was the head of the department from 2004 to 2010. He currently teaches courses in spectrum management and communication and navigation satellite systems as a full professor. Dr. Garcia Sanchez's research interests focus on radio systems, and include studies of indoor and outdoor radio channels, channel sounding and modeling for narrow- and wideband applications, interference detection and analysis, design of impairment mitigation techniques, and radio systems design. These results are applied to point-to-multipoint radio links, mobile communications, and wireless networks, at microwave and millimeter wave frequencies.

1 Next-Generation MIMO Satellite Systems

From Channel Modeling to System Performance Evaluation

Konstantinos P. Liolis and Pantelis-Daniel Arapoglou

CONTENTS

1.1 INTRODUCTION

Satellite communications (SatCom) are currently undergoing a strong expansion to follow the dramatic increase in demand for higher capacity, improved quality of service (QoS), and ubiquitous connectivity. Characteristic examples of emerging paradigms in SatCom are the hybrid satellite/

terrestrial transmission systems based on the ETSI (European Telecommunications Standardization Institute), DVB-SH (Digital Video Broadcasting—Satellite to Handheld) (DVB-SH, 2008), and DVB-NGH (Digital Video Broadcasting—Next-Generation Handheld) (DVB-NGH, 2013) standards that provide rich multimedia broadcast content to mobile users. Although benefiting from their large geographic coverage, satellite networks have some limitations compared to terrestrial networks, which make them suitable mainly for serving sparsely populated areas. Under these limitations, new advanced technologies at the physical layer and system level are urgently required to boost the performance of SatCom and follow the capacity growth trends of terrestrial wireless communications.

In this course, the utilization of multiple-input multiple-output (MIMO) technology is a very promising candidate. In the last few years, single-user (SU) and multiuser (MU) MIMO transmission systems have received significant attention from both the research community and the wireless industry due to their impressive potential capacity gains with respect to the conventional single-input single-output (SISO) transmission systems (Mietzner et al., 2009). The appealing gains obtained by SU MIMO techniques in terrestrial cellular and WiFi networks generate further interest in investigating the applicability of the same principle in satellite networks (Arapoglou et al., 2011a). However, the fundamental differences between the terrestrial and the satellite channels make such applicability a nontrivial and nonstraightforward task but instead, a rather challenging research subject. These differences are mainly related to the requirement for having line-of-sight (LOS) reception of satellite signals (due to the limited power arriving on the ground) and to the absence of scatterers in the vicinity of the satellite, which eliminate multipath-fading profiles over the space segment and lead to an inherent rank deficiency of the MIMO channel matrix. Further details on such key differences, which justify the increasing interest of the research community in MIMO SatCom, are highlighted in Horvath et al. (2007), Liolis et al. (2007), and Arapoglou et al. (2011a); and Petropoulou et al. (2014).

DVB-SH is a satellite-driven standard first published in 2007 (DVB-SH, 2008) that does not foresee any type of MIMO application. Nevertheless, this air interface has been adopted in the majority of research and development (R&D) works as the baseline system configuration, which is then extended to introduce MIMO applications. On the other hand, DVB-NGH was published in 2013 (DVB-NGH, 2013) and includes a sheer terrestrial base profile, a sheer terrestrial MIMO profile, a hybrid satellite/terrestrial profile, and a hybrid satellite/terrestrial MIMO profile, the last three being optional (Jokela et al., 2012).

Both DVB-SH and DVB-NGH standards in general refer to satellite digital multimedia broadcasting (SDMB) systems. SDMB systems are used for the provision of digital mobile broadcasting at L (1/2 GHz) or S (2/4 GHz) frequency bands by means of geostationary or highly elliptical orbit satellites and a complementary ground component (CGC) to cover urban areas (see Figure 1.1). The typical applications envisaged are audio/video broadcasting and software updates for mobile platforms, an example being the commercially successful U.S. Sirius XM Radio system (DiPierro et al., 2010). The broadcasting mission of SDMB systems can be complemented by some interactive return link capability for messaging services (Scalise et al., 2013), leading to interactive mobile satellite systems (Gallinaro et al., 2014).

Regarding the type of coverage, for European scenarios, a few linguistic beams reusing the system bandwidth in, for example, a three-color frequency reuse scheme is a suitable approach (Gallinaro et al., 2014) to customize the digital content to the specific language adopted in each region covered. This multibeam approach allows to focus the satellite power better and reuse the frequency among the beams when there is enough isolation. The satellite also feeds the CGC, which is typically deployed in densely populated urban areas, at Ku-band. The CGC repeaters typically convert the Ku-band downlink CGC feeder link signal into an S- or L-band DVB-SH terrestrial signal, either in the same (single-frequency network, SFN) or in a different (multiple-frequency network, MFN) frequency band. An alternative coverage paradigm is the single beam continental U.S. coverage adopted by Sirius XM, where there is no need to service different languages over the coverage area.

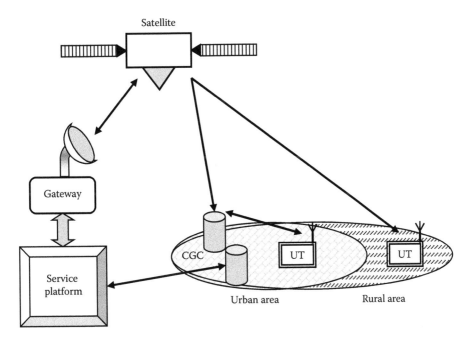

FIGURE 1.1 DVB-SH system architecture.

Fundamental in building any MIMO system are the MIMO channel degrees of freedom and diversity it offers. The possible diversity sources that can be exploited in a satellite environment to form an MIMO matrix channel are identified in Karagiannidis et al. (2007). This chapter focuses on two types of diversity, namely polarization and spatial, which result in two respective MIMO satellite system configurations for L/S-band SDMB systems: (i) Scenario A: single-satellite/dual-polarization diversity configuration (see Figure 1.2) and (ii) Scenario B: dual-satellite diversity/ single-polarization configuration (see Figure 1.3). For these two MIMO satellite system configurations, a novel unifying statistical model of the underlying MIMO land mobile satellite (LMS) fading channels is described, which extends the model in Liolis et al. (2010), Liolis (2011), and has not

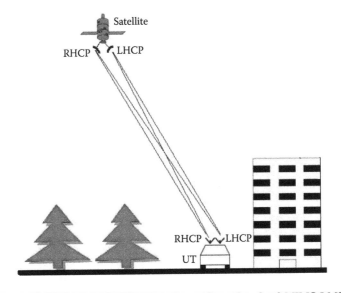

FIGURE 1.2 Single-satellite/dual-polarization diversity configuration: 2 × 2 MIMO LMS channel.

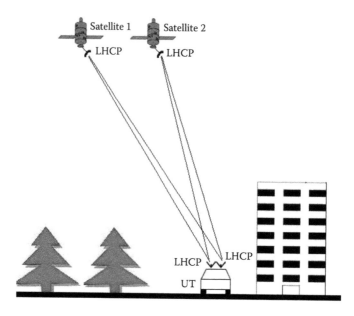

FIGURE 1.3   Dual-satellite diversity/single-polarization configuration: 2×2 MIMO LMS channel.

been presented before in the literature. This generalized 2×2 MIMO LMS channel model unifies both cases of dual-polarization diversity and dual-satellite diversity, is fully parameterized, and can be easily fine-tuned upon the availability of experimental data. In addition, a useful step-by-step recipe of the corresponding MIMO LMS time-series generation for use in computer simulation is provided.

Beyond propagation channel-modeling aspects, the chapter also addresses performance evaluation aspects of satellite systems employing MIMO signal-processing techniques particularly well aligned with the proposed MIMO LMS channel. In this regard, the outage capacity statistics of MIMO LMS-fading channels well suited to SDMB systems are evaluated through computer simulations. Moreover, useful numerical results obtained through end-to-end system computer simulations are provided, which assess the gain of MIMO techniques over dual-polarized satellite systems.

The outcomes of recent European Space Agency (ESA) cofunded R&D activities, focusing on MIMO satellite channel measurements and hardware (HW) demonstration, are also overviewed. For completeness of this chapter, other open or ongoing research avenues of MIMO over satellite are briefly reviewed in the last section. The chapter concludes with general comments and directions on the applicability of MIMO technology over next-generation satellite systems.

1.2   MIMO SATELLITE SYSTEM CONFIGURATIONS

Two specific MIMO satellite system configurations for L/S-band SDMB systems are considered hereinafter:

- Scenario A: Single-satellite/dual-polarization diversity configuration (see Figure 1.2)
- Scenario B: Dual-satellite diversity/single-polarization configuration (see Figure 1.3)

1.2.1   SCENARIO A: SINGLE-SATELLITE/DUAL-POLARIZATION DIVERSITY CONFIGURATION

In Figure 1.2, both types of coverage (linguistic beam and single beam) lend themselves to a dual-polarization per beam (DPPB) MIMO architecture (Liolis et al., 2010; Arapoglou et al., 2011b), which consists of a single geosynchronous equatorial orbit (GEO) satellite with two right/left-hand circularly polarized transmit antenna (RHCP/LHCP) set up as a DPPB payload and a user terminal (UT) with two colocated circularly polarized receive antennas. A vehicular type of UT is assumed.

The choice of circular polarization is related to the legacy coming from SISO mobile satellite systems to avoid the effect of Faraday rotation (depolarization). Nevertheless, as both polarizations are being transmitted, Faraday rotation can be removed at the receiver by operating in the polarization domain. Note that at the satellite antenna side, typically the antenna feed provides two linear polarization ports. The two circular polarizations can then be simply obtained using two different orthomode transducer topologies able to simultaneously generate the RHCP/LHCP inputs (ESA TAS-E, 2012). For a given overall radio frequency (RF) payload power, the transmission in two distinct polarizations within the same beam has the further advantage of reducing by 3 dB the power handling in the high-power part of the payload.

To perform a fair comparison between conventional SISO and dual-polarization MIMO systems, it has been assumed that the same overall RF satellite power will be present in a single- or dual-polarization system. While the satellite RF power is the same using DPPB, it allows doubling the available bandwidth per beam. From a system point of view and as explained in Arapoglou et al. (2010), adopting a DPPB architecture leads to an increased interbeam interference that depends on the number of beams and the frequency reuse scheme; this limits the operational signal-to-noise-plus-interference ratio (SNIR). The design of a multibeam (linguistic) payload for European coverage using DPPB has been performed in ESA TAS-E (2012). In Byman et al. (2015), typical S-band SDMB link budgets for both a European multibeam linguistic beam and a single beam (continental U.S.-like) cases are provided. For the multibeam (linguistic) case, the LOS edge-of-coverage SNIR for a vehicular type of terminal amounts to about 11.5 dB. In the single-beam case, thanks to the absence of another beam interference, the SNIR increases to 20 dB.

1.2.2 Scenario B: Dual-Satellite Diversity/Single-Polarization Configuration

In the joint use of two satellites (satellite or orbital diversity), the diversity is obtained by having two satellites separated by angular separation (see indicatively Lutz, 1996; Vazquez Castro et al., 2002; Heuberger, 2008; DiPierro et al., 2010; Pérez-Neira et al., 2011). In Figure 1.3, both satellites transmit in single polarization (i.e., either RHCP or LCHP) resulting either in a distributed 2 × 1 SISO configuration (single-antenna UT) or a distributed 2 × 2 MIMO configuration (dual-antenna UT, i.e., with antenna diversity). A vehicular type of UT is assumed.

A critical aspect of this scenario is the separation of the two satellites in the GEO arc in relation to the elevation angle toward them and the correlation properties of the channels formed by the two Earth–space paths. However, the blocking point for applying dual-satellite MIMO in SFN SDMB systems is the relative delay in the arrival of the signals originating from the two satellites, which is expected to be very large for any meaningful (from a channel correlation point of view) range of angular separations and, due to the broadcasting nature of the service, cannot be precompensated at the transmitter side. The result is that achieving synchronization and decoding the two streams becomes unfeasible in SFN mode. In case of MFN, a dual-satellite MIMO configuration has not been investigated so far, mainly because it requires two satellites in a simultaneous view, which is economically considered a less-attractive option. Results from a feasibility analysis on the use of orthogonal frequency division multiplexing (OFDM) as a potential countermeasure against the inherent issue of the relative delay in satellite diversity are provided (see Figure 1.13 later in chapter).

1.3 OVERVIEW OF RELEVANT STATE-OF-THE-ART WORK IN MIMO LMS CHANNEL MODELING

Before introducing the MIMO LMS channel model in the following section, it is important that we set the landscape of past modeling approaches referring to dual-polarization MIMO. Early developments included the dual-polarization channel model in Sellathurai et al. (2006) to support the corresponding signal-processing schemes. The very first set of MIMO LMS measurements was conducted in Guildford, UK, at 2.45 GHz and presented in King (2007) and King and Stavrou (2007). To emulate the satellite scenario, an artificial terrestrial platform acting as the satellite transmitter was installed on a hilltop transmitting to a mobile vehicle (van) acting as a mobile terminal. Large- and small-scale first- and second-order statistics were recorded, along with correlation statistics over the delay and spatial/polarization domains. The data collected also enabled the development of a statistical MIMO LMS channel model.

On the basis of the same principles as King (2007) and King and Stavrou (2007), a flexible statistical MIMO LMS channel was introduced in Liolis et al. (2010). The main assumptions of this model, which is in wide use and will be further elaborated in the following section, are

- Log-normal distribution is assumed for the large-scale fading.
- Small-scale fading is generated separately and added to the large-scale fading.
- A Markov or semi-Markov process is used for modeling the change of channel state (e.g., good/bad state).
- For each environment, different statistics apply.

A consolidation of the statistical approach in MIMO LMS channel modeling is attempted in Carrie et al. (2013).

A second set of dual-polarization MIMO LMS measurement results came from the ESA MIMOSA project (Eberlein et al., 2011; ESA MIMOSA, 2012), which covers two measurement campaigns: the first measurement campaign was performed in August 2010 at Erlangen and Lake Constance and its goal was to allow statistical analysis of the dual-polarized 2×2 MIMO channel from a real S-band satellite. A complementary measurement campaign (channel sounder-based field trial) took place in Berlin in March 2011, using a high tower for satellite transmission emulation aiming at covering the detailed evaluation of selected scenarios with respect to channel frequency selectivity and angle of arrival.

In Cheffena et al. (2012), an MIMO LMS channel model is parameterized by means of a physical model based on the multiple-scattering theory, which accounts for the signal attenuation and scattering by trees. Moreover, finite-difference time-domain electromagnetic computations are performed to characterize the scattering pattern of an isolated tree, and to calculate the MIMO-shadowing correlation matrix. In King et al. (2012), the procedure for implementing a simple empirical–stochastic-based model for the dual circular polar MIMO LMS channel is presented along with results to validate the model at low elevation. Given the simplicity of generating a Markov chain and correlated small- and large-scale fading, it is proposed as being highly appropriate for conformance testing for satellite MIMO applications.

The vast majority of the state-of-the-art MIMO LMS channel models listed above is based on a statistical narrowband approach. Instead, a completely different variant of MIMO LMS channel modeling is presented in Burkhardt et al. (2014) referred to as Quadriga model and based on the WINNER II class of channel models (EC FP6 IST, 2007). The WINNER II is a terrestrial wideband model relying on a mixed statistical and deterministic approach that in Quadriga is adapted to the satellite case allowing for antenna effects to be taken into account through angle-of-arrival modeling.

1.4 UNIFYING STOCHASTIC MIMO LMS CHANNEL MODEL

For the two MIMO satellite system configurations introduced, a novel unifying statistical model of the underlying MIMO LMS-fading channels is hereinafter proposed, which has not been published before. The modeling philosophy can be summarized as follows: *even if limited (or no) experimental MIMO LMS channel models are available, statistical MIMO LMS channel models can be built by relying on solid modeling principles that allow for consolidating the model parameters whenever a plethora of datasets become available.* In this regard, the proposed MIMO LMS channel model takes into account the existing vast literature on SISO LMS and MIMO wireless channel modeling as well as the few experimental results available in the MIMO LMS context as appropriate. However, the more experimental datasets become available, the better the parameterization can become.

Emphasis is put on the characterization of the downlink channel, that is, from the satellite(s) to the UT. Owing to the local environment in the vicinity of the mobile UT (i.e., mostly adjacent buildings and vegetation), the LOS link between the satellite(s) and the UT might be clear, partially or even fully obstructed that gives rise to multipath, shadowing, and blockage effects. The resulting fading channel is assumed to be narrowband (i.e., frequency nonselective) since the multipath echoes are not significantly spread in time (Loo, 1985; Pérez-Fontán et al., 2001; Prieto-Cerdeira et al., 2010).

In the following, the speed of the mobile UT is denoted by v (m/s) whereas the satellite elevation angle is denoted by θ (deg). Under the assumptions stated above, the MIMO LMS channel assumed in both Scenario A and Scenario B is modeled by a 2×2 MIMO fading channel matrix $\mathbf{H} = [h_{ij}]$ ($i,j = 1,2$), where h_{ij} represent the complex fading components of the four SISO LMS subchannels formed between the transmit and receive sides. Note that h_{ij} incorporate both the large-scale fading effects (i.e., those accounting for direct LOS shadowing) and the small-scale fading effects (i.e., those accounting for a diffuse multipath). For both Scenario A and Scenario B, a common unifying modeling methodology is followed comprising five main steps:

- Modeling of SISO LMS subchannel effects
- Modeling of cross-channel discrimination effects
- Modeling of spatial correlation of large-scale fading components
- Modeling of spatial correlation of small-scale fading components
- Modeling of temporal correlation

1.4.1 Modeling of SISO LMS Subchannel Effects

For the modeling of the envelope $|h_{ij}|$ ($i,j = 1,2$), the vast literature of SISO LMS channel modeling is employed. In this chapter, the Loo distribution (Loo, 1985) is assumed for $|h_{ij}|$ because it has been fundamental to the modeling of the SISO LMS channel (Pérez-Fontán et al., 2001) and has been extensively used and validated in the frame of DVB-SH standardization (DVB-SH IG, 2008).

Under this assumption, the 2×2 MIMO LMS-fading channel matrix $\mathbf{H}$ is expanded as

$$\mathbf{H} = [h_{ij}] = [\bar{h}_{ij}] + [\tilde{h}_{ij}] = \bar{\mathbf{H}} + \tilde{\mathbf{H}} \quad (i,j = 1,2) \tag{1.1}$$

where

$$h_{ij} = |h_{ij}| \exp(j\phi_{ij}) = |\bar{h}_{ij}| \exp(j\bar{\phi}_{ij}) + |\tilde{h}_{ij}| \exp(j\tilde{\phi}_{ij}) \quad (i,j = 1,2) \tag{1.2}$$

$\bar{\phi}_{ij}$, $\tilde{\phi}_{ij}$ are uniformly distributed over $[0,2\pi)$, the envelope $|\bar{h}_{ij}|$ of the large-scale fading components is assumed *lognormally* distributed with parameters (α, ψ), and the envelope $|\tilde{h}_{ij}|$ of the small-scale fading components is assumed *Rayleigh* distributed with parameter MP. Thus, the Loo probability density function (PDF) of the overall envelope $|h_{ij}|$ ($i,j = 1,2$) is given by (Loo, 1985)

$$p(|h_{ij}|) = \frac{|h_{ij}|}{b_0\sqrt{2\pi d_0}} \int_0^\infty \frac{1}{z} \exp\left[-\frac{(\ln z - \mu)^2}{2d_0} - \frac{|h_{ij}|^2 + z^2}{2b_0}\right] I_0\left(\frac{|h_{ij}| z}{b_0}\right) dz \tag{1.3}$$

where $\alpha = 20 \log_{10}[\exp(\mu)]$ is the mean and $\psi = 20 \log_{10}[\exp(\sqrt{d_0})]$ is the standard deviation of the lognormally distributed large-scale fading components, $MP = 10 \log_{10}(2b_0)$ is the average power of the Rayleigh-distributed small-scale fading components (α, ψ, and MP are all expressed in dB, relative to LOS), and $I_0(\cdot)$ is the modified Bessel function of the first kind and zero order.

The Loo statistical parameter triplet (α, ψ, MP) refers to the experimental dataset that is originally presented in Pérez-Fontán et al. (2001) and further properly revised in Prieto-Cerdeira et al. (2010). The specific choice of these parameters is pertinent to each operational scenario assumed. That is, different (α, ψ, MP) are obtained for different frequency bands: L (1/2 GHz) and S (2/4 GHz); for different user environments: intermediate/heavy/light tree shadowed, urban, suburban, and open rural; and for different elevation angles: $\theta = 40$–$80°$ (S-band) and $\theta = 10$–$70°$ (L-band). Specifically, two sets of narrowband experimental data were taken into account in Pérez-Fontán et al. (2001); Prieto-Cerdeira et al. (2010): one at S-band and another one at L-band. Both datasets were measured in moving-vehicle conditions. In this respect, note that the originally derived SISO LMS channel model in Pérez-Fontán et al. (2001) assumes a three-state Markov model with, namely, "clear LOS," "moderate shadowing," and "deep shadowing" states, whereas the revised model in Prieto-Cerdeira et al. (2010), which is adopted here, assumes a two-state semi-Markov model with "GOOD" and "BAD" states. According to the semi-Markov-state model (Bråten and Tjelta, 2002), the nonfade duration distribution, that is, the duration of the "GOOD" state is given by a power law distribution and that of the "BAD" states is given by a lognormal distribution. The methodology to obtain the Loo parameters (α, ψ, MP) based on tabulated experimental datasets is detailed in Prieto-Cerdeira et al. (2010).

Unlike the SISO LMS case where the characterization of statistical distribution for $|h_{ij}|$ $(i,j = 1,2)$ would suffice, in the MIMO LMS case addressed here, there are fundamental differences mainly due to the presence of multiple antenna elements both at the transmit and receive sides. The modeling of these differences is dealt with next for both cases of Scenario A and Scenario B.

1.4.2 Modeling of Cross-Channel Discrimination Effects

1.4.2.1 Scenario A

In the case of polarization diversity, the cross-channel discrimination is referred to as cross-polar discrimination (XPD). The XPD of the large-scale fading components $\bar{h}_{ij}(i,j = 1,2)$ is related only to the XPD of the antenna, denoted by XPD_{ant}, whereas the XPD of the small-scale fading components $\tilde{h}_{ij}$ $(i,j = 1,2)$ is related to both XPD_{ant} and the cross-polar coupling (XPC) of the propagation environment, denoted by XPC_{env}. Note that the antenna that is mainly critical in the presented analysis is that of the UT (i.e., reception side), whose XPD in most practical configurations of satellite networks is not greater than 15 dB. On the contrary, the XPD of the satellite antenna (i.e., transmission side) is assumed to approximate ∞ due to its very large value in practice. Thus, XPD_{ant} will denote hereinafter only the UT antenna XPD, which critically affects the analysis.

The power imbalance between the co-polar and cross-polar components in the cases of $0°/90°$ and $\pm 45°$ polarization diversity configurations was investigated in detail in Oestges et al. (2004) and Sørensen et al. (1998) in the context of fixed, wireless, and mobile radio channels, respectively, through experimental results and, in particular, through the branch power ratio defined (in dB) as

$$\text{BPR} = 10 \log_{10}(E[|h_{11}|^2]/E[|h_{22}|^2]) \tag{1.4}$$

where h_{11} and h_{22} refer to copolar components and $E[\cdot]$ denotes the expectation operator. In most examined scenarios, specifically those related to the $\pm 45°$ polarization diversity case, it was found that BPR = 0 dB, that is, symmetry can be assumed. Thus, for the R/L-HCP circular polarization diversity of interest, symmetry is also assumed. This assumption is in line with experimental results obtained in the context of LMS channels and reported in King and Stavrou (2007), which further indicate such power balance. On the basis of this symmetry assumption, the power of the small- and large-scale fading components is given by

$$E[|\bar{h}_{ij}|^2] = \begin{cases} (\psi^2 + \alpha^2) \cdot (1 - \beta_{ant}) & i = j \\ (\psi^2 + \alpha^2) \cdot \beta_{ant} & i \neq j \end{cases} \qquad (1.5)$$

$$E[|\tilde{h}_{ij}|^2] = \begin{cases} MP \cdot (1 - \gamma) & i = j \\ MP \cdot \gamma & i \neq j \end{cases} \qquad (1.6)$$

where $i,j = 1,2$, (α, ψ, MP) are expressed in linear scale (i.e., not in dB), $\beta_{ant} \in [0,1]$ depends only on XPD_{ant}, and $\gamma \in [0,1]$ depends on both XPD_{ant} and XPC_{env}.

Concerning the relationship between the XPD modeling factors β_{ant} and γ in Equations 1.5 and 1.6 as well as the actual measurable parameters XPD_{ant} and XPC_{env}, one gets

$$XPD_{ant} = 10\log_{10}(E[|\bar{h}_{ii}|]^2 / E[|\bar{h}_{ij}|^2]) = 10\log_{10}[(1 - \beta_{ant})/\beta_{ant}] \qquad (1.7)$$

$$XPC_{env} = 10\log_{10}[(1 - \gamma_{env})/\gamma_{env}] \qquad (1.8)$$

$$\gamma = \beta_{ant}(1 - \gamma_{env}) + (1 - \beta_{ant})\gamma_{env} \qquad (1.9)$$

γ_{env} is related to the XPC of the propagation environment, XPC_{env}. Several works such as Lempiäinen and Laiho-Steffens (1998), Sellathurai et al. (2006), Brown et al. (2007), King (2007), and King and Stavrou (2007) propose specific values for parameters XPD_{ant}, XPC_{env} based on extensive measurement campaigns, which are also taken into account in this chapter for Scenario A. Typical values of these parameters are provided in Table 1.1.

1.4.2.2   Scenario B

In the case of satellite diversity, similar to the cross-polarization discrimination effect, the attenuation experienced from the different satellite depending on the receive antenna radiation pattern is taken into account. If $\epsilon \in [0,1]$ is a parameter depending on the antenna radiation pattern and denoting the cross-channel attenuation effect due to satellite discrimination, each receiver antenna is assumed to receive the signal from one satellite and attenuate the signal from the other satellite by $\sqrt{\epsilon/(1-\epsilon)}$. Thus, assuming symmetry in the problem, the cross-satellite antenna attenuation factor XSD_{ant} affecting the large-scale fading components $\bar{h}_{ij}$ $(i,j = 1,2)$ is modeled (in dB) as

$$XSD_{ant} = 10\log_{10}(E[|\bar{h}_{ii}|^2]/E[|\bar{h}_{ij}|^2]) = 10\log_{10}[(1 - \epsilon)/\epsilon] \qquad (1.10)$$

As an illustration, for hemispherical antenna radiation patterns, $\epsilon = 0.5$ and thus $XSD_{ant} = 0$ dB.

Regarding the small-scale fading components $\tilde{h}_{ij}$ $(i,j = 1,2)$ in Scenario B, similar to the cross-polarization discrimination effect in Scenario A, the attenuation experienced from the satellite depending on the receiver antenna radiation pattern is taken into account. Thus, assuming a similar symmetry in the problem and by simply interchanging ϵ and γ in Equations 1.5 and 1.6, similar expressions can be derived for $E[|\bar{h}_{ij}|^2]$ and $E[|\tilde{h}_{ij}|^2]$ in the case of satellite diversity, as well.

1.4.3   MODELING OF SPATIAL CORRELATION OF LARGE-SCALE FADING COMPONENTS

Owing to the huge Earth–space distance and the colocation of multiple antenna elements at the UT, the large-scale fading components $\bar{h}_{ij}$ $(i,j = 1,2)$ undergo a strong spatial correlation. If $\bar{C}$ denotes

TABLE 1.1

Parameters Assumed for MIMO LMS Channel Simulations in Scenario A (Polarization Diversity)

Parameter	Open Rural Environment	Suburban Environment	Urban Environment	Reference
Operating frequency, f	2.2 GHz (S-band)	2.2 GHz (S-band)	2.2 GHz (S-band)	DVB-SH (2008)
Satellite orbit	GEO	GEO	GEO	DVB-SH IG (2008)
Polarization	RHCP and LHCP	RHCP and LHCP	RHCP and LHCP	DVB-SH IG (2008)
Mobile UT speed, v	50 km/h	50 km/h	50 km/h	DVB-SH IG (2008)
Satellite elevation angle, θ	40°	40°	40°	DVB-SH IG (2008)
XPC of the environment, XPC_{env}	15 dB	6 dB	5 dB	Lempiäinen and Laiho-Steffens (1998), Sellathurai et al. (2006), and Mansor et al. (2010)
XPD of UT antenna, XPD_{ant}	15 dB	15 dB	15 dB	Lempiäinen and Laiho-Steffens (1998), Sellathurai et al. (2006), and Mansor et al. (2010)
Loo statistical parameter triplet (α, ψ, MP)	Each time a new state is reached, (α, ψ, MP) are drawn from the corresponding joint distribution	Each time a new state is reached, (α, ψ, MP) are drawn from the corresponding joint distribution	Each time a new state is reached, (α, ψ, MP) are drawn from the corresponding joint distribution	Prieto-Cerdeira et al. (2010)
Polarization correlation coefficient of small-scale fading components at Tx, $\tilde{\rho}_{tx}$	0.4	0.5	0.5	Lempiäinen and Laiho-Steffens (1998), Sellathurai et al. (2006), and Mansor et al. (2010)
Polarization correlation coefficient of small-scale fading components at Rx, $\tilde{\rho}_{rx}$	0.5	0.5	0.5	Lempiäinen and Laiho-Steffens (1998), Sellathurai et al. (2006), and Mansor et al. (2010)
Polarization correlation matrix of large-scale fading components, $\bar{\mathbf{C}}$	$\begin{bmatrix} 1 & 0.86 & 0.85 & 0.90 \\ 0.86 & 1 & 0.91 & 0.87 \\ 0.85 & 0.91 & 1 & 0.88 \\ 0.90 & 0.87 & 0.88 & 1 \end{bmatrix}$	$\begin{bmatrix} 1 & 0.76 & 0.76 & 0.83 \\ 0.76 & 1 & 0.83 & 0.75 \\ 0.76 & 0.83 & 1 & 0.78 \\ 0.83 & 0.75 & 0.78 & 1 \end{bmatrix}$	$\begin{bmatrix} 1 & 0.86 & 0.86 & 0.92 \\ 0.86 & 1 & 0.89 & 0.85 \\ 0.86 & 0.89 & 1 & 0.93 \\ 0.92 & 0.85 & 0.93 & 1 \end{bmatrix}$	King (2007)
Interstate temporal variations	First-order two-state Markov chain model with the respective absolute state and state transitions probability matrices $\mathbf{W}_{MIMO} = \mathbf{W}_{SISO}$ and $\mathbf{P}_{MIMO} = \mathbf{P}_{SISO}$	First-order two-state Markov chain model with the respective absolute state and state transitions probability matrices $\mathbf{W}_{MIMO} = \mathbf{W}_{SISO}$ and $\mathbf{P}_{MIMO} = \mathbf{P}_{SISO}$	First-order two-state Markov chain model with the respective absolute state and state transitions probability matrices $\mathbf{W}_{MIMO} = \mathbf{W}_{SISO}$ and $\mathbf{P}_{MIMO} = \mathbf{P}_{SISO}$	Prieto-Cerdeira et al. (2010)

the 4×4 positive semidefinite Hermitian covariance matrix for the large-scale fading components, a 2×2 channel matrix $\bar{\mathbf{H}}_{w,corr}$ with spatially correlated, identically distributed, and circularly symmetric complex Gaussian elements of zero mean and unit variance is generated as

$$vec(\bar{\mathbf{H}}_{w,corr}) = \bar{\mathbf{C}}^{1/2} \cdot vec(\bar{\mathbf{H}}_w) \tag{1.11}$$

where $vec(\cdot)$ denotes the operator that stacks a matrix into a vector column wise, and $\bar{\mathbf{H}}_w$ is the 2×2 channel matrix with spatially uncorrelated, identically distributed, and circularly symmetric complex Gaussian elements of zero mean and unit variance. After appropriately incorporating the mean α (dB) and standard deviation ψ (dB) obtained from Prieto-Cerdeira et al. (2010) in the Gaussian channel matrix $\bar{\mathbf{H}}_{w,corr}$ and then exponentiating to generate the lognormal channel matrix, the spatially correlated 2×2 MIMO channel matrix $\bar{\mathbf{H}}$ accounting for the large-scale fading components comes up in the form

$$vec(\bar{\mathbf{H}}) = 10^{[vec(\bar{\mathbf{H}}_{w,corr}) \cdot (\psi/20) + (\alpha/20)]} \tag{1.12}$$

The covariance matrix $\bar{\mathbf{C}}$ in Equation 1.11 is different for the two considered cases of polarization diversity (Scenario A) and satellite diversity (Scenario B).

1.4.3.1 Scenario A

In the case of polarization diversity, experimental results characterizing $\bar{\mathbf{C}}$ in the MIMO LMS context are reported in King (2007). Although these results refer to low satellite elevation angles, the polarization correlation coefficients affecting the large-scale fading components get relatively high values (i.e., close to 1) for all user environments examined. This indicates that even in the more general and practical case of higher-elevation angles, the polarization correlation coefficients will also get similarly high values and thus the same matrices $\bar{\mathbf{C}}$ are also considered in this chapter (see Table 1.1).

At this point, note that in Carrie et al. (2013), it is claimed that the covariance matrix $\bar{\mathbf{C}}$ proposed in King (2007) can be used to validate channel models but cannot be used as input parameters to state-oriented channels models. To this end, it is shown that the covariance matrix $\bar{\mathbf{C}}$ is closely connected to the states parameters and a procedure is proposed to extract $\bar{\mathbf{C}}$ from an experimental dataset that is compatible with state-oriented channel models (Carrie et al., 2013).

1.4.3.2 Scenario B

In the case of satellite diversity, there are scarce experimental results available in the literature (Milojevic et al., 2009), especially characterizing $\bar{\mathbf{C}}$ in the MIMO LMS context. Nonetheless, based on the vast literature on SISO LMS channel modeling, here, it is proposed to model the shadowing correlation matrix $\bar{\mathbf{C}}$ in the MIMO LMS case of interest as

$$\bar{\mathbf{C}} = \begin{bmatrix} 1 & 1 & \bar{\rho}_w & \bar{\rho}_w \\ 1 & 1 & \bar{\rho}_w & \bar{\rho}_w \\ \bar{\rho}_w & \bar{\rho}_w & 1 & 1 \\ \bar{\rho}_w & \bar{\rho}_w & 1 & 1 \end{bmatrix} \tag{1.13}$$

where $\bar{\rho}_w$ is the spatial correlation coefficient introduced in the large-scale fading components $\bar{h}_{ij}$ $(i,j = 1,2)$ between two different satellites and the same receiver antenna. Note that due to the huge Earth–space distance and the relative smaller antenna elements' separation distance, Equation 1.13 further assumes that the spatial correlation introduced in the links between two receiver antennas and the same satellite equals to 1 whereas that between the links from different satellites and different receiver antennas equals to $\bar{\rho}_w$ considering that the satellite correlation dominates.

However, note that $\overline{\mathbf{C}}$ in Equation 1.11 models the spatial correlation introduced among Gaussian channel inputs. Thus, to calculate the actual (measured) correlation coefficient $\bar{\rho}_{LN}$ characterizing the lognormally distributed large-scale fading components, the following transformation applies (Liolis et al., 2007):

$$\bar{\rho}_{LN} = [\exp(\bar{\rho}_w \psi_1 \psi_2) - 1]/\sqrt{(\exp(\psi_1^2) - 1)(\exp(\psi_2^2) - 1)} \tag{1.14}$$

where ψ_i $(i = 1,2)$ are the standard deviations of the lognormally distributed large-scale fading components $\bar{h}_{ij}$ $(i,j = 1,2)$. That is, Equation 1.14 takes into account the general case of $\psi_1 \neq \psi_2$ that corresponds to the case where the two satellites are characterized by different elevation angles (*unbalanced* satellite diversity). The shadowing correlation coefficient $\bar{\rho}_{LN}$ in satellite diversity configurations has been studied extensively based on experimental measurement campaigns for different user environments, for example, in Vazquez Castro et al. (2002), Heuberger (2008), and Lacoste et al. (2011), where it has been shown to depend on the mobile user environment, the azimuth separation of the two satellites, and their elevation angles. Also note that $\bar{\rho}_{LN}$ has been estimated in the relevant literature as averaged over all channel states (i.e., possible combination of "GOOD" and "BAD" channel states). As no information is currently available for characterizing the spatial correlation introduced per channel state, the same correlation coefficient $\bar{\rho}_{LN}$ is assumed here for all channel states.

1.4.4 Modeling of Spatial Correlation of Small-Scale Fading Components

Owing to the angular spread $\delta\theta$ of the multipath components and the colocation of multiple antenna elements at the UT, the small-scale fading components h_{ij} $(i,j = 1,2)$ also suffer from spatial correlation, that is,

$$vec(\tilde{\mathbf{H}}) = \tilde{\mathbf{C}}^{1/2} \cdot vec(\tilde{\mathbf{H}}_w) \tag{1.15}$$

where $\tilde{\mathbf{H}}_w$ is the 2×2 channel matrix with independent identically distributed zero-mean circularly symmetric complex Gaussian elements of variance MP obtained from Prieto-Cerdeira et al. (2010), and $\tilde{\mathbf{C}}$ is the 4×4 positive semidefinite Hermitian covariance matrix for the small-scale fading components based on the Kronecker product approach (Chuah et al., 2002)

$$\tilde{\mathbf{C}} = \tilde{\mathbf{R}}_{tx}^T \otimes \tilde{\mathbf{R}}_{rx} \tag{1.16}$$

where $\otimes$ denotes the Kronecker product operator, the superscript T denotes matrix transposition, and $\tilde{\mathbf{R}}_{tx}$, $\tilde{\mathbf{R}}_{rx}$ are the 2×2 positive semidefinite Hermitian, covariance matrices of the transmit and receive sides, respectively, regarding the small-scale fading components. On the basis of Equations 1.15 and 1.16, the spatially correlated small-scale fading component $\tilde{\mathbf{H}}$ is given by (Chuah et al., 2002)

$$\tilde{\mathbf{H}} = \tilde{\mathbf{R}}_{rx}^{1/2} \cdot \tilde{\mathbf{H}}_w \cdot \tilde{\mathbf{R}}_{tx}^{1/2} \tag{1.17}$$

To further model the covariance matrices $\tilde{\mathbf{R}}_{tx}$, $\tilde{\mathbf{R}}_{rx}$, the cases of polarization diversity (Scenario A) and satellite diversity (Scenario B) are treated differently.

1.4.4.1 Scenario A

In the case of polarization diversity, based on Equations 1.5, 1.6, and 1.17, $\tilde{\mathbf{R}}_{tx}$, $\tilde{\mathbf{R}}_{rx}$ are given by

$$\tilde{\mathbf{R}}_{rx} = E[\tilde{\mathbf{H}}\tilde{\mathbf{H}}^H] = \frac{1}{2}\begin{bmatrix} 1 & 2MP\sqrt{(1-\gamma)\gamma}\tilde{\rho}_{rx} \\ 2MP\sqrt{(1-\gamma)\gamma}\tilde{\rho}_{rx} & 1 \end{bmatrix} \qquad (1.18)$$

$$\tilde{\mathbf{R}}_{tx} = E[\tilde{\mathbf{H}}^H\tilde{\mathbf{H}}] = \frac{1}{2}\begin{bmatrix} 1 & 2MP\sqrt{(1-\gamma)\gamma}\tilde{\rho}_{tx} \\ 2MP\sqrt{(1-\gamma)\gamma}\tilde{\rho}_{tx} & 1 \end{bmatrix} \qquad (1.19)$$

where the superscript H denotes matrix-conjugate transposition and $\tilde{\rho}_{rx}$, $\tilde{\rho}_{tx}$ are the spatial correlation coefficients introduced in the small-scale fading components at the receive and transmit sides, respectively. Typical values for the parameters γ, $\tilde{\rho}_{rx}$, and $\tilde{\rho}_{tx}$ can be found in the relevant literature, for example, in Lempiäinen and Laiho-Steffens (1998), Sellathurai et al. (2006), King (2007), and King and Stavrou (2007), according to the corresponding scenario assumed.

1.4.4.2 Scenario B

By simply interchanging ε and γ in Equations 1.18 and 1.19, the relevant covariance matrices $\tilde{\mathbf{R}}_{tx}$, $\tilde{\mathbf{R}}_{rx}$ in Equation 1.17 can be derived for the case of satellite diversity, as well. The relevant small-scale fading spatial correlation coefficients $\tilde{\rho}_{rx}$, $\tilde{\rho}_{tx}$ depend on the angular spread $\delta\theta$ of the multipath components, which in turn depends on the user environment, on the elevation angle, as well as on the antenna element separation distance Δ at the receive and transmit sides, respectively.

In the context of MIMO LMS channels employing satellite diversity, available results for the correlation coefficients $\tilde{\rho}_{rx}$, $\tilde{\rho}_{tx}$ are reported in King et al. (2005) and King (2007) that, however, refer to relatively low separation distances between satellites where the assumption of the rays arriving approximately parallel at the mobile UT holds. Nevertheless, small-scale fading correlation coefficients have been widely studied in the conventional MIMO terrestrial wireless communications literature (Mietzner et al., 2009). In the MIMO LMS case of interest though, the distribution of the multipath power along the elevation angle is also relevant (Shafi et al., 2006). Thus, in the absence of currently available valid experimental results for the small-scale fading correlation coefficients in the case of Scenario B, educated intuitive estimation on the values of $\tilde{\rho}_{rx}$, $\tilde{\rho}_{tx}$ is employed (see Table 1.2). To this end, in the case where the satellite separation distance is sufficiently large (e.g., 30° [Heuberger, 2008; DiPierro et al., 2010]), $\tilde{\rho}_{tx}$ will get relatively low values for all relevant user environments assumed because of the absence of scatterers in the vicinity of each satellite. Regarding $\tilde{\rho}_{rx}$, in urban environments, it is expected to be higher with respect to suburban and open rural environments since the latter have usually more widely spaced clusters of scatterers, that is, $\delta\theta$ is lower in urban user environments with respect to suburban and open rural environments.

1.4.5 Modeling of Temporal Correlation

To model the temporal correlation induced as the UT moves at a certain speed v in the LMS environment, the following two different scales of channel temporal variations are considered: *interstate* and the *intrastate* channel temporal variations.

1.4.5.1 Interstate Temporal Variations

In the SISO LMS case, the temporal evolution of the large-scale fading components is characterized by a stochastic process $c(t)$ given by

$$c(t) = \begin{cases} 0 & \text{BAD} \\ 1 & \text{GOOD} \end{cases} \qquad (1.20)$$

TABLE 1.2

Parameters Assumed for MIMO LMS Channel Simulations in Scenario B (Satellite Diversity)

Parameter	Open Rural Environment	Suburban Environment	Urban Environment	Reference
Operating frequency, f	2.2 GHz (S-band)	2.2 GHz (S-band)	2.2 GHz (S-band)	DVB-SH (2008)
Satellite orbit	GEO	GEO	GEO	DVB-SH IG (2008)
Polarization	RHCP	RHCP	RHCP	DVB-SH IG (2008)
Mobile UT speed, v	50 km/h	50 km/h	50 km/h	DVB-SH IG (2008)
Satellite elevation angle, θ	40°	40°	40°	DVB-SH IG (2008)
Satellite separation angle	30°	30°	30°	Heuberger (2008), DiPierro et al. (2010)
Cross-channel attenuation parameter, ε	0.5	0.5	0.5	Hemispherical antenna pattern
Loo statistical parameter triplet (α, ψ, MP)	Each time a new state is reached, (α, ψ, MP) are drawn from the corresponding joint distribution	Each time a new state is reached, (α, ψ, MP) are drawn from the corresponding joint distribution	Each time a new state is reached, (α, ψ, MP) are drawn from the corresponding joint distribution	Prieto-Cerdeira et al. (2010)
Spatial correlation coefficient of small-scale fading components at Tx, $\tilde{\rho}_{tx}$	0	0	0	
Spatial correlation coefficient of small-scale fading components at Rx, $\tilde{\rho}_{rx}$	0.1	0.2	0.3	
Spatial correlation coefficient of large-scale fading components, $\bar{\rho}_{LN}$	0.25	0.3	0.7	Heuberger (2008), Vazquez Castro et al. (2012)
Interstate temporal variations	First-order four-state Markov chain model with Loo distribution per state w/($\bar{\rho}_{LN} = 0.25$)	First-order four-state Markov chain model with Loo distribution per state w/($\bar{\rho}_{LN} = 0.3$)	First-order four-state Markov chain model with Loo distribution per state w/($\bar{\rho}_{LN} = 0.7$)	Lutz (1996), Prieto-Cerdeira et al. (2010)

which is modeled by a first-order two-state Markov chain model with a given absolute state probability matrix $\mathbf{W}_{SISO}$ and state transitions probability matrix $\mathbf{P}_{SISO}$ for the specific user environment considered (Prieto-Cerdeira et al., 2010).

1.4.5.1.1 Scenario A
In the 2×2 MIMO LMS case of polarization diversity, as described above, the polarization correlation coefficient gets relatively high values, which further indicates that all the SISO LMS sublinks between the satellite and the UT follow the *same* sequences of "GOOD" and "BAD" channel states. Thus, the resulting Markov chain model for the MIMO LMS case of interest is also a two-state one with absolute state and state transitions probability matrices being the same as in the SISO LMS case.

1.4.5.1.2 Scenario B
In the 2×2 MIMO LMS case of satellite diversity, to model the interstate channel temporal variations, a four-state Markov chain model is employed here such as that proposed in Lutz (1996). To generate this four-state correlated model, the only parameters needed are the single-link two-state transition probabilities, which are obtained from Prieto-Cerdeira et al. (2010) assuming a first-order Markov model; and the desired correlation coefficient $\bar{\rho}_{LN}$ (Lutz, 1996; Vazquez Castro et al., 2002; Heuberger, 2008). Note that to characterize this four-state Markov model, 12 independent transition probabilities would have to be determined, which would imply to make pairs of channel measurements for each kind of environment, for different pairs of elevation and azimuth angles. The approach taken into account here instead is to characterize a two-state Markov model for each SISO LMS link and then to combine them to obtain the four-state Markov model by means of the correlation coefficient $\bar{\rho}_{LN}$ (Lutz, 1996). Thus, given the two-state Markov model transition probabilities $\mathbf{P}_{SISO,i}(b_i, g_i)$ $(i = 1,2)$ characterizing each SISO LMS link, the four-state Markov model transition probability matrix becomes

$$\mathbf{P}_{MIMO} = \mathbf{P}_u + \mathbf{C}_{temp} \tag{1.21}$$

where

$$\mathbf{P}_u = \begin{bmatrix} (1-g_1)(1-g_2) & (1-g_1)g_2 & g_1(1-g_2) & g_1g_2 \\ (1-g_1)b_2 & (1-g_1)(1-b_2) & g_1b_2 & g_1(1-b_2) \\ b_1(1-g_2) & b_1g_2 & (1-b_1)(1-g_2) & (1-b_1)g_2 \\ b_1b_2 & b_1(1-b_2) & (1-b_1)b_2 & (1-b_1)(1-b_2) \end{bmatrix} \tag{1.22}$$

and $\mathbf{C}_{temp} \in \mathbb{R}^{4\times4}$ whose coefficients are built through $\bar{\rho}_{LN}$ as analytically described in Lutz (1996).

1.4.5.2 Intrastate Temporal Variations
Regarding the intrastate channel temporal variations, within each channel state, as the mobile UT moves at a certain speed v, the large-scale fading components undergo a temporal correlation. To model a specific coherence distance, there is a low-pass filtering process introduced in the large-scale fading components as detailed in Gudmundson (1991). Although originally developed for terrestrial mobile radio systems, the model in Gudmundson (1991) has also been successfully validated through experimental results for LMS systems (King, 2007). In this regard, for each state, a Gaussian random sequence of samples with zero mean and unit variance is first generated. One sample per transmission block of duration T is assumed. These Gaussian samples are then low-pass filtered to introduce the temporal correlation, using the following low-pass infinite impulse response (IIR) filter of one coefficient

$$y_n = x_n + A \cdot y_{n-1} \tag{1.23}$$

where $A = \exp(-vT/r_c)$, T is the sampling time being equal to the transmission block, and r_c is the coherence distance. The filtered samples are then scaled by $(1 - A^2)$ to restore the statistics of the samples before the filtering. Finally, based on Equations 1.11 and 1.12, the uncorrelated Gaussian samples generated (incorporated in matrix $\bar{\mathbf{H}}_w$) are transformed into spatially correlated lognormal samples (incorporated in matrix $\bar{\mathbf{H}}$) for a given covariance matrix $\bar{\mathbf{C}}$ and statistical parameters (α, ψ).

Regarding the small-scale fading components, to model their induced temporal correlation within each state, a simplified approach to decrease central processing unit (CPU) simulation time is followed. To this end, the generated multipath channel samples are assumed constant during a transmission block T (i.e., full temporal correlation within each block) but change independently from block to block. T is assumed equal to the channel coherence time, that is, the channel is sampled only once per block, which significantly reduces the CPU simulation time. The channel coherence time depends on the mobile UT speed, the operating frequency, and the specific user environment assumed. Typical values for the channel coherence time can be found in Gudmundson (1991) and King (2007), and refer to 2.5–10 ms depending on the specific S-band channel scenario under consideration. Note that the assumed block-fading model suffices to conduct statistical outage analyses (Ozarow et al., 1994) in delay-limited systems (Hanly and Tse, 1995), as is the case in DVB-SH systems with a relatively short forward error correction (FEC) code (DVB-SH, 2008) and where the LMS channel is slowly fading. In the more general case where this assumption of transmission block T does not hold, other more sophisticated approaches modeling the Doppler spectrum with a low-pass low-order Butterworth filter (Prieto-Cerdeira et al., 2010) can be followed.

1.5   UNIFYING MIMO LMS CHANNEL SIMULATOR STEP-BY-STEP METHODOLOGY

The following step-by-step procedure outlines a useful methodology for computer simulation and time-series generation of the derived MIMO LMS channels. The channel time-series generator can be easily integrated in a comprehensive end-to-end system simulator for in-depth performance assessment of advanced MIMO-enabled SDMB systems employing advanced space/frequency/ polarization time-coding techniques (Bhavani Shankar et al., 2012) (see Figure 1.4). Note that the proposed MIMO LMS channel simulator step-by-step methodology unifies both cases of polarization diversity (Scenario A) and satellite diversity (Scenario B).

Step 1: Definition of Markov chain and generation of random walk
- Obtain state frame length, absolute state probability matrix $\mathbf{W}_{SISO}$, and state transition probability matrix $\mathbf{P}_{SISO}$ for the two-state Markov chain model as described in Prieto-Cerdeira et al. (2010) for the specific user environment assumed.

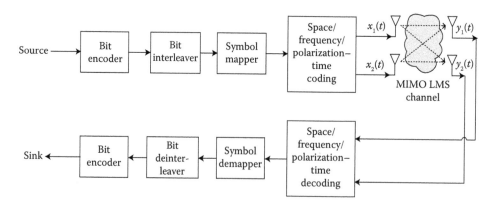

FIGURE 1.4   Block diagram for end-to-end simulation of MIMO satellite system.

- In the case of polarization diversity (Scenario A), assume a two-state Markov chain model similar to the one described in Prieto-Cerdeira et al. (2010) for the SISO LMS case. In the case of satellite diversity (Scenario B), follow the approach analytically described in Lutz (1996) and outlined above, based on which, a four-state Markov chain is created by means of the correlation coefficient $\bar{\rho}_{LN}$ defined in Equation 1.14 between the two respective SISO LMS links. This results in the four-state Markov model transition probability matrix $\mathbf{P}_{MIMO}$.
- Let $P(i,j)$ denote the transition probability from state i to state j, S_t the current state, S_{t+1} the next state, and N the total number of possible states, that is, $N = 2$ (Scenario A—polarization diversity) and $N = 4$ (Scenario B—satellite diversity). Given the current state S_t, the next state S_{t+1} can be generated as follows: Generate $U \sim Uniform(0,1)$; set $i = 1$ and test the condition $U \leq \sum_{i=1}^{i} P(S_t, i)$; if true, then the new state is $S_{t+1} = i$; otherwise, test the condition for $i = i + 1$.

Step 2: Calculation of Loo statistical parameter set (α, ψ, MP)
- Each time a new channel state is reached, a Loo statistical parameter triplet (α, ψ, MP) is drawn from the corresponding joint distribution and for the specific user environment assumed, as described in detail in Prieto-Cerdeira et al. (2010).

Step 3: Generation of large-scale fading components
- For each state and for each polarization or spatial dimension, a Gaussian random sequence of samples with zero mean and unit variance is generated. One sample per transmission block of duration T is assumed.
- These samples are then low-pass filtered to introduce the temporal correlation, using the one-coefficient low-pass IIR filter of Equation 1.23.
- The filtered samples are then scaled by $(1 - A^2)$ to restore the statistics of the samples before the filtering.
- The spatial correlation among the Gaussian-generated samples of each dimension is introduced using the covariance matrix $\bar{\mathbf{C}}$ and the related operation in Equation 1.11, which results in a Gaussian spatially correlated 2×2 matrix with zero mean and unit variance.
- According to the lognormal statistical parameters (α, ψ) drawn in Step 2, these Gaussian spatially correlated samples generated are then exponentiated accordingly as shown in Equation 1.12 to generate the spatially correlated lognormally distributed 2×2 channel matrix $\bar{\mathbf{H}}$.
- The cross-channel discrimination (cross-polarization discrimination in Scenario A or cross-channel attenuation in Scenario B) effects with respect to the cochannel component are then introduced for the specific user environment assumed through the related operations in Equation 1.5, which results in the channel matrix $\bar{\mathbf{H}}$ of Equation 1.1.

Step 4: Generation of small-scale fading components
- Generate a 2×2 channel matrix $\tilde{\mathbf{H}}_w$ assuming independent identically distributed zero-mean circularly symmetric complex Gaussian elements with variance MP.
- Within each channel state, assume that the generated multipath channel samples are constant during a transmission block T but change independently from block to block. T is assumed equal to the channel coherence time, that is, the channel is sampled only once per block.
- The spatial correlation among elements of $\tilde{\mathbf{H}}_w$ is introduced using the multipath covariance matrices $\tilde{\mathbf{R}}_{tx}$, $\tilde{\mathbf{R}}_{rx}$ and the operations in Equations 1.15 and 1.16, which results in the spatially correlated small-scale fading matrix $\tilde{\mathbf{H}}$.
- The cross-channel discrimination (cross-polarization discrimination in Scenario A or cross-channel attenuation in Scenario B) effects with respect to the cochannel component are then introduced for the specific user environment assumed through the related operations in Equation 1.6, which results in the channel matrix $\tilde{\mathbf{H}}$ of Equation 1.1.

Step 5: Combination of large- and small-scale fading components
- Combine the generated large- and small-scale fading components by simply adding the signals in each MIMO channel at each time sample as indicated in Equation 1.1, that is, $\mathbf{H} = \overline{\mathbf{H}} + \tilde{\mathbf{H}}$. Note that the produced time series are normalized with respect to the LOS power level since the Loo statistical parameters (α, ψ, MP) in Prieto-Cerdeira et al. (2010) are normalized, as well.

1.6 MIMO SATELLITE SYSTEM PERFORMANCE EVALUATION

Next, useful numerical results on the MIMO satellite systems performance are provided based on the proposed MIMO LMS channel models for both cases of polarization diversity (Scenario A) and satellite diversity (Scenario B). To this end, specific MIMO LMS channel scenarios are assumed for each case whose proposed parameters pertinent to the analysis are given in Tables 1.1 and 1.2, respectively. Moreover, the outage capacity $C_{out,q}$ (i.e., the information rate guaranteed for $(1 - q)100\%$ of the channel realizations (Ozarow et al., 1994)) of an MIMO LMS channel is evaluated through Monte Carlo simulations over 10,000 channel realizations using the proposed MIMO LMS channel simulator. Note that since broadcasting systems are of interest, the channel $\mathbf{H}$ is considered perfectly known only to the UT receiver (via training and tracking) while the transmitter is assumed to have no channel knowledge. In addition, the performance of specific MIMO transmission techniques over MIMO LMS channels is evaluated through detailed computer simulations based on the proposed channel model for Scenario A.

1.6.1 Scenario A

First, the time series generated for the 2×2 MIMO LMS channel in the case of polarization diversity assuming operation in an open rural, suburban, and urban environment is illustrated in Figure 1.5. Input parameters assumed are the ones reported in Table 1.1. A sequence of "GOOD" and "BAD" channel states along the UT-traveled distance can be seen. Within each user environment, note that the level of all "GOOD" states is not the same. This also applies for all "BAD" states and it is because every time a new channel state is reached along the UT-traveled distance, a new *Loo* statistical parameter triplet (α, ψ, MP) is drawn from the corresponding joint distribution. The increase in the amount of multipath fading when moving from an open rural to an urban environment can also be observed. Moreover, it can be seen that the channel XPD between the copolar h_{ii} ($i = j = 1,2$) and the cross-polar h_{ij} ($i \neq j = 1,2$) components is not constant along the UT-traveled distance, which is mainly due to the XPC_{env}. The temporal and polarization correlation between each of the components of the channel samples is also effectively incorporated.

In Figure 1.6, the 1% outage capacity of a 2×2 MIMO LMS channel assuming operation in an open rural environment is plotted versus SNR to investigate the effect of the terminal antenna XPD. As XPD_{ant} increases, the cross-polar interferences become weaker, the MIMO channel becomes diagonal, and the outage capacity achieved increases.

In Figure 1.7, the 1% outage capacity of a 2×2 MIMO LMS channel assuming operation in a suburban environment is plotted versus SNR to investigate the effect of the satellite elevation angle. As θ decreases, the signal is blocked with higher probability; the "BAD" state becomes more dominant and thus the outage capacity achieved by the respective MIMO case decreases.

In Figure 1.8, the 1% outage capacity of a 2×2 MIMO LMS channel assuming operation in an urban environment is plotted versus SNR to investigate the effect of the polarization correlation between the small-scale fading components. As $\tilde{\rho}_{rx}$ (or $\tilde{\rho}_{tx}$) decreases, the outage capacity achieved by each considered MIMO case increases. Comparing Figures 1.6 and 1.7 to Figure 1.8, the latter indicates that the outage capacity achieved in an urban environment is much lower than that in open and suburban environments, as expected. However, in all cases simulated above, including the case of the harsh urban environment where the satellite signal is usually blocked, there is a significant

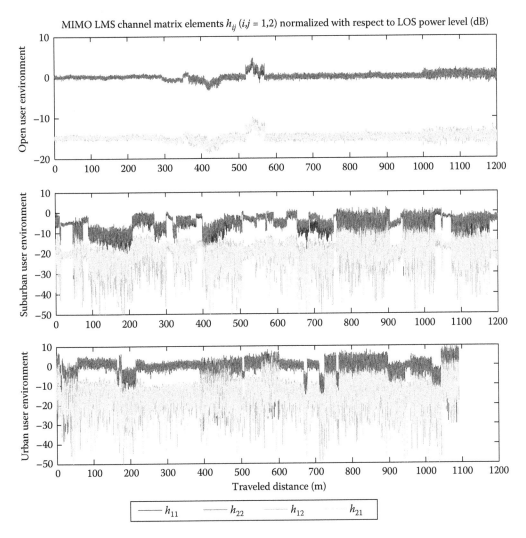

FIGURE 1.5 Time series generated for a 2 × 2 MIMO LMS channel (Scenario A—polarization diversity): open rural, suburban, and urban environment.

capacity gain evident between the respective MIMO LMS and SISO LMS performance curves. Summarizing the results, in harsher multipath environments, the absolute MIMO performance (capacity) deteriorates, but the relative gain with respect to SISO performance increases.

Beyond outage capacity results, the MIMO LMS channel model described in this chapter can be used to derive system performance evaluation results via computer simulations. Indeed, as briefly reviewed next, the model in Liolis et al. (2010) along with some datasets from the MIMOSA campaign have been used in past publications to assess the gain offered by MIMO in a purely satellite setup corresponding to Scenario A. According to MIMO theory, an MIMO technique is designed to offer either multiplexing gain or diversity gain. To increase the multiplexing gain, a spatial multiplexing (SM) technique is usually adopted or another more sophisticated space–frequency–time coding, such as Alamouti or the Golden code (Bhavani Shankar et al., 2012). These schemes were tested in several independent simulation platforms, all based on the DVB-SH waveform (Cioni et al., 2010; Arapoglou et al., 2011c, 2012; Kyröläinen et al., 2014). Specifically, Cioni et al. (2010) and Arapoglou et al. (2011c) present simulation results in a DVB-SH framework of various MIMO schemes for OFDM and time division multiplexing (TDM), respectively, while

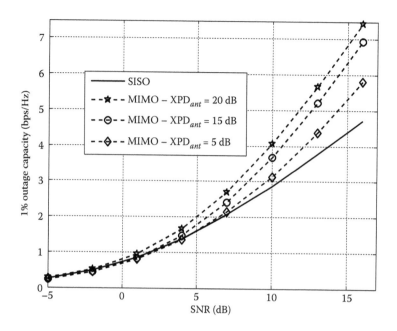

FIGURE 1.6 (Scenario A—polarization diversity) 1% outage capacity of a 2 × 2 MIMO LMS channel versus SNR in an open rural environment. The effect of antenna cross-polarization discrimination, XPD$_{ant}$.

Arapoglou et al. (2012) investigates the impact of interleaver length, cross-polarization isolation, imperfect channel estimation, and nonlinear amplification on MIMO performance. It seems that key properties of the dual-polarization LMS channel as analyzed throughout the chapter favor the simpler SM-like transmission schemes. To complement this statement, Table 1.3 provides an extract of the simulation results obtained in Kyröläinen et al. (2014), where it is observed that

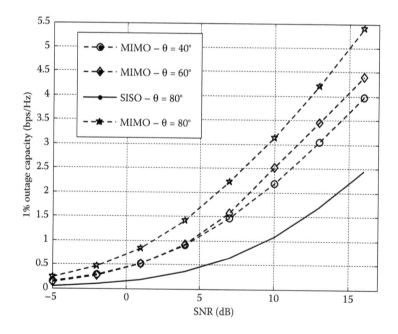

FIGURE 1.7 (Scenario A—polarization diversity) 1% outage capacity of a 2 × 2 MIMO LMS channel versus SNR in a suburban environment. The effect of satellite elevation angle, θ.

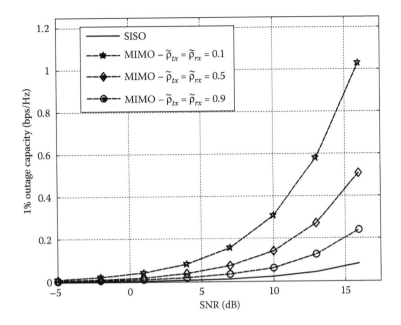

FIGURE 1.8 (Scenario A—polarization diversity) 1% outage capacity of a 2×2 MIMO LMS channel versus SNR in an urban environment. The effect of polarization correlation of small-scale fading components, $\tilde{\rho}_{rx}$, $\tilde{\rho}_{tx}$. The special case of $\tilde{\rho}_{rx} = \tilde{\rho}_{tx}$ is assumed and only the "BAD" state is considered, as more relevant in urban environments.

for typical system and channel parameters, SM can result in gains up to 3 dB over SISO at equal spectral efficiency levels. The results in Table 1.3 have been collected for an intermediate tree-shadowed (ITS) parameterization of the MIMO LMS channel in Liolis et al. (2010) corresponding to a UT speed of 60 km/h and for the longest possible depth of the interleaver (10 s). They refer to the same error second ratio ESR5(20) level, which is fulfilled in a time interval of 20 s if there is at most 1 s in error.

1.6.2 Scenario B

In Figure 1.9, the time series generated for the 2×2 MIMO LMS channel in the case of satellite diversity assuming operation in an open rural, suburban, and urban environment is provided. Input parameters assumed are the ones reported in Table 1.2. The sequence of ("GOOD," "GOOD"), ("GOOD," "BAD"), ("BAD," "GOOD"), and ("BAD," "BAD") channel states along the UT- traveled

TABLE 1.3
Gain of MIMO over SISO and $2 \times$ SISO in a Single-Satellite Dual-Polarization DVB-SH Configuration[a]

DVB-SH-A MIMO Scheme Gain over SISO 16QAM 1/3				
SM QPSK 1/3	Golden QPSK 1/3	Alamouti SFBC 16QAM 1/3	Alamouti STBC 16QAM 1/3	$2 \times$ SISO QPSK 1/3
2.9 dB	2.6 dB	2.4 dB	1.7 dB	0.5 dB

[a] The $2 \times$ SISO scheme in Table 1.3 corresponds to an independent encoding and decoding of the two polarizations (Arapoglou et al., 2011b). SFBC and STBC correspond to the space–frequency and space–time variants of Alamouti.

MIMO LMS channel matrix elements h_{ij} ($i,j = 1,2$) normalized with respect to LOS power level (dB)

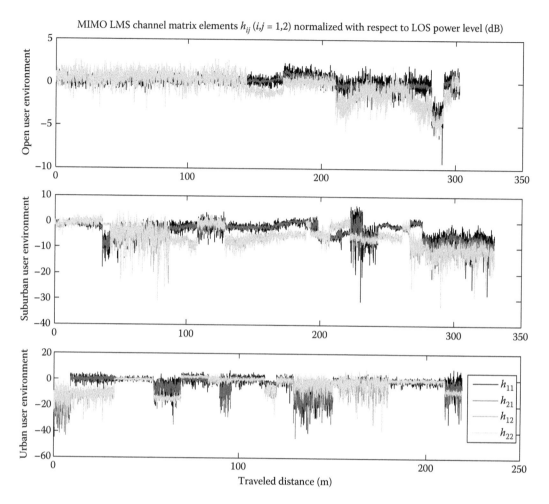

FIGURE 1.9 Time series generated for a 2×2 MIMO LMS channel (Scenario B—satellite diversity): open rural, suburban, and urban environment.

distance can be observed. Moreover, similar observations to those made for Figure 1.5 apply in this case, as well. However, note that since both satellites transmit on the same polarization and the receiver antennas are assumed hemispherical, there is no significant cross-channel attenuation between the diagonal and cross-diagonal elements of the channel matrix **H**.

In Figure 1.10, the 1% outage capacity of a 2×2 MIMO LMS channel assuming satellite diversity and operation in an open rural environment is plotted versus SNR. In particular, the effect of cross-satellite discrimination due to antenna radiation pattern, XSD_{ant} is investigated. Since the parameter XSD_{ant} in Scenario B (satellite diversity) plays a similar role to that of XPD_{ant} in Scenario A (polarization diversity), the results observed in Figure 1.10 regarding the effect of XSD_{ant} on the outage capacity achieved are similar to those in Figure 1.6.

In Figure 1.11, the 1% outage capacity of a 2×2 MIMO LMS channel assuming satellite diversity and operation in a suburban environment is plotted versus SNR. In particular, the effect of the shadowing correlation coefficient is investigated. As can be seen, as $\bar{\rho}_{LN}$ increases, the outage capacity achieved decreases.

In Figure 1.12, the 1% outage capacity of a 2×2 MIMO LMS channel assuming satellite diversity and operation in an urban environment is plotted versus SNR. In particular, the effect of the spatial correlation coefficient $\tilde{\rho}_{rx}$ of small-scale fading components at the receiver side is investigated

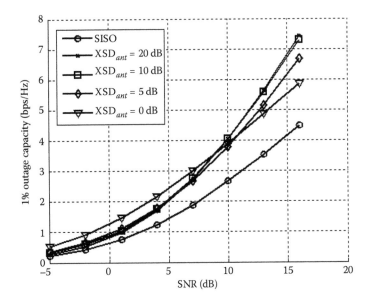

FIGURE 1.10 (Scenario B—satellite diversity) 1% outage capacity of a 2×2 MIMO LMS channel in an open rural environment. The effect of cross-satellite discrimination due to the antenna radiation pattern, XSD_{ant} (or else cross-channel attenuation parameter, ε).

whereas $\tilde{\rho}_{tx} = 0$ is assumed. Given that only the multipath components are affected by these spatial correlation parameters, both satellite links are forced to be in the relevant "BAD" state. Of notable interest is the small difference between the two MIMO LMS curves for $\tilde{\rho}_{rx} = 0.5$ and $\tilde{\rho}_{rx} = 0.9$. The results observed in Figure 1.12 are similar to those observed in the relevant Figure 1.8 in the case of Scenario A (polarization diversity).

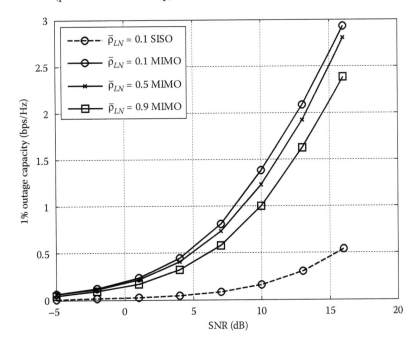

FIGURE 1.11 (Scenario B—satellite diversity) 1% outage capacity of a 2×2 MIMO LMS channel in a suburban environment. The effect of a spatial correlation coefficient of large-scale fading components, $\bar{\rho}_{LN}$.

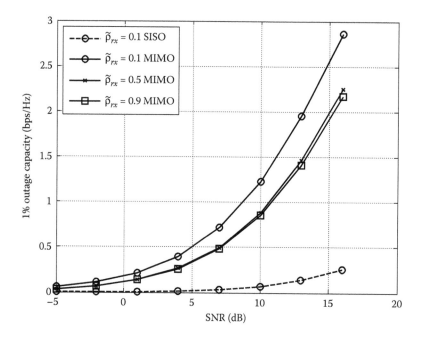

FIGURE 1.12 (Scenario B—satellite diversity) 1% outage capacity of a 2×2 MIMO LMS channel in an urban environment. The effect of a spatial correlation coefficient of small-scale fading components at the receiver, $\tilde{\rho}_{rx}$.

In Figure 1.13, results from a feasibility analysis on the use of OFDM as a potential countermeasure against the inherent issue of the relative delay in satellite diversity are provided. As discussed above, a critical aspect of Scenario B and the blocking point for applying dual-satellite MIMO in SDMB systems is the relative delay in the arrival of the signals originating from the two satellites, which is expected to be very large for any meaningful (from a channel correlation point of view) range of angular separations. The result of this is that achieving synchronization and decoding the two streams becomes unfeasible in SFN mode. To assess the feasibility of OFDM use in satellite diversity MIMO systems, we consider both the cyclic prefix (CP) duration and the maximum relative propagation delay. The guard interval length depends on the subcarrier number. In the DVB-SH standard, for each mode, four values are possible for the CP duration: (i) 8 k mode: (CP Max/Min) 224/28 μs; (ii) 4 k mode: (CP Max/Min) 112/14 μs; (iii) 2 k mode: (CP Max/Min) 56/7 μs; and (iv) 1 k mode: (CP Max/Min) 28/3.5 μs. To this end, Figure 1.13 shows the worst-case relative delay seen by the UT. We assume a coverage area with a 2000-km diameter and assume that satellites are synchronized so that a UT in the center of the coverage area receives both signals with zero relative delay. Then, we evaluate the relative delay as seen by two UTs at the edges of the coverage area as the angular separation increases. One satellite is assumed to be located at 10° east while the second satellite moves eastward. A beam centered in Berlin is assumed. As can be seen, for angular separations less than approximately 35°, it is possible to use OFDM, while for larger angular separations, the relative delay exceeds the maximum guard interval. That is, for certain angular separations, band sharing between two satellites in OFDM mode is not feasible as the difference in propagation delay from two satellites exceeds the DVB-SH-compliant OFDM signal CP length.

1.7 OVERVIEW OF MIMO SATELLITE TECHNOLOGY HW DEMONSTRATION

Intensive ESA R&D efforts over the last decade devoted to developing the MIMO technology culminated in 2013 with the completion of the MIMO HW demonstrator developed by a project team led by Elektrobit as part of the ESA ARTES 5.1 program (ESA MIMOHW, 2013; Byman et al., 2015).

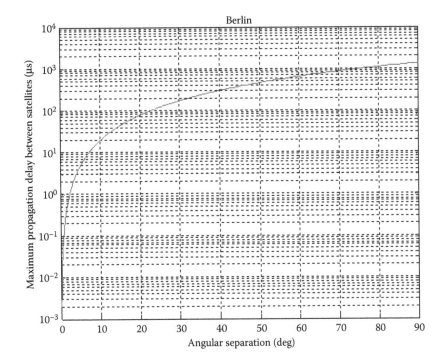

FIGURE 1.13 (Scenario B—satellite diversity) maximum relative delay between two UTs located at the edges of a coverage area. The beam centered in Berlin.

The activity resulted in a comprehensive laboratory HW demonstrator (test bed) for a hybrid satellite/terrestrial S-band mobile digital broadcasting system. The physical layer is based on an enhanced version of the DVB-SH standard exploiting dual-polarization MIMO technology. This complete digital MIMO HW demonstrator, the first of its kind, allows in-depth verification and optimization of the MIMO techniques applied to satellite-broadcasting networks, as well as complementing and confirming the theoretical or simulation-based findings published before the HW was developed.

This novel technology requires light modifications to the DVB-SH standard, particularly since the DVB-SH standard does not specify the pilot patterns for multistream transmission. Thus, the transmitter unit for the OFDM mode slightly diverges from the standard to allocate pilot sequences that allow MIMO channel estimation at the receiver. Overall, the changes to the DVB-SH standard are very minimal. This MIMO test bed was run to perform a comprehensive system test campaign to evaluate a number of performance aspects of a dual-polarization DVB-SH system (Byman et al., 2015) using the measured MIMOSA channel samples from Burkhardt et al. (2014) and the channel model in Liolis et al. (2010) elaborated in this chapter. The key findings out of a huge volume of laboratory measurements are summarized in Kyröläinen et al. (2014), and a sample of the results is reproduced here.

Figure 1.14 depicts the results for MIMO (SM), 2 × SISO, and SISO in a spectral efficiency versus LOS SNR plot, which was run for a mixture of propagation environments (open, suburban, and tree shadowed) referred to as MIX, an output of the MIMOSA measurement campaign (Burkhardt et al., 2014). The first conclusion is that MIMO always performs better than SISO and 2 × SISO, with the improvement becoming higher as the SNR increases. There are two ways of interpreting this improvement: either by fixing the spectral efficiency and quantifying the power saving from employing MIMO or by assuming an operational SNR and then evaluating the corresponding improvement in data rate. As an illustration, with SISO technology based on DVB-SH, a typical system configuration would be the one based on QPSK 1/2 (1 bit/symbol). For that level of spectral efficiency, MIMO yields at least 3 dB of valuable on-board power savings compared to SISO.

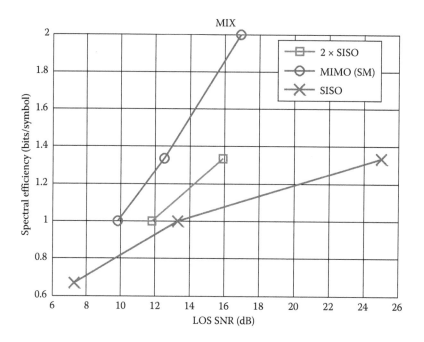

FIGURE 1.14   Satellite-only performance based on the MIX-measured channel.

Note that the same ESR5(20) will in general correspond to a different frame error rate (FER), a fact that explains the differences in slope between the SISO, 2 × SISO, and MIMO curves. In general, it is remarked that the MIMO gain in an LMS DPPB case for a given ESR5(20) performance target increases with the LOS SNR. Recalling the link budget results from Byman et al. (2015), one can conclude that the single-beam satellite configuration operating at around 20 dB of SNIR has the highest potential for DPPB MIMO.

1.8   ADDITIONAL MIMO SATELLITE AVENUES OF R&D

After a decade of R&D efforts, the mobile satellite-broadcasting MIMO technology is quite mature and has reached the level of a faithful laboratory demonstration based on real satellite measurements. Dual-polarization satellite MIMO and its terrestrial complementary component allows to increase the spectral efficiency of hybrid digital mobile-broadcasting networks with a limited increase of power compared to conventional SISO solutions. Alternatively, dual-polarization MIMO allows reducing the satellite equivalent isotropically radiated power (EIRP) for achieving the same throughput of current systems. These improvements are feasible with an affordable increase in receiver complexity.

In recent years, however, two additional areas of satellite R&D related to MIMO technology have emerged as potentially very promising, namely SU MIMO for interactive satellite services and MU MIMO in multibeam-fixed broadband-interactive systems. As these areas are strongly linked to satellite MIMO technology, it is worth summarizing the developments in each.

1.8.1   SU MIMO for Mobile Satellite Interactive Services

The exploitation of DPPB may also be considered for a mobile satellite interactive system (MSS). Nowadays, L-band multibeam MSS systems are typically based on high-power geostationary satellite platforms and a large single-antenna reflector (typically >9 m) with a feed array RF front end and a transparent digital onboard processor (OBP) generating several hundreds of user beams (see, e.g., Inmarsat's I4 constellation of GEO satellites). The current state-of-the-art L-band payload supports a single polarization/beam with a four- (or higher) color frequency reuse scheme. Doubling the number of polarizations/beams will require a major payload mass and power increase as the number of RF chains has to be doubled. However, a DPPB approach may be adopted only on specific "hot spots" over the coverage area, thus reducing the stress on the satellite platform resources.

The study of potential benefits from dual-polarization MIMO for interactive satellite systems is a topic still not well covered by the literature. Some indication of the potential dual-polarization MIMO gains achievable has been obtained in the context of the ESA ARTES 1 Contract (2014), where a MIMO bit-interleaved coded modulation (BICM) transmission system model is adopted. The selected MIMO approach is very similar to the one previously described for digital satellite broadcasting. The main difference of the air interface is that in interactive systems, the time interleaver size is limited by the latency constraints. The simulated air interface features QPSK modulation with Turbo FEC code rate ranging from 0.34 to 0.87 and an 80-ms time interleaver. For performing the simulations, the maritime parameterization of the dual-polarization MIMO channel model in Sellathurai et al. (2006) has been employed. In the maritime case, dual-polarization MIMO allows to double the throughput with approximately a 3-dB power increase instead of 4–6 dB required by a conventional single-polarization SISO system (e.g., through increasing the code rate). The MIMO gain is critically dependent on the system C/I due to cochannel intrasystem interference and the operating environment. Preliminary results indicate that the MIMO gain reduces for lower C/I values and for the ITS type of channels.

Moreover, note that satellite MIMO for propagation environments other than the vehicular LMS environment considered in this chapter (e.g., maritime, aeronautical, and railway environments) are under further investigation also awaiting for relevant commercial developments.

1.8.2 MU MIMO (Precoding) in Multibeam-Fixed Broadband-Interactive Systems

The content of the chapter as well as the references listed hitherto refer to SU MIMO applied over mobile satellite systems. In contrast, there is a growing interest from academia and standardization on MU MIMO applied over fixed-satellite systems (Arapoglou et al., 2011a). Because of the limited fixed-satellite service (FSS) spectrum available to broadband systems operating predominantly at Ka-band (20/30 GHz), increasing the frequency reuse amid the multiple-spot beams of high-throughput satellite (HTS) systems is the main avenue for substantially improving capacity and reducing the offered cost per bit. Nevertheless, increasing the frequency reuse leads to a high increase in intrasystem interference between the cochannel beams, which renders the use of an additional spectrum futile.

To address the issue of high interbeam interference in aggressive-frequency reuse multibeam configurations, joint processing of the signals intended to the different beams can be carried out at the transmitter (usually the gateway or hub) for interference management. This processing, referred to under the generic term *precoding* or *beamforming* and falling under the category of MU MIMO signal-processing techniques, "reverts" the impact of the satellite RF channel and interferences. This way, the additional spectrum can be exploited and a much higher system capacity can be delivered compared to existing systems. A precondition for an efficient forward-link precoding is that the UTs (receivers) provide high-quality reports of their channel conditions (amplitude and phase) back to the gateway (transmitter) that is responsible for deriving the appropriate precoding matrix.

The studies on precoding in a fixed-satellite system context, starting from the early works in Caire et al. (2005) up to the more recent ones (Christopoulos et al., 2012; Zheng et al., 2012) have mainly focused on evaluating various linear and nonlinear precoding techniques over the multibeam satellite channel. It turns out that simple linear techniques already grasp the largest part of the potential MU gains with manageable complexity and deliver improvements that at least double the throughput of existing systems (Christopoulos et al., 2012; Zheng et al., 2012).

Despite the extremely promising results reported by these studies, the actual application of pre-coding techniques in the forward link of a DVB-S2- (digital video broadcasting—satellite second generation) based broadband-interactive system has been hindered by a number of practical limitations, the most important one being the lack of appropriate signaling and framing support. Nevertheless, the DVB recently issued the DVB-S2 extension (DVB-S2X) with an optional specification that provides the necessary framing and signaling support to interference management techniques (DVB-S2X, 2014). Thereby, all elements to support precoding in terms of a physical layer have been put in place and the technique is awaiting a proof of concept.

1.9 SUMMARY AND CONCLUSIONS

This chapter addresses the potential of applying MIMO in next-generation SatCom systems, covering from propagation channel modeling to system performance evaluation and standardization aspects. It mainly focuses on L/S-band SDMB systems and considered two specific MIMO diversity configurations: (i) single-satellite/dual-polarization diversity scenario and (ii) dual-satellite diversity/single-polarization scenario. For both scenarios, a novel unifying statistical model of the underlying MIMO LMS-fading channels is presented, which unifies the inherently different cases of polarization diversity and satellite diversity, is fully parameterized, and can be easily fine-tuned upon the availability of experimental data.

Furthermore, outage capacity statistics of the assumed MIMO LMS channels are provided in both cases of polarization diversity and satellite diversity. On the basis of these simulation results, the effects of several critical channel-modeling parameters, such as the spatial and temporal correlation, shadowing, XPD, antenna radiation pattern, elevation angle, and user environment, are numerically quantified. In all simulated cases, the observed capacity gains between the respective MIMO and SISO LMS-fading channels are considerable. Moreover, the end-to-end system-level performance of MIMO DVB-SH dual-polarized satellite system has been assessed and useful numerical results are provided. In all simulated cases, DVB-SH standard has been adopted as the baseline satellite system configuration, which has been extended to introduce MIMO applications. Actually, it turns out that this extension from SISO DVB-SH to MIMO DVB-SH is feasible with minor modifications to the existing standard, namely a change in the pilot pattern for channel estimation. It is also shown that dual-polarized MIMO LMS channels favor simple MIMO SM transmission schemes where typical gains up to 3 dB over SISO at equal spectral efficiency levels are demonstrated.

Finally, different ongoing research avenues of MIMO applicability over next-generation satellite systems are highlighted and two recently emerged and potentially very promising areas of MIMO satellite R&D are overviewed, namely SU MIMO for interactive mobile satellite services and MU MIMO in multibeam-fixed broadband-interactive systems. Currently, L- and S-band satellite operators in Europe and elsewhere are considering the option of adding this exciting new technology in their next-generation constellations.

REFERENCES

Arapoglou, P.-D., M. Bertinelli, A. Bolea Alamanac, and R. De Gaudenzi, To MIMO or not to MIMO in mobile satellite broadcasting systems, *IEEE Transactions of Wireless Communication*, 10(9), 2807–2811, 2011c.

Arapoglou, P.-D., P. Burzigotti, A. Bolea Alamanac, and R. De Gaudenzi, Capacity potential of mobile satellite broadcasting systems employing dual polarization per beam, in *5th Advanced Satellite Multimedia System Conference 11th Signal Processing Space Communication Workshop*, ASMS/SPSC 2010, pp. 213–220, Cagliari, Italy, September 2010.

Arapoglou, P.-D., P. Burzigotti, A. Bolea Alamanac, and R. De Gaudenzi, Practical MIMO aspects in dual polarization per beam mobile satellite broadcasting, *International Journal of Satellite Communication System Network*, 30(2), 76–87, 2012.

Arapoglou, P.-D., K.P. Liolis, M. Bertinelli, A.D. Panagopoulos, P.G. Cottis, and R. De Gaudenzi, MIMO over satellite: A review, *IEEE Communications Surveys and Tutorials*, 13(1), 27–51, 2011a.

Arapoglou, P.-D., M. Zamkotsian, and P.G. Cottis, Dual polarization MIMO in LMS broadcasting systems: Possible benefits and challenges, *International Journal of Satellite Communication Network*, 29(4), 349–366, 2011b.

Bhavani Shankar, M.R., P.-D. Arapoglou, and B. Ottersten, Space–frequency coding for dual polarized hybrid mobile satellite systems, *IEEE Transactions on Wireless Communication*, 11(8), 2806–2814, 2012.

Bråten, L.E. and T. Tjelta, Semi-Markov multistate modelling of the land mobile propagation channel for geostationary satellites, *IEEE Transactions of Antennas Propagation*, 50(12), 1795–1802, 2002.

Brown, T.W.C., S.R. Saunders, S. Stavrou, and M. Fiacco, Characterization of polarization diversity at the mobile, *IEEE Transactions of Vehicular Technology*, 56(5), 2440–2447, 2007.

Burkhardt, F., E. Eberlein, S. Jaeckel, G. Sommerkorn, and R. Prieto-Cerdeira, MIMOSA—A dual approach to detailed LMS channel modeling, *International Journal of Satellite Communication Network*, 32(4), 309–328, 2014.

Byman, A., A. Hulkkonen, P.-D. Arapoglou, M. Bertinelli, and R. De Gaudenzi, MIMO for mobile satellite digital broadcasting: From theory to practice, *IEEE Transactions on Vehicular Technology*, PP(99), 1–1, 2015. doi: 10.1109/TVT.2015.2462757.

Caire, G., M. Debbah, L. Cottatellucci, R. De Gaudenzi, R. Rinaldo, R. Mueller, and G. Gallinaro, Perspectives of adopting interference mitigation techniques in the context of broadband multimedia satellite systems, in *23rd AIAA International Communication Satellite System Conference*, ICSSC 2005, Rome, Italy, September 1–5, 2005.

Carrie, G., F. Pérez-Fontán, F. Lacoste, and J. Lemorton, A generative MIMO channel model encompassing single satellite and satellite diversity cases, *Space Communications*, 22(2–4), 133–144, 2013.

Cheffena, M., F. Pérez-Fontán, F. Lacoste, E. Corbel, H. Mametsa, and G. Carrie, Land mobile satellite dual polarized MIMO channel along roadside trees: Modeling and performance evaluation, *IEEE Transactions of Antennas Propagation*, 60(2), 597–605, 2012.

Christopoulos, D., S. Chatzinotas, G. Zheng, J. Grotz, and B. Ottersten, Linear and non-linear techniques for multibeam joint processing in satellite communications, *EURASIP Journal of Wireless Communication and Network*, 1, 1–13, 2012.

Chuah, C.-N., D.N.C. Tse, J.M. Kahn, and R.A. Valenzuela, Capacity scaling in MIMO wireless systems under correlated fading, *IEEE Transactions of Information Theory*, 48(3), 637–650, 2002.

Cioni, S., A. Vanelli-Coralli, G.E. Corazza, P. Burzigotti, and P.-D. Arapoglou, Analysis and performance of MIMO–OFDM in mobile satellite broadcasting systems, in *IEEE Global Communication Conference*, GLOBECOM 2010, Miami, USA, December 2010.

DiPierro, S., R. Akturan, and R. Michalski, Sirius XM satellite radio system overview and services, in *5th Advanced Satellite Multimedia System Conference 11th Signal Processing Space Communication Workshop*, ASMS/SPSC 2010, pp. 506–511, Cagliari, Italy, September 2010.

DVB-NGH, ETSI EN 303 105 V1.1.1, Digital Video Broadcasting (DVB); Next generation broadcasting system to handheld, physical layer specification (DVB-NGH), 2013.

DVB-SH, ETSI EN 302 583 V1.1.1 2008–03, Digital Video Broadcasting (DVB); Framing structure, channel coding and modulation for satellite services to handheld devices (SH) below 3 GHz, 2008.

DVB-SH IG, ETSI TS 102 584 V1.1.1 2008–12, Digital Video Broadcasting (DVB); Guidelines for implementation for satellite services to handheld devices (SH) below 3 GHz, 2008.

DVB-S2X, ETSI EN 302 307 Part II: Digital Video Broadcasting (DVB); Second generation framing structure, channel coding and modulation systems for broadcasting, interactive services, news gathering and other broadband satellite applications; DVB-S2—Extensions (DVB-S2X), 2014.

Eberlein, E., F. Burkhardt, C. Wagner, A. Heuberger, D. Arndt, and R. Prieto-Cerdeira, Statistical evaluation of the MIMO gain for LMS channels, EUCAP 2011, *European Conference Antennas Radio Propagation*, pp. 2695–2699, Rome, Italy, April 11–15, 2011.

EC FP6 IST, WINNER Phase II model homepage: http://www.ist-winner.org/phase_2_model.html, 2007.

ESA ARTES 1 Contract No. 4000106528/12/NL/NR, Next generation waveforms for improved spectral efficiency, Prime contractor: DLR, Final Report, 2014.

ESA MIMOHW, ESA ARTES 5.1 Contract No. 4000100894 MIMO Hardware (MIMOHW) Demonstrator, Prime contractor: Elektrobit, Final Report, 2013.

ESA MIMOSA, ESA ARTES 5.1 Contract No. 4000100936 Characterisation of the MIMO Channel for Mobile Systems (MIMOSA), Prime contractor: Fraunhofer, Final Report, 2012.

ESA TAS-E, ESA ARTES 5.1, S-band high-power reconfigurable front-end demonstrator, TAS-E, Final Report, 2012.

Gallinaro, G., E. Tirrò, F. Di Cecca, M. Migliorelli, N. Gatti, and S. Cioni, Next generation interactive S-band mobile systems: Challenges and solutions, *International Journal of Satellite Communication System Network*, 32(4), 247–262, 2014.

Gudmundson, M., Correlation model for shadow fading in mobile radio systems, *IEE Electrical Letters*, 27(23), 2145–2146, 1991.

Hanly, S.V. and D.N. Tse, The multi-access fading channel: Shannon and delay limited capacities, in *Proceedings of 33rd Allerton Conference*, pp. 786–795, Monticello, IL, 1995.

Heuberger, A., Fade correlation and diversity effects in satellite broadcasting to mobile users in S-band, *International Journal of Satellite Communication Network*, 26(5), 359–379, 2008.

Horvath, P., G.K. Karagiannidis, P.R. King, S. Stavrou, and I. Frigyes, Investigations in satellite MIMO channel modeling: Accent on polarization, *EURASIP Journal of Wireless Communication Network*, 2007.

Jokela, T., P.-D. Arapoglou, C. Hollanti, B.M.R. Shankar, and V. Tapio, Hybrid satellite–terrestrial MIMO for mobile digital broadcasting, in *Next Generation Mobile Multimedia Broadcasting*, David Gómez-Barquero (ed.), CRC Press, 2013, pp. 713–748. Print ISBN: 978-1-4398-9866-6 eBook ISBN: 978-1-4398-9869-7. DOI: 10.1201/b14186-30.

Karagiannidis, G.K. et al., Diversity techniques and fade mitigation, in *Digital Satellite Communications*, G.E. Corazza (ed.), Springer, 2007, pp. 313–365, ISBN: 978-0-387-25634-4.

King, P.R., Modeling and measurement of the land mobile satellite MIMO radio propagation channel, PhD thesis, University of Surrey, UK, June 2007.

King, P.R., T.W.C. Brown, A. Kyrgiazos, and B.G. Evans, Empirical–stochastic LMS–MIMO channel model implementation and validation, *IEEE Transactions on Antennas and Propagation*, 60, 606–614, 2012.

King, P.R., and S. Stavrou, Low elevation wideband land mobile satellite MIMO channel characteristics, *IEEE Transactions of Wireless Communication*, 6(7), 2712–2720, 2007.

Kyröläinen, J., A. Hulkkonen, J. Ylitalo, A. Byman, B.M.R. Shankar, P.-D. Arapoglou, and J. Grotz, Applicability of MIMO to satellite communications, *International Journal of Satellite Communication Network*, 32(4), 343–357, 2014.

Lacoste, F. et al., Hybrid single frequency network propagation channel sounding and antenna diversity measurements, *International Journal of Satellite Communication Network*, 29(1), 7–21, 2011.

Lempiäinen, J.A. and J.K. Laiho-Steffens, The performance of polarization diversity schemes at a base station in small/micro cell at 1800 MHz, *IEEE Transactions of Vehicular Technology*, 47(3), 1087–1092, 1998.

Liolis, K.P., Statistical analysis, modelling and simulation of MIMO satellite communications channels, PhD thesis, National Technical University of Athens, Greece, December 2011 (in Greek).

Liolis, K.P., J. Gómez-Vilardebó, E. Casini, and A. Pérez-Neira, Statistical modeling of dual-polarized MIMO land mobile satellite channels, *IEEE Transactions on Communications*, 58(11), 3077–3083, 2010.

Liolis, K.P., A.D. Panagopoulos, and P.G. Cottis, Multi-satellite MIMO communications at Ku band and above: Investigations on spatial multiplexing for capacity improvement and selection diversity for interference mitigation, *EURASIP Journal Wireless Communications and Networking, Special Issue on Satellite Communications*, 2007, 1–12, 2007. doi:10.1155/2007/59608.

Loo, C., A statistical model for a land mobile satellite link, *IEEE Transactions of Vehicular Technology*, 34(3), 122–127, 1985.

Lutz, E., A Markov model for correlated land mobile satellite channels, *International Journal of Satellite Communication*, 14, 333–339, 1996.

Mansor, M.F.B., T.W.C. Brown, and B.G. Evans, Satellite MIMO measurements with co-located quadrifilar helix antennas at the receiver terminal, *IEEE Antennas Wireless Propagation Letters*, 9, 712–715, 2010.

Mietzner, J., R. Schober, L. Lampe, W.H. Gerstacker, and P.A. Hoeher, Multiple-antenna techniques for wireless communications—A comprehensive literature survey, *IEEE Communication Surveys Tutorials*, 11(2), 87–105, 2009.

Milojevic, M., M. Haardt, E. Eberlein, and A. Heuberger, Channel modeling for multiple satellite broadcasting systems, *IEEE Transactions Broadcasting*, 55(4), 705–718, 2009.

Oestges, C., V. Erceg, and A. Paulraj, Propagation modeling of MIMO multipolarized fixed wireless channels, *IEEE Transactions of Vehicular Technology*, 53(3), 644–654, 2004.

Ozarow, L.H., S. Shamai (Shitz), and A.D. Wyner, Information theoretic considerations for cellular mobile radio, *IEEE Transactions of Vehicular Technology*, 43, 359–378, 1994.

Pérez-Fontán, F., M.-A. Vazquez Castro, C. Enjamio Cabado, J. Pita García, and E. Kubista, Statistical modeling of the LMS channel, *IEEE Transactions of Vehicular Technology*, 50(6), 1549–1567, 2001.

Pérez-Neira, A., C. Ibars, J. Serra, A. del Coso, J. Gómez-Vilardebó, M. Caus, and K.P. Liolis, MIMO channel modeling and transmission techniques for multi-satellite and hybrid satellite–terrestrial mobile networks, *Elsevier's Physical Communication*, 4(2), 127–139, 2011.

Petropoulou, P., E.T. Michailidis, A.D. Panagopoulos, and A.G. Kanatas, Radio propagation channel measurements for multi-antenna satellite communication systems: A survey, *IEEE Antennas and Propagation Magazine*, 56(6), 102–122, 2014.

Prieto-Cerdeira, R., F. Pérez-Fontán, P. Burzigotti, A. Bolea Alamañac, and I. Sanchez Lago, Versatile two-state land mobile satellite channel model with first application to DVB-SH analysis, *International Journal of Satellite Communication Network*, 28(5–6), 291–315, 2010.

Scalise, S., C. Parraga Niebla, R. De Gaudenzi, O. Del Rio Herrero, D. Finocchiaro, and A. Arcidiacono, S-MIM: A novel radio interface for efficient messaging services over satellite, *IEEE Communication Magazine*, 51(3), 119–125, 2013.

Sellathurai, M., P. Guinand, and J. Lodge, Space–time coding in mobile satellite communications using dual-polarized channels, *IEEE Transactions of Vehicular Technology*, 55(1), 188–199, 2006.

Shafi, M., Z. Min, P.J. Smith, A.L. Moustakas, and A.F. Molisch, The impact of elevation angle on MIMO capacity, in *Proceedings of IEEE International Conference Communication*, ICC 2006, Istanbul, Turkey, June 2006.

Sørensen, T.B., A.Ø. Nielsen, P.E. Mogensen, M. Tolstrup, and K. Steffensen, Performance of two-branch polarization antenna diversity in an operational GSM network, in *Proceedings of IEEE VTC'98*, pp. 741–746, Ottawa, Canada, May 1998.

Vazquez Castro, M., F. Pérez-Fontán, and S.R. Saunders, Shadowing correlation assessment and modeling for satellite diversity in urban environments, *International Journal of Satellite Communication*, 20, 151–166, 2002.

Zheng, G., S. Chatzinotas, and B. Ottersten, Generic optimization of linear precoding in multibeam satellite systems, *IEEE Transactions of Wireless Communication*, 11(6), 2308–2320, 2012.

2 Propagation Phenomena and Modeling for Fixed Satellite Systems
Evaluation of Fade Mitigation Techniques

Athanasios D. Panagopoulos

CONTENTS

2.1 INTRODUCTION

In modern satellite communication systems, the reliable design is constrained by the radio propagation effects, the interference, and the noise that are inherently present in all radio systems (Panagopoulos, 2014). New and rate-demanding satellite applications have been evolved and have led to the employment of higher-frequency bands. The current satellite communications use geosynchronous Earth orbit (GEO) satellites and fixed-satellite services usually employ (Panagopoulos et al., 2004a) Ku- (12/14 GHz) and Ka- (20/30 GHz) bands. These frequency bands are used for direct-to-user (DTU) applications, for broadcasting satellite applications, for feeder links, and satellite backhaul networks. Next-generation low Earth orbit (LEO) satellite systems will also operate at higher frequencies, that is, Ka-band due to a high available bandwidth and the congestion of conventional frequency bands, such as X-band and Ku-band (Rosello et al., 2012). These frequency bands for nongeostationary Earth orbit (NGEO) satellites may be used either for Earth-observation links or telecommunication links (Rosello et al., 2012; Toptsidis et al., 2012). However, there are numerous activities mostly from the European Space Agency (ESA) for the exploitation of Q/V (33–50 GHz) and W (75–110 GHz) bands. Q/V band is proposed either for feeder links or for DTU applications, while W band is studied for feeder links and backhauling applications.

For the installation of the satellite communication systems, the impact of the phenomena must be accurately quantified through experimental data, analytical theoretical models, and empirical or semiempirical models (Crane, 2003). In this chapter, we focus on the nonionized atmosphere. These effects take into account (ITU-R, P. 618-11, 2013) the loss of the signal (pointing error) due to beam divergence of the ground terminal antenna, the absorption, the scattering, and the depolarization by gases (water and ice droplets in precipitation and clouds), in a reduction of the effective antenna gain, due to phase decorrelation across the antenna aperture, caused by irregularities in the refractive-index structure and fast fluctuation of the signal due to turbulence (wet and dry scintillation).

Consequently, the loss due to propagation phenomena induced on an Earth–space slant path, relatively qualified to the free-space loss, is the sum of various influences such as attenuation due to atmospheric gases; attenuation due to rain and clouds and other precipitation (e.g., snow); focusing and defocusing of the beam; reduction to the antenna gain due to wave-front incoherence; scintillation and multipath effects; and attenuation due to sand and dust storms (Crane, 2003).

The fade margin namely the system gain that ensures the specified quality of service (QoS) of the satellite link must be much greater to compensate the transmission and propagation impairments for satellite communications operating at frequencies much more than 10 GHz. The larger fade margins are not feasible of technical and economic reasons. The cost of the ground station would increase dramatically. The fade margins should satisfy at least the availability specifications proposed and predicted in ITU-R, P. 618-11 (2013) model for the estimation of the total attenuation. The total attenuation (the effect of multiple sources of simultaneously occurring atmospheric attenuation) should be considered for the evaluation of the fade margins for satellite links operating at frequencies above about 18 GHz, and especially if they operate with low-elevation angles.

Consequently, to design modern satellite communication networks effectively and efficiently and with a view to operating under a low fade margin and satisfy the strict QoS requirements, appropriate fade mitigation techniques (FMTs) have been proposed (COST 255, 2002; Crane, 2003; Panagopoulos et al., 2004a; Castanet et al., 2007).

Summing up in this chapter, we present the propagation phenomena through an ionized atmosphere (very briefly) that play a significant role for frequencies below 3 GHz; we present the tropospheric effects that are the most important ones for the spectrum above 10 GHz. In addition, the prediction models that are well accepted in the satellite communication society are presented. The proposed FMTs are categorized with respect to their applications and studied with emphasis on modeling, prediction methods, and experimental work done in past and future trends. Finally, the outcomes of recent ESA cofunded research and development (R&D) activities, focusing on channel measurements using ALPHASAT (Koudelka, 2011; Codispoti et al., 2012) are also overviewed.

For completeness of this chapter, other open or ongoing research issues of fixed-satellite systems (FSS) are briefly reviewed in the last section. The chapter concludes with general comments and directions.

2.2 PROPAGATION PHENOMENA AND MODELING

2.2.1 Frequencies of Operation below 3 GHz (Ionospheric Effects)

When the signal propagates through the ionosphere, the following effects may occur to a slant path (ITU-R, P. 618-11, 2013): rotation of the polarization (Faraday rotation) due to the interaction of the transmitted electromagnetic wave with the ionized medium in the Earth's magnetic field along the path; group delay and phase advance of the signal due to the total electron content (TEC) accumulated along the path; fast fluctuation of amplitude and phase (scintillations) of the signal due to small-scale irregular structures in the ionosphere; a change in the expected direction of arrival due to refraction; and Doppler effects due to nonlinear polarization rotations and time delays. For the satellite system's design, the impact of the ionospheric propagation can be summarized (ITU-R, P. 618-11, 2013): the TEC accumulated along the slant path through the ionosphere causes rotation of the polarization (Faraday rotation), time delay of the transmitted signal, and a change in the apparent direction of arrival due to refraction; ionospheric irregularities that further cause excess and random rotations and time delays; and dispersion or group velocity distortion of the satellite system carriers. The frequency dependence of the most important ionospheric effects can be found in Table 2.1.

2.2.2 Frequencies of Operation above 10 GHz (Tropospheric Effects)

In this section, the most significant tropospheric phenomena affecting the FSS operating above 10 GHz (see Figure 2.1) are presented briefly along with the well-accepted models in the related literature. The most important tropospheric propagation effects with a small explanation of their physical cause are tabulated in Table 2.2.

2.2.2.1 Atmospheric Gaseous Attenuation

The signal attenuation due to atmospheric gases is caused by absorption. It depends on the frequency, elevation angle, altitude above sea level, and water vapor density (absolute humidity). The impact of atmospheric gaseous attenuation increases for frequencies above 10 GHz, especially for low-elevation angles. In the Recommendation ITU R P.676-10 (ITU R P.676-10, 2013), a full methodology for the

---

TABLE 2.1
Frequency Dependence of the Most Important Ionospheric Effects

Ionospheric Effect	Frequency Dependence
Faraday rotation	$1/f^2$
Propagation delay	$1/f^2$
Refraction	$1/f^2$
Variation in the direction of arrival (rms)	$1/f^2$
Absorption	$\approx 1/f^2$
Dispersion	$1/f^3$

Source: Adapted from ITU-R. P. 618-11, *International Telecommunication Union*, Geneva, 2013.

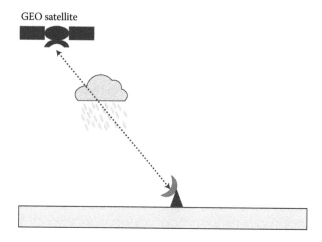

FIGURE 2.1 Fixed-satellite slant path affected by tropospheric propagation impairments.

calculation of gaseous attenuation can be found based on millimeter-wave propagation model (MPM). In the literature, the MPM was proposed by Liebe (1989). In ITU R P.676-10 (2013), an approximate method for frequencies up to 350 GHz can also be found. Vertical profiles of pressure, temperature, and water vapor density, are inputs in the model and can be obtained from annual reference models of ITU-R. P. 835 (ITU-R. P. 835-5, 2012). In a simpler model, the approximation algorithms as well as surface meteorological data and maps of integrated water vapor content for various time percentages derived by ITU-R. P. 836-4 (2009) for the calculation of the gaseous absorption are used.

2.2.2.2 Atmospheric Attenuation Due to Hydrometeor Precipitation

When the satellite signal (either uplink or downlink) is propagating through rain, snow, hail, or ice droplets, it suffers from power loss due to hydrometeor scattering (see Figure 2.1 and Table 2.2). The most dominant fading mechanism is rain attenuation due to rainfall rate that mostly affects the heavy-rain climatic regions. For frequencies up to 30 GHz, the hydrometeor absorption is the dominant phenomenon. For frequencies above 30 GHz, the hydrometeor scattering is a greater limiting factor. The power loss of the signal in dB is analogous to the square of the frequency, which is due to the combined effect of hydrometeor scattering and absorption. Rain attenuation mainly depends on the rainfall rate and the raindrop size distribution and is greater for lower-elevation angles.

For the calculation of the specific rain attenuation (A_o, dB/km), the power law model (Olsen et al., 1978) is the most accepted one and provides a very simple expression in terms of rainfall rate (R, mm/h)

$$A_o(\text{dB/km}) = aR^b \tag{2.1}$$

TABLE 2.2

Tropospheric Propagation Impairments Affecting Satellite Communication Systems

Propagation Impairment	Physical Origin
Radiowave signal attenuation, sky noise increase	Absorption due to gases and scatter due to hydrometeors
Signal depolarization	Phase shift and differential attenuation due to raindrop shape and ice crystals
Tropospheric signal scintillation	Refractive-index variations in the troposphere
Refraction and atmospheric multipath	Tropospheric density variations
Propagation delays and delay variations	Free-space propagation and time variations due to the troposphere
Intersystem interference	Spatial inhomogeneity of the propagation medium—troposcatter

where a and b are two parameters of specific attenuation that depend on the raindrop size distribution, the elevation angle, frequency, and polarization (ITU-R. P. 838-3, 2005). The rain attenuation along the radiopath can be calculated as

$$A(dB) = \int_0^L A_o(x)dx \qquad (2.2)$$

where L is the slant path that is affected by rain and is calculated using the $0°$ isotherm level. The most accepted raindrop size distributions are the Marshall–Palmer model (Marshall and Palmer, 1948) for temperate climatic regions and the Ajayi–Olsen model (Ajayi and Olsen, 1985) more appropriate for subtropical and climatic regions. In Kanellopoulos et al. (2005), the MAS (method of auxiliary sources) is employed for the calculation of the forward-scattering amplitude from a Pruppacher–Pitter raindrop (Pruppacher and Pitter, 1970) and the results have been applied for the evaluation of the exceedance probability and is compared with experimental data with very encouraging results (Georgiadou et al., 2006).

In the literature, there are numerous propagation models for the prediction of rain attenuation that can be classified into (i) empirical, (ii) semiempirical, (iii) statistical, and (iv) physical–mathematical models. The most accepted empirical model is the ITU-R model that is given in the updated Recommendation ITU-R P. 618-11. It is a step-by-step algorithmic model that results in the calculation of the exceedance cumulative distribution function (CDF) of rain attenuation given by a specific location that is the ground terminal, the electrical, and geometrical characteristics of the satellite link as well as the rain height found in ITU-R P. 839-4 (2013). The rainfall rate value, which is exceeded for 0.01% of time that is needed, can be found either from local measurement or using data from ITU-R rainmaps (ITU-R P. 837-6, 2012). In Figure 2.2, the exceedance probability of rainfall rate from ITU-R rainmaps for three different climatic regions is presented. The differences between the climatic regions are obvious.

A variety of models exists for the prediction of rain attenuation. For the statistical models, it is assumed that the rainfall rate and rain attenuation follow the same probability distributions. Afterward, employing a proper spatial coefficient model either for rainfall rate or specific attenuation that is usually isotropic, the statistical terms of rain attenuation are analytically found in terms of the statistical

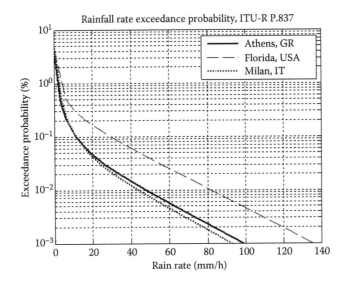

FIGURE 2.2 Rainfall rate exceedance probability versus rainfall rate for various climatic regions.

parameters of rainfall rate. The most well-accepted statistical models in the related radiowave propagation literature are based on lognormal distribution (Lin, 1975; Kanellopoulos and Koukoulas, 1987; Kanellopoulos et al., 2000a; Panagopoulos and Kanellopoulos, 2003a,b; Bertorelli and Paraboni, 2005), gamma distribution (Morita and Higutti, 1976; Panagopoulos and Kanellopoulos, 2002a), Weibull distribution (Livieratos et al., 2000; Panagopoulos et al., 2005), and very recently on inverse Gaussian distribution (Kourogiorgas and Panagopoulos, 2013a,b; Kourogiorgas et al., 2013a). The outage probabilities of the previously briefly described statistical models adopting an unconditional distribution (including rainy and nonrainy time) for rain attenuation are the following:

Unconditional lognormal distribution:

$$P[A \geq A_{thr}] = 0.5\,\mathrm{erfc}\left(\frac{\ln(A_{thr}/A_m)}{\sqrt{2}S_A}\right) \tag{2.3}$$

Unconditional gamma distribution:

$$P[A \geq A_{thr}] = 1 - \frac{\gamma(v_A, \beta A_{thr})}{\Gamma(v_A)} \tag{2.4}$$

Unconditional Weibull distribution:

$$P[A \geq A_{thr}] = \exp(-w_A(A_{thr})^{m_A}) \tag{2.5}$$

Unconditional inverse Gaussian distribution:

$$P[A \geq A_{thr}] = 1 - \left\{ Q\left(\sqrt{\frac{\lambda_A}{A_{thr}}}\left(1 - \frac{A_{thr}}{\mu_A}\right)\right) + e^{2(\lambda_A/\mu_A)}Q\left(\sqrt{\frac{\lambda_A}{A_{thr}}}\left(1 + \frac{A_{thr}}{\mu_A}\right)\right) \right\} \tag{2.6}$$

Details for the above expressions can be found in the corresponding papers. A great advantage of the statistical models is that with the proper modifications (an appropriate spatial correlation coefficient), they can also be used for the prediction of point-to-point rain attenuation statistics. A simple model based on regression-fitting analysis on a statistical model that has been produced by Weibull distribution is presented in Panagopoulos et al. (2005). Other long-term rain attenuation prediction models are those that are based on the simulation of the raincells such as EXCELL (Capsoni et al., 1987), multi-EXCELL (Luini and Capsoni, 2011), HYCELL (Feral et al., 2003a,b), and the Leitao–Watson model (Leitao and Watson, 1986).

In Figure 2.3, the rain attenuation exceedance probability curves using the ITU-R P. 618-11 Recommendation model for a hypothetical satellite link in Athens, Greece with Hellas Sat 2, for frequencies 14, 30, and 50 GHz, are presented. It simulates uplink frequencies of Ku, Ka, and V bands. It is obvious that for greater frequencies, the attenuation increases leading to very large fade margins.

2.2.2.3   Attenuation Due to Clouds

Cloud attenuation is important for the evaluation of the satellite links performance for higher-frequency bands above 10 GHz. The impact of cloud attenuation increases with the increase of frequency. For operation at Ku band (12/14 GHz), it is considered negligible but for operation at Ka, Q/V, and W bands, the cloud attenuation should be properly incorporated in the satellite link budget analysis (Panagopoulos et al., 2004a). Cloud and gaseous absorption determine the satellite system's performance during nonrainy periods.

The liquid water content is the physical cause of cloud attenuation. Several models have been developed in the literature (Salonen and Uppala, 1991; Luini and Capsoni, 2014a,b). Figure 2.4

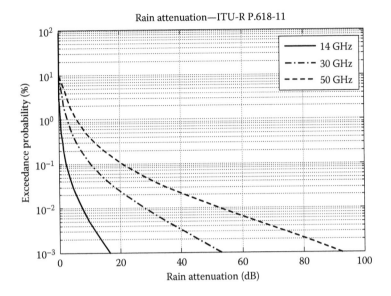

FIGURE 2.3   Rain attenuation exceedance probability, Athens, Greece, uplink, and Hellas Sat (39°E).

depicts the application of ITU-T P.840-6 model (ITU-R. P. 840-6, 2013) for the prediction of cloud attenuation for three downlink frequencies for a hypothetical link in Athens using Hellas Sat 2. In Resteghini et al. (2014), another model that has global applicability is presented and is based on the average properties of four cloud types (Stratus, Nimbostratus, Cumulus, and Cumulonimbus) and their occurrence probabilities. Very recently an engineering model has been presented for the prediction of long-term cloud attenuation statistics using a time-series synthesizer (Lyras et al., 2015) based on stochastic differential equations (SDEs). It has been employed for optical frequencies but similarly, it can be used for RF frequencies.

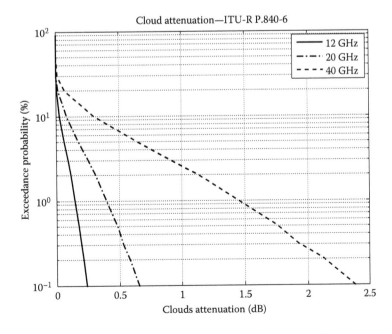

FIGURE 2.4   Cloud attenuation exceedance probability, Athens, Greece, downlink, and Hellas Sat (39°E).

2.2.2.4 Attenuation Due to Melting Layer

The layer of the atmosphere that is above the effective rain height (0° isotherm level) where the snow and the ice are melting into rain precipitation is called a melting layer. It has been shown for slant paths with low-elevation angles and during light rain, it plays a significant role in the measured attenuation. A prediction model for the attenuation due to a melting layer that verifies the above experimental observation can be found in Zhang et al. (1994).

2.2.2.5 Sky Noise Increase

With the increase of the attenuation, we also observe the increase of the noise, something that is expected due to the scattering phenomena from the precipitation hydrometeors. The following formula can be used for the calculation of the sky noise temperature taking into account the dominant fading mechanism rain attenuation (Kourogiorgas et al., 2015a)

$$T_{sky} = T_{cosmic} 10^{-A_R/10} + T_{equiv}(1 - 10^{-A_R/10}) \tag{2.7}$$

where T_{cosmic} is equal to 3°K and for the cosmic noise and T_{equiv} is the equivalent noise temperature of the receiver under clear sky conditions considering the impact of the water vapor and the oxygen and can be computed with the following formula:

$$T_{equiv} = 275(1 - 10^{-(A_{WV} + A_{OX})/10}) \tag{2.8}$$

For the accurate evaluation of the signal-to-noise ratio (SNR), the above expressions should be considered to incorporate the statistical behavior of the propagation phenomena.

2.2.2.6 Radiowave Signal Depolarization

Frequency reuse satellite systems have been proposed from the 1970s with a view to increasing the capacity of space telecommunication systems. However, this method is restricted due to the depolarization effects on the atmospheric propagation paths. Differential phase shift and differential attenuation that produced due to the nonspherical nature of the scatterers along the propagation paths, are the main causes of depolarization. The nonspherical scatterers are the raindrops and the ice crystals. The single-polarized satellite links are not affected by these phenomena. The depolarization of the signal results in cross-polar interference; that is, part of the transmitted power in one polarization interferes with its orthogonal polarization. Metrics of the depolarization effects are the cross-polarization discrimination (XPD) and the cross-polarization isolation. In ITU-R P. 618-11, there is a well-accepted empirical method for the calculation of long-term statistics of hydrometeor-induced cross-polarization, XPD, in terms of copolar rain attenuation that is evaluated for the same percentage of time. XPD depends on polarization tilt angle, the frequency of operation and the elevation angles, and finally the distribution of the canting angles. The ice crystals depolarization factor can be calculated in terms of XPD due to rain and the required exceeded time percentage. In Kanellopoulos and Panagopoulos (2001), the impact of ice crystals depolarization to the radio interference between adjacent satellite systems has been considered.

Very recently, the dual-polarized satellite link has been evaluated as a multiple-input multiple-output (MIMO) technique. A new channel model based on SDEs for the evaluation of dual-polarization satellite channel has also been introduced and the outage capacity of the dual-polarized channel has been presented (Kourogiorgas et al., 2015b). The main capacity is increased with the use of dual-polarization MIMO compared to a single-polarization system. The improvement is almost twofold for most of time percentages due to the use of a double bandwidth (second polarization) with respect to single input and single output (SISO). That is why the MIMO capacity is very similar when using two single-polarization SISO (2xSISO).

2.2.2.7 Tropospheric Scintillation

The phenomenon of fast fluctuations of the received signal is due to the space–time variations of the magnitude and the profile of the refractive index that cause variations in the angle of the arrival of the received signal. The fluctuations increase with the frequency and depend on the slant path length and decrease with the increase of the antenna beamwidth. The scintillation can be categorized as a wet scintillation during a rain event or dry scintillation when we have clear sky. A model for the calculation of the scintillation depth is presented in ITU-R P. 618-11 where the scintillation does not depend only on the electrical and the geometrical characteristics of the satellite link but also on the wet term of radio refractivity. The model has been tested mostly for the Ku band and from the model, it can be seen that with the increase in the elevation angle, the scintillation depth is decreasing, while with the increase in the frequency of operation, the scintillation depth is also increasing. Other models for the estimation of the impact of the received signal can be found elsewhere (Mousley and Vilar, 1982; Vasseur, 1999; Kourogiorgas and Panagopoulos, 2013a,b).

2.2.2.8 Total Attenuation-Combined Propagation Effects

In the literature, various models have been proposed for the prediction of the exceedance probability of each attenuation component. The main tropospheric propagation phenomena and the corresponding ITU-R recommendations for their prediction of the attenuation components are presented in Table 2.3. However, the proposal of a unified model that would combine the impact from all the tropospheric phenomena and would predict the exceedance probability of total attenuation is not an easy subject. The main problem is to find the interdependencies between the propagation effects. For example, a melting layer is associated with low-intensity rain, while gaseous absorption increases when there are rainfall events and this is due to the increase of the water vapor content in the atmosphere. If we want to make a reliable design and keep the outage probability very low, all the rest propagation phenomena should be considered for the calculation of the corresponding fade margins. Especially when the availability that must be guaranteed is low, apart from rain attenuation, all the other effects must be taken into account. There are various methods for combining these effects considering if the effects are fully correlated or uncorrelated (COST 255, 2002).

A simple approach is that all the attenuation effects may be considered correlated and therefore, the total attenuation A_{tot} (dB) can be calculated with the coherent summation

$$A_{tot} = A_G + A_C + A_R + A_{ML} + A_S \qquad (2.9)$$

TABLE 2.3
ITU-R Models for Tropospheric Propagation Prediction

Propagation Effect—Attenuation Component	ITU-R Recommendation	Model Parameters and Range of Validation	Range of Validity
Oxygen attenuation	P.676-10 (09/2013)	Surface/ground level or profile temperature and pressure	Up to 350 GHz
Water vapor attenuation	P.676-10 (09/2013)	Pressure, temperature, and water vapor density at ground level or profile IWVC pdf	Up to 350 GHz
Cloud attenuation	P.840-6 (09/2013)	ILWC pdf	Up to 200 GHz
Rain attenuation	P.618-11 (09/2013)	Rain height	Up to 55 GHz
Long-term rain attenuation frequency scaling	P.618-11 (09/2013)	Attenuation, frequency	From 7 to 55 GHz
Scintillation	P.618-11 (09/2013)	Wet term of radio refractivity, N_{wet}	4–20 GHz
Rain and ice crystals depolarization	P.618-11 (2013)	Attenuation	6–55 GHz
Depolarization scaling			4–30 GHz

where A_G (dB) is the gaseous attenuation, A_C (dB) is the cloud attenuation, A_R (dB) is the rain attenuation, A_{ML} (dB) is the melting layer attenuation, and A_S (dB) is the attenuation due to scintillation.

Another approach is the calculation of the total attenuation through summation of the components of total attenuation for the same certain exceedance percentage of time (probability level) $p\%$. That means that if we have two attenuation elements with the corresponding values $A_I(p\%)$ (dB), $A_{II}(p\%)$ (dB), for a percentage of time $p\%$ the total attenuation is given as

$$A_{tot}(p\%) = A_I(p\%) + A_{II}(p\%) \tag{2.10}$$

Moreover, another methodology for the calculation of the cumulative distribution of total attenuation considering two uncorrelated phenomena, is the convolution method. That means the cumulative distribution of the total attenuation is calculated from

$$P_t(A_{tot}) = \int_{-\infty}^{A_{tot}} p_1(A) P_2(A_{tot} - A) dA \tag{2.11}$$

where $p_1(A)$ is the probability density function (PDF) of induced attenuation due to propagation phenomenon 1 and $P_2(A)$ is the complementary cumulative distribution function (CCDF) of the induced attenuation due to propagation phenomenon 2.

In addition, the exceedance probabilities of two propagation phenomena can be summed if they can be considered disjoint. A characteristic example is the case that the CCDF of attenuation in rainy and nonrainy time is available, and the CCDF of total attenuation could be calculated as

$$P(A_{tot} > A_{thr}) = P(A_{no\_rain} > A) + P(A_{rain} > A_{thr}) \tag{2.12}$$

Another methodology for the calculation of the total attenuation is the root-square summation according to the following expression:

$$A_{tot}(p\%) = \sqrt{A_I^2(p\%) + A_{II}^2(p\%)} \tag{2.13}$$

that underestimates the predicted total attenuation value in comparison to the summation of equiprobable values.

For the calculation of total attenuation in terms of time series, there is another methodology that is applied if the attenuation values are known for the given time stamps t_j. That means the total attenuation can be computed through the following expression:

$$A_{tot}(t_j) = \sum_{k=1}^{K} A_k(t_j) \tag{2.14}$$

From the above time series, the total attenuation first- and second-order statistics can be calculated.

Finally, another method for the calculation of total attenuation effects considering that its components are partially uncorrelated and already proposed in ITU-R P. 618 is given through the formula:

$$A_T(P) = A_G(P) + \sqrt{(A_R(P) + A_C(P))^2 + A_S^2(P)} \tag{2.15}$$

where A_T, A_G, A_R, A_C, and A_S are the total attenuation, the gaseous attenuation, the rain and cloud attenuation, and the scintillation fade depth, respectively, at a given probability. In Equation 2.15, the total attenuation has been computed following ITU-R. P. 618-11 for a satellite downlink from Hellas Sat (39°E) to a ground terminal located in Athens, Greece, operating at Ka and Q bands, respectively. The deterioration of the propagation phenomena and their impact on the total attenuation due to the increase of the frequency of operation is obvious in Figure 2.5.

2.2.2.9 Aggravation of Intersystem Interference Due to Propagation Phenomena

Interference may occur between two radio systems that are sharing the same frequency band (spectrum coexistence). The impact of harmful interference may become greater considering that the radio is adjacent. The intersystem interference may occur between a satellite system and terrestrial systems or between satellite systems where their satellites (space segment) are in adjacent orbital positions. Under clear sky conditions, the intersystem interference is mainly caused by the signal leakage of the side lobes of the communication antennas. In the general case, intersystem interference may cause all the forms of interference: cochannel and copolar interference and cross-polar interference. The interference calculations are important for the cognitive satellite and cognitive hybrid satellite terrestrial systems (Vassaki et al., 2013, 2015). The quantitative metrics of intersystem interference are mainly the carrier-to-interference ratio CIR(C/I), the interference-to-noise ratio INR(I/N), the carrier-to-noise plus interference ratio CNIR (C/N + I), and in the more general case the carrier-to-noise plus total interference ratio CNTIR (C/N + TI) including depolarization effects. The above metrics are taking their nominal values under clear sky conditions. However, from the propagation impairments perspective, the interference is deteriorated due to the spatial inhomogeneity of the propagation phenomena. Most specifically, the above metrics can be calculated under rain conditions and it is important to evaluate their statistical behavior. For example, in Figure 2.6, we consider a dual-cognitive satellite communication scenario that both satellite systems operate at the same frequency bands. We consider the wanted link from GEO satellite that in the cognitive terminology is called either incumbent or primary and the interfering link in the cognitive terminology is called either a cognitive or secondary link. If we consider that we have a cognitive satellite scenario (Sharma et al., 2013a,b), the two ground stations are coordinated (cognition of the spectrum sharing). The wanted (incumbent) link is the WS-GT$_1$, and the cognitive one is the IS-GT$_2$. There are two interfering links: (a) WS-GT$_2$, which interferes the communications of the cognitive user and (b) IS-GT$_1$, which interferes with the incumbent link. In Figure 2.6, the

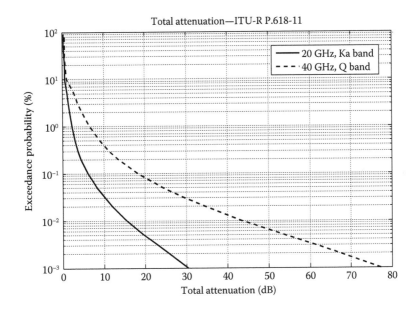

FIGURE 2.5 Total attenuation exceedance probability, Athens, Greece, downlink, and Hellas Sat (39°E).

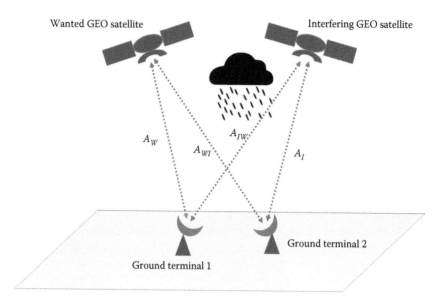

FIGURE 2.6   Interference satellite scenario: Cognitive approach.

variables of rain attenuation induced on the four satellite slant paths are also shown: (i) A_W (dB) is the rain attenuation induced on the desired link of an incumbent user (WS-GT$_1$), (ii) A_I (dB) is the rain attenuation induced on the desired link of the cognitive user (IS-GT$_2$), (iii) A_{IW} (dB) is the rain attenuation induced on the interfering link of an incumbent user (IS-GT$_1$), and (iv) A_{WI} (dB) is the rain attenuation induced on the interfering link of the cognitive user (WS-GT$_2$). All these are considered random variables.

As described previously, to assess the performance of the wanted link, the CIR statistics are needed to estimate the effect of the power level of interference to the signal power. The clear sky CIR is calculated using simple transmission theory taking into account the antenna diagrams of the satellite transponder and of the ground terminals.

The CIR in dB scale for the incumbent users also considering rain attenuation is

$$\text{CIR}_p\big|_{\text{dB}} = \text{SIR}_{CS,\text{dB}} - A_W + A_{IW} \tag{2.16}$$

where CIR_{CS} is the signal-to-interference ratio under clear sky conditions at dB values and A_p and A_{sp} are the rain attenuation random variables in dB that correspond to rain attenuation values induced in the links as this was explained before. Similarly for the cognitive user, the SIR in dB values is

$$\text{CIR}_s\big|_{\text{dB}} = \text{CIR}_{CS,\text{dB}} - A_I + A_{WI} \tag{2.17}$$

Equations 2.16 and 2.17 show the dependence of SIR under rainfall conditions, on the differential rain attenuation (Kourogiorgas and Panagopoulos, 2014; Panagopoulos et al., 2014). For the evaluation of the impact of interference, we are interested in calculating the following outage probabilities:

$$P_{out,p} = P\left[\text{CIR}_p \leq \text{CIR}_{th}, A_W \leq A_{th,W}\right] \tag{2.18}$$

$$P_{out,s} = P\left[\text{CIR}_s \leq \text{CIR}_{th}, A_I \leq A_{th,I}\right] \tag{2.19}$$

The probabilities of $(A_j \leq A_{th,j}, j = W, I)$ refer to the cases that the wanted satellite link or cognitive user is available with $A_{th,j}$ that are the thresholds for the two links for the various services provided. Similarly, one can calculate CNIR under rain fades taking into consideration the basic transmission theory, the antenna side lobes gain, and then calculate the outage probability in terms of CNIR:

$$P_{out} = P\left[\text{CNIR} \le \text{CNIR}_{th}\right] \tag{2.20}$$

In ITU-R P. 1815 (2007), a model for the prediction of the joint differential rain attenuation statistics between a satellite and two locations on the surface of the Earth is presented employing the statistical model presented in ITU-R P. 618-11 (2013) for site diversity (SD) performance prediction, that means the slant paths are considered parallel. Moreover (ITU-R, P. 619-1, 1992), a simple empirical model for the calculation of differential rain attenuation statistics is presented considering converging adjacent slant paths. Other models in the literature that incorporate the aggravation of the satellite system due to differential rain attenuation are Kanellopoulos et al. (2000a,b), Panagopoulos and Kanellopoulos (2002a), and Panagopoulos et al. (2002, 2005), considering various climatic conditions and general geometrical configurations.

2.3   FADE MITIGATION TECHNIQUES

All the propagation phenomena discussed in the previous sections exhibit space–time variations and differ from the deterministic system losses that can be evaluated in a static approach of the satellite link availability evaluations. Some tropospheric phenomena show greater space–time variations than others and for this reason we must incorporate them in the analysis stochastically. The propagation effects influence significantly for less than 1% of the annual time and for this reason we have to accurately calculate the extra-required fade margin to satisfy the prescribed availability and the QoS specifications. The availability is defined as the time percentage during a year that a specific figure of merit such as bit-error ratio (BER) is lower than a certain threshold beyond which an outage of the satellite link occurs.

$$P_{avail} = P[\text{BER} \le \text{BER}_{th}] \tag{2.21}$$

The availability can also be defined with a similar probability using other figures of merit, such as carrier-to-noise ratio (CNR), CNIR, packet error ratio (PER), frame error rate (FER), etc. The choice of the figure of merit depends on the satellite service and on the data that are available to the satellite radio designers. Another metric is the satellite link reliability that is given through the annual percentage of time that BER is greater than a specific BER threshold:

$$P_{sp} = P\left[\text{BER} > \text{BER}_{th}\right] \tag{2.22}$$

The above probabilistic metrics can also be described in terms of total attenuation and the corresponding exceedance probabilities can also be defined. The static approach for the satellite link design is to choose a percentage of time for availability and then the fade margin is defined as the difference in dB between the precipitation total attenuation leading to an outage and the clear sky attenuation.

To satisfy the QoS requirements and with a view to exploit higher millimeter-wave frequencies, high-availability systems must be designed in which a small fraction of time is significant for system design and due to the fact that total attenuation induced into the system can take high values for this small fraction of time, the application of a high fixed-power margin to deal with total attenuation (especially rain attenuation) does not give the optimum and efficient engineering solution, as this extra power will remain unexploited for the greatest time percentage of the year. Consequently,

FMTs must be proposed to protect the system from atmospheric attenuation and to operate at smaller fade margins. The FMTs based on different design approaches regarding the satellite signal impairments due to tropospheric propagation, are categorized in the following three major classifications (Castanet et al., 1998; Panagopoulos et al., 2004a): (i) power control techniques, (ii) link adaptation techniques, and (iii) diversity techniques.

2.3.1 GENERAL CONCEPT AND APPLICATION OF FMTs

In this section, the general concept and some comments for the application of FMTs in the satellite communication systems will be given. First of all, the FMTs under consideration follow the same four steps (see Figure 2.7): (i) measurement of propagation conditions by monitoring the satellite link quality, (ii) the information is sent in the satellite gateway where the decision for the FMT is made, (iii) short-term prediction of attenuation, and (iv) decision for the appropriate FMT and reconfiguration of the transmission parameters.

Considering the procedure described in Figure 2.7, it can be easily deduced that for the two of the major points of the procedure, it is really difficult to have an accurate solution. The first step, which is the attenuation measurement, is usually linked, for example, with the corresponding measurement of BER at the output of the receiver and then the E_b/N_0 threshold. This indirect method needs much extra measurements to estimate the fade level. Moreover, the third step is also very crucial for the FMT performance since the short-term prediction of the total atmospheric attenuation is one of the most difficult tasks due to the random nature of the various physical phenomena. There are numerous research projects and papers on this subject. A whole chapter in this book is devoted to time-series synthesizers of propagation phenomena that are required for the evaluation and the implementation of the FMTs. More specifically for the third step of the above-described procedure, the dynamics or the second-order statistics (fade duration, fade slope, power spectrum, autocorrelation function, etc.) are usually employed to design prediction algorithms. An important issue is to effectively decompose the dynamic characteristics of the propagation phenomena in the frequency domain using filters (low-pass filters for the gaseous absorption, midterm frequencies for clouds and rain, and a higher-frequency filter for scintillation). As explained previously, the phenomenon that plays the most important role in the satellite link reliability is rain attenuation; consequently, in this section, we present two well-established and well-accepted models and new metric definitions on rain attenuation dynamic characteristics that can be easily extended and applied to various FMTs.

First, we present a metric for second-order statistics of rain attenuation, the hitting-time statistics. Hitting-time distribution describes the diffusion processes, which are modeled with SDEs. Rain attenuation phenomena are well described using SDEs. Hitting-time distribution of rain attenuation random process has been presented in Kanellopoulos et al. (2007). In Kanellopoulos et al. (2007),

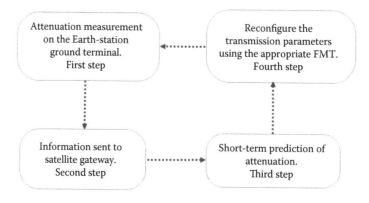

FIGURE 2.7 FMTs as a control loop in four steps.

the hitting-time distribution has been employed for the accurate determination of the dynamic input parameter of Maseng–Bakken (M–B) rain attenuation model (Maseng and Bakken, 1981).

For the rain attenuation stochastic process $A(t)$ (dB), the random variable called hitting-time τ_h is defined as the time needed for the stochastic variable to reach a rain attenuation specific value A_{th} in finite time, given that at the starting point ($t = t_0$), the random process has an initial value A_0. The definition of the hitting time as well as the fade duration metrics are depicted in Figure 2.8.

A more general definition for the hitting-time distribution is given in Karatzas and Shreve (1991) and is considered as the time needed for the stochastic variable to reach one of the two thresholds A_{min} or A_{max} in finite time, given that at the starting point, the random process has an initial value A_0, with $A_{min} \leq A_0 \leq A_{max}$. The complementary cumulative density function of hitting-time random variable, can be obtained from the expression below

$$P\left[\tau_h \geq \tau \,|\, A(t_0) = A_0\right] = M(A_{min}, A_{max}, A_0, \infty) - M(A_{min}, A_{max}, A_0, \tau) \tag{2.23}$$

where M can be obtained if the inverse Laplace transform (ILT) is applied to the function U of

$$U(A_{min}, A_{max}, A_0, \lambda) = \frac{u(A_{min}, A_{max}, A_0, \lambda)}{\lambda} + \frac{M(A_{min}, A_{max}, A_0, 0)}{\lambda} \tag{2.24}$$

Here, it must be noticed that the ILT of the second term of the right-hand side of Equation 2.24 is $M(A_{min}, A_{max}, A_0, \lambda)$ and it vanishes due to the subtraction in Equation 2.23. Therefore, this term is ignored. Consequently, to compute the CCDF of hitting time for rain attenuation, the function $u(\ )$ is defined through M–B model. For the calculations of rain attenuation hitting-time distribution, we consider $A_{max} \rightarrow \infty$ and $A_{th} = A_{min}$ and in this way, we calculate the hitting time as described in Figure 2.8. After algebra and using the assumptions of M–B model, we can calculate the function $u(\ )$:

$$u(X_{th}, X_0, \lambda) = \frac{u_2(X_0)}{u_2(X_{th})} \tag{2.25}$$

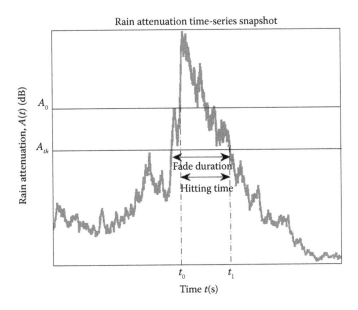

FIGURE 2.8 Definition of hitting time on a rain attenuation event.

where

$$X = \ln\left(A\middle/A_m\right)/S_a \tag{2.26}$$

and

$$u_2(x) = \frac{\sqrt{\pi}}{\Gamma\left(\frac{\lambda}{2d_A}+\frac{1}{2}\right)}\Phi\left(\frac{\lambda}{2d_A},\frac{1}{2};\frac{x^2}{2}\right) - \frac{\sqrt{2\pi}}{\Gamma\left(\frac{\lambda}{2d_A}\right)} x \cdot \Phi\left(\frac{\lambda}{2d_A}+\frac{1}{2},\frac{3}{2};\frac{x^2}{2}\right) \tag{2.27}$$

where Γ is the gamma function and Φ is the Kummer's function of the first kind and $d_A(s^{-1})$ is the dynamic parameter of the slant path rain attenuation.

Hitting-time distribution is an important metric for diffusion processes that can be employed from the radio communication engineers to characterize the channel models. For rain attenuation modeling, the hitting-time statistics reflect the dynamic properties of the satellite channel that is modeled with SDEs. In the next section, the advantages and methods for using hitting-time distribution to advanced FMTs and how they can be employed in future satellite communication systems and evaluation of the system are pointed out.

At this point, we present a two-dimensional stochastic model that correlates the induced rain attenuation values in the time domain (Karagiannis et al., 2013). It may be used for various FMTs since it provides a useful tool for the generation of simulated frequency–time-correlated time series. It can be used, for example, for efficiently implementing an up-link power control scheme or as a general framework for the effective choice of power control algorithms taking into account rain attenuation values at different frequencies.

We assume the rain attenuation stochastic processes $a^d(t)$, $a^u(t)$ (dB) induced on the slant paths (downlink and uplink) that generally operate at different frequencies f_d, f_u, respectively, and follow the system of SDEs:

$$\begin{cases} da_t^d = a_t^d\left[\frac{s_{11}^2+s_{12}^2}{2}-\beta_d\ln\left(\frac{a_t^d}{a_m^d}\right)\right]dt + a_t^d s_{11}dW_t^1 + a_t^d s_{12}dW_t^2 \\[4mm] da_t^u = a_t^u\left[\frac{s_{21}^2+s_{22}^2}{2}-\beta_u\ln\left(\frac{a_t^u}{a_m^u}\right)\right]dt + a_t^u s_{21}dW_t^1 + a_t^u s_{22}dW_t^2 \end{cases} \tag{2.28}$$

The subscripts d and u denote downlink and uplink and the random variables of rain attenuation follow the bivariate lognormal distribution and equivalently the random variables $\ln a^d(t)$, $\ln a^u(t)$ follow the bivariate Gaussian distribution and the static parameters are $\sigma_a^d, \ln a_m^d, \sigma_a^u, \ln a_m^u$ (Papoulis and Pillai, 2002). W_t^1, W_t^2 are independent Wiener processes known as Brownian motion processes (Karlin and Taylor, 1975). Using the following well-known nonlinear transformations:

$$X_t^u = \ln\left(a_t^u/a_m^u\right), \quad X_t^d = \ln\left(a_t^d/a_m^d\right) \tag{2.29}$$

the system in expression (2.28) of the linear system of SDEs becomes of Ornstein–Uhlenbeck kind (Karatzas and Shreve, 1991)

$$\begin{cases} dX_t^d = -\beta_d X_t^d dt + s_{11}dW_t^1 + s_{12}dW_t^2 \\ dX_t^u = -\beta_u X_t^u dt + s_{21}dW_t^1 + s_{22}dW_t^2 \end{cases} \tag{2.30}$$

In the above equations β_u, β_d are the dynamic parameters of rain attenuation of the uplink and the downlink, respectively, that are basically different since they are referred to different frequencies (Panagopoulos and Kanellopoulos, 2003a,b; Kanellopoulos et al., 2007). Moreover, the parameters s_{ij}, $1 \le i, j \le 2$ (matrix S elements) are the parameters that combine and correlate the independent Wiener processes for the derivation of the correlated white noise in the two links. The solution of the above system in Equation 2.30 is given as the vector stochastic process

$\mathbf{X}_t = \left[X_t^d, X_t^u \right]^T$, from Karlin and Taylor (1981) and Karatzas and Shreve (1991)

$$\mathbf{X}_t = e^{t \cdot \mathbf{B}} \cdot \mathbf{X}_0 + e^{t \cdot \mathbf{B}} \cdot \int_0^t e^{-s \cdot \mathbf{B}} \cdot \mathbf{S} \cdot d\mathbf{W}_s \tag{2.31}$$

where $e^{t \cdot \mathbf{B}} = \sum_{n=0}^{\infty} (t^n/n!)\mathbf{B}^n$ and $\mathbf{W}_t = \left[W_t^1, W_t^2 \right]^T$ are the two-dimensional Wiener process and for the matrixes, we have used these symbols

$$\mathbf{B} = \begin{bmatrix} -\beta_d & 0 \\ 0 & -\beta_u \end{bmatrix} \tag{2.32}$$

and και $\mathbf{S} = [s_{ij}]_{1 \le i, j \le 2}$. More details for the calculation of the elements s_{ij}, $1 \le i, j \le 2$ through the covariance matrix can be found in Karagiannis et al. (2012, 2013).

From these models, two important final close formulas can be calculated that are very important for the optimization of the application of FMTs. The first one is the transition probability of the reduced random variable in Equation 2.29 of the uplink considering a specific time lag t, given the value of rain attenuation on a specific link:

$$p\left(x_t^u \mid x_0^d\right) = \frac{1}{\sqrt{2\pi\left(1-\rho_{n_{du}}^2 e^{-2\beta_u t}\right)}\sigma_a^u} e^{-\left(x_t^u - \rho_{n_{du}}\left(\sigma_a^u/\sigma_a^d\right)x_0^d e^{-\beta_u t}\right)^2 \big/ 2\left(\sigma_a^u\right)^2\left(1-\rho_{n_{du}}^2 e^{-2\beta_u t}\right)} \tag{2.33}$$

The rest parameters in the right-above equation can be also found in Karagiannis et al. (2013). The second one is the expected value of rain attenuation at a specific time lag t considering the rain attenuation on the second link is given:

$$E\left[a_t^u \mid a_0^d\right] = a_m^u \cdot \exp\left(\rho_{n_{du}} \frac{\sigma_a^u}{\sigma_a^d}\left[\ln\left(\frac{a_0^d}{a_m^d}\right)\right] \cdot e^{-\beta_u t} + \frac{1}{2}\left(\sigma_a^u\right)^2 \cdot \left(1-\rho_{n_{du}}^2 e^{-2\beta_u t}\right)\right) \tag{2.34}$$

The error of the prediction is defined as the difference of the real measured value and the predicted value from the model that means

$$e = a_t^u - E\left[a_t^u \mid a_0^d\right] \tag{2.35}$$

This value is also a random variable with zero-mean value since the conditional expectation is a conditional estimator and finally, its variance can be obtained

$$E[e^2] = \left(a_m^u\right)^2 e^{\left(\sigma_a^u\right)^2} \left(e^{\left(\sigma_a^u\right)^2} - e^{\left(\sigma_a^u\right)^2 \rho_{n_{du}}^2 \cdot e^{-\beta_u t}}\right) \tag{2.36}$$

Some general guidelines and two important models that can be used for the optimum design and evaluation of FMTs have been presented in this section.

2.3.2 POWER CONTROL TECHNIQUES

In the category of power control techniques, we generally consider the variation of EIRP (effective isotropic radiated power) in dBW that is defined as the product of the antenna gain and the transmitted power. That means here, we present EIRP control that is considered as either the variation of the carrier power or the antenna gain to compensate the stochastic variations of the propagation losses. The adjustment can be implemented either at the Earth stations (then we have uplink power control—UPLC), or onboard satellite transponder (downlink power control—DLPC). Another technique that belongs to this category is the spot beam shaping (SBS) technique, where the antenna diagram is properly adjusted to dynamically compensate the variations of the atmospheric attenuation.

The power control techniques, in principle, can be designed and implemented in two ways: (i) open-loop power control where the transmitted power is controlled based on the measurements of the received power either from a pilot signal at a different frequency or from the information signal itself and (ii) the closed-loop power control where the transmitted power is adjusted after receiving feedback information from the receiver. In practice, when it is applied to real satellite systems, the closed-loop system must take into account the propagation delay due to the round trip time between the Earth stations. For this reason, when the closed loop is applied, hybrid techniques including prediction algorithms are considered. For geostationary satellite systems, this delay almost makes the application of a closed control loop power control since the most aggravating tropospheric phenomena have generally short durations (Panagopoulos et al., 2004a). For this reason, for the rest of the section, we will consider only the open-loop principle for the application in satellite communication systems.

The ULPC is achieved by adjusting the transmitted power of an Earth terminal with a view to keeping the power flux density at the satellite's transponder input at a certain predefined level. At this point, we have to differentiate the two types of transponders. For a regenerative satellite, the UPLC system takes into account only the uplink, while for the transparent transponders, the UPLC system of the Earth station aims to compensate both the uplink and the downlink power variations due to atmospheric phenomena. The two main drawbacks of the ULPC technique is the causal of adjacent channel interference due to the increased power transmitted through the sidelobes and the adjacent satellite interference. The adjacent satellite interference occurs due to the increased power that is received by the satellites that are in a closed orbit to the receiving satellite.

The DLPC system is a technique that will be implemented in the satellite transponder and unlike the ULPC, it is much more difficult due to the satellite size and the satellite weight constraints. In a DLPC system, adjacent channel interference may occur, the intermodulation interference and the intersystem interference. Intermodulation interference is caused by the nonlinear amplification of the multiple carriers, while intersystem interference is caused by the interference of the Earth–space system to terrestrial networks due to the increased power.

Another EIRP control technique is the SBS that is achieved by the adjustment of the antenna gain. A GEO satellite that covers a wide region may provide services using a spot beam that the antenna beam width is reduced and the gain is increased. In this technique short-term weather predictions are needed that may be derived from satellite images or numerical weather products. Using these predictions, the antenna gain may be changed in the region that is mostly affected by the atmospheric phenomena at the specific time instance. Given the meteorological conditions, the optimization of the spot beams may be static or dynamic (Paraboni et al., 2009a,b).

Generally speaking, the power control techniques when they are used to improve the availability of a link may cause interference to adjacent links and for this reason, the whole analysis should be considered as a cognitive system taking into account satellite and terrestrial networks and a unified and holistic interference calculation problem with a probabilistic solution should be given (Panagopoulos et al., 2013).

In the power control techniques, the methodology of hitting time and the dynamic frequency-scaling model may be used (Kourogiorgas et al., 2012). First of all, in a power control system, the transmitted power of the signal is modified according to the current state of the rain attenuation value. For example, if the rain attenuation value increases above a certain threshold, the transmitted power will also be increased to deal with rain attenuation. To design a system in which the energy efficiency as well as the availability are considered, the employment of the hitting-time distribution for the various power thresholds should lead to more efficient results. In this application, with hitting-time statistics, we have an accurate way to have a distribution function for the time needed to reach a certain value of rain attenuation, which may be the new power control threshold. The hitting-time statistics may be applied for both uplink and DLPC schemes. To measure the distribution of the time for a passage from one threshold to another, the metric of hitting time must be used. Moreover, in the case in which the FMT control loop is not able to measure the rain attenuation values for a short time, the hitting-time statistics must be used to define the time needed to use a certain power to compensate rain attenuation or the time needed to change to a higher or lower transmitted power to keep the system available. Generally speaking, transmitted power should follow rain attenuation channel hitting-time distribution to optimize the FMT control loop performance.

2.3.3 Link Adaptation Techniques

Techniques that are belonging to this category focus on properly modifying the manner that the signals are transmitted either by the Earth station terminals or by satellite transponders, whenever the link quality is deteriorated. The different types of signal processing should be available in more than one node of the satellite network. They can be classified into three categories: adaptive modulation (AM), adaptive coding (AC), and data rate reduction (DRR). The first two categories in the modern satellite systems (DVB-S2-based systems) (ETSI STANDARD DVB-S2, 2005, 2006) are almost considered simultaneously for defining the levels of their application and for this reason, they have ACM—adaptive coding and modulation technique. The DVB-S2 has introduced the possibility of adapting transmission amid a set of modulation and coding schemes to significantly enhance the satellite link throughput.

Historically, the satellite links have been designed to a constant coding and modulation strategy that means a single modulation and coding scheme have been selected and are not changed during the satellite communication.

2.3.3.1 Adaptive Modulation

AM decreases the required $E_b/N_0 \rightarrow$ SNR (dB) for a certain bit-error probability, by reducing the spectral efficiency (bps/Hz) when the SNR at the input of the demodulator decreases due to propagation effects (Filip and Vilar, 1990). That means during heavy rainfall events, the AM techniques exchange spectral efficiency for power requirements, while in the power control techniques, there is an adjustment in the power transmission. In Figure 2.9, we can see the implementation of the AM technique (e.g., considering M-PSK modulation schemes), that means when the attenuation increases and the SNR decreases and to avoid the outage, we are moving to lower-modulation schemes. Consequently, when the atmospheric attenuation is low, the satellite link can operate at higher-modulation schemes and moves to more robust ones when the atmospheric attenuation increases. The technique is preferably implemented as a closed FMT control loop where the receiving Earth station communicates with the transmitting station through a terrestrial return link in low rates (Castanet et al., 1998).

For open-loop AM techniques, a modem that will be able to adapt to the changing modulation formats must be available. This is the current state-of-the-art DVB-S2 and DVB-RCS protocols.

2.3.3.2 Adaptive Coding

AC is another FMT that may be used by the satellite communication networks. The coding is employed in the satellite links to detect and correct the errors in bits and this is done by adding

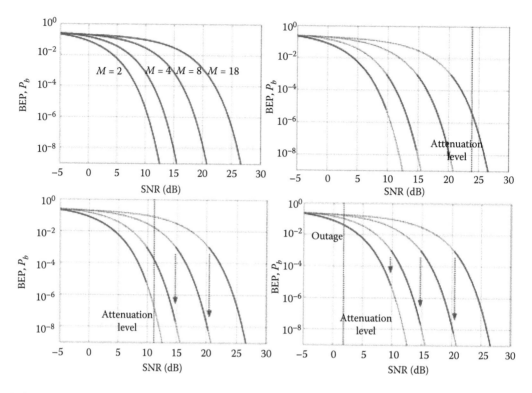

FIGURE 2.9 Implementation of an adaptive modulation technique.

redundancy to the information signal. As the redundancy increases, the error probability is decreased but, at the same time, the bandwidth required also increases. Consequently, the error correction coding can be seen as a trade-off between bandwidth and the power requirements to achieve a certain error probability. The decoding process takes place at the satellite receiver without using any feedback by the transmitters. This type of error correction is known as FEC (forward error correction). In another type of error correction coding called ARQ (automatic repeat request), error detection and correction is achieved by retransmitting the erroneous blocks of bits. ARQ is widely implemented in satellite packet communications under the main limitation of the large propagation delays involved and that we have delay-tolerant services.

The error-correcting codes have been initially designed for use against randomly spaced errors, that is, those caused by thermal noise or multipath phenomena. However, after the emergence of satellite links operating above 10 GHz, where fading caused by precipitation is the main reason for signal degradation, errors occur in bursts and not independently. Therefore, the availability of a satellite link can be preserved by varying the rate of codes more resistant to bursts (Panagopoulos et al., 2004a). A technique known as interleaving is effective at minimizing the effect of burst errors by spreading each message in time. The idea behind this technique is to apply coding to the columns of a shift register arranged as a matrix and then transmit the coded word row wise, so that after descrambling, the errors are spread and can be considered as independent. However, interleaving proves to be efficient only against very short fades, particularly against scintillation. Whenever the link suffers from severe propagation impairments, more efficient coding schemes that may be employed within the scope of AC may originate from concatenated codes, that is, combinations of block codes with convolutional codes. For the rest of the time, a less-complex coding scheme may be used. An example of concatenated codes widely used in practice to combat error bursts is to combine the Reed Solomon outer code and use convolutional coding as the inner code together with Viterbi decoding at the receiver.

2.3.3.3 Adaptable Coding and Modulation Strategies

As already discussed previously, the exploitation of changing coding and modulation formats to increase availability and throughput takes place simultaneously in the current communication systems. The first degree of adaptability is a plan for static–variable coding and modulation schemes. The modulation and coding schemes are preprogrammed off-line so as to be modified in the corresponding thresholds. The static–variable and coding scheme finds a very good application for the LEO satellite downlinks for Earth-observation applications (Toptsidis et al., 2012). The preprogrammed modulation and coding schemes are designed according to the varying link geometry and the long-term distribution of atmospheric attenuation in various elevation angles.

Another strategy is the dynamic-coding and modulation strategy and is another step toward achieving the optimal channel capacity. In this strategy, again, the actual propagation phenomena are not taken into account, but a factor that separates clear sky and rainy conditions is considered. The main motivation behind this differentiation is that the most aggravating atmospheric impairments such as deep fades occur for small percentages of time. That means two sets of modulation and coding schemes are preprogrammed off-line: the first set covering the higher modulation and coding schemes corresponding to a fade margin only because of clear sky effects and a second set that contains the lower modulation and coding schemes that are consistent with the prediction of total attenuation including rain effects. From the system point of view, we have moved from an open-loop system to a close-loop system. More elegant solutions also use information from meteorological satellites (satellite imagery forecasting) known as nowcasting or even deployed meteorological sensors (e.g., rain gauge meters) in a wider area.

The highest degree of adaptability is considered in the ACM strategy, which fully exploits all the system degrees of freedom. In this technique, the modulation and coding schemes are fully adapted to the actual propagation phenomena induced on the satellite link. The ACM scheme in its implementation is similar to a power control technique, since the optimum rain attenuation thresholds for changing ACM schemes have to be evaluated. The average hitting time between two rain attenuation threshold values can be defined using the hitting-time statistics. Consequently, the thresholds of rain attenuation will be defined using the hitting-time statistics to design an optimal system that exploits the various modulation and coding schemes. The average time that the satellite link stays (uses) on the modulation and coding level is analytically calculated. The FMT control loops for an ACM scheme in case that channel measurements with a small sampling time are not feasible, the hitting-time statistics must be used to define the time needed to change the scheme as this is actually the time between two threshold values of rain attenuation.

2.3.3.4 Data Rate Reduction

In this signal-processing technique, the information data rate is reduced whenever the FMT control loop predicts a deep atmospheric fade. This technique has been designed during the OLYMPUS experiment (Castanet et al., 2002) for the simulation of a video-conferencing system. The method is very similar to the AM and coding strategies since this procedure results in data spreading using a pseudorandom sequence. The gain obtained from this DRR in terms of the margin over the required threshold varies from 3 to 9 dB. This procedure results in data spreading with higher-processing gains corresponding to higher-spreading factors.

2.3.4 Diversity Techniques

Diversity techniques are a countermeasure against propagation phenomena. They used to be called diversity protection schemes. They constitute the most efficient FMTs, since rain-induced attenuation that is the dominant factor deteriorating the availability and performance of a satellite link operating above 10 GHz, shows significant space–time variations and these variations are fully exploited by the diversity techniques. The set of diversity techniques consists of SD, orbital diversity

(OD) or satellite diversity (SatD), frequency diversity (FD), and time diversity (TD). The first two techniques take advantage of the spatial inhomogeneous structure of the rainfall medium, whereas FD and TD are based on the spectral and the temporal dependence of rain, respectively. There are two well-known metrics that are widely used to describe diversity performance: the diversity improvement I and the diversity gain G (Panagopoulos et al., 2004a). The improvement metric is defined as the ratio of the single-site time percentage and the multiple-site time percentage, at which a given attenuation level is exceeded. On the other hand, diversity gain is defined as the difference between the single-site attenuation threshold and multiple-site attenuation threshold exceeded for a given time percentage (Panagopoulos et al., 2004a). Equivalently, these metrics can also be defined in terms of the corresponding SNR values.

There are many signal processing that may be implemented for the combination of the signals considering them in various degrees of freedom (space, angle, frequency, and time). Here, we present the most popular ones: the selection combining (SC) and maximal ratio combining (MRC) diversity schemes for two receiving terminals. In the SC scheme, the link with the highest SNR is selected from the receiver to communicate, while in MRC scheme, the receiver combines the two signals received at the two antennas of the receiver, that the SNR of the received signal is the sum of the SNR of the two signals. The latter summation holds considering the SNR in linear terms. Another difference between the MRC and SC diversity techniques is that for the former technique (MRC), the sum of the SNR of the two signals received at the antennas on the receiver side takes place after cophasing these two signals at the receiver side, while in the latter one, there is no need for cophase of the two signals. The received SNR_r in linear terms, that is, watts, can be calculated as (Kourogiorgas et al., 2013)

$$\text{SNR}_r = \begin{cases} \max(\text{SNR}_1, \text{SNR}_2), & \text{SC} \\ \text{SNR}_1 + \text{SNR}_2, & \text{MRC} \end{cases} \tag{2.37}$$

with SNR_1 that is the SNR of the signal received by the first ground terminal and SNR_2 is the SNR of the received signal in the second ground terminal. An equivalent method to describe the performance of the SC receiver is the usage of atmospheric attenuation values, and consequently the joint total atmospheric attenuation A_d (dB)

$$A_d = \min(A_1, A_2) \tag{2.38}$$

2.3.4.1 Site Diversity

In the SD technique, two or more ground terminals are communicating with the same satellite and the Earth–space path with the least attenuation, that is, a higher SNR is selected resulting in an SC scheme; for example, as explained above, SD takes advantage of the rainfall rate inhomogeneity with distance between the two or more radio paths (see Figure 2.10).

Therefore, the probability of simultaneous rain attenuation occurring on two or more network routes is less than the probability of rain attenuation occurring on each individual path. In SD, the joint cumulative distribution is of great importance, since the outage of an SD system can be calculated. Various models have been developed for the prediction and calculation of the three above-mentioned metrics: joint exceedance probability, improvement factor, and diversity gain. More particularly, there has been a segmentation of these models into empirical and physical based. In empirical models, the following are included: Hodge model (Hodge, 1982): in this model, the diversity gain is computed by modeling the effect of the frequency, the baseline distance, the elevation angle, and the baseline orientation in the diversity gain. It is an empirical model since its parameters have been derived after a regression-fitting analysis to the available experiment. This model has a limitation that the baseline distance must be less than 20 km and greater than 1.7 km. National Technical University of Athens (NTUA)-simplified model (Panagopoulos et al., 2005): an empirical

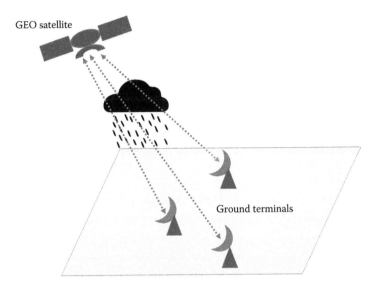

FIGURE 2.10 Configuration of an SD system.

formula is proposed for the computation of the diversity gain, based on Hodge methodology. This model is presented here:

$$G_{SD} = G_{A_S} \cdot G_D \cdot G_\theta \cdot G_f \cdot G_\Delta$$

$$G_{A_S} = 8.19 A_S^{0.0004} + 0.1809 A_S - 8.2612$$

$$G_D = \ln(3.6101D)$$ (2.39)

$$G_\theta = 1.2347(1 - \theta^{-0.356})$$

$$G_\Delta = \exp(-0.0006f)$$

In the above relationship, $G_{A_S}, G_D, G_\theta, G_f, G_\Delta$ are the factors expressing the dependence of the SD gain on the single-site attenuation A_S (dB), the site separation distance D (km), the common elevation angle of both slant paths θ (degrees), the frequency of operation f (GHz), and the orientation of the baseline between the two Earth stations Δ (degrees). As far as the attenuation threshold A_S ($p\%$) of an average year is concerned, one can resort to either the ITU-R model or to local experimental data. The above model has been derived from the application of an extended regression-fitting analysis to the results that are derived from the analytical model in Panagopoulos et al. (2004a).

Some extra empirical models are in Allnutt and Rogers (1982), Goldhirsh (1982), and Dissanayake and Lin (2000). However, more sophisticated SD models have also been developed that are based on the physical statistical properties of the medium: ITU-R 618.11 Section 2.2.4.2 model that is the Paraboni–Barbaliscia model (Paraboni and Barbaliscia, 2002), where the cumulative CDF of rain attenuation for an unbalanced system is calculated assuming that rain attenuation follows the lognormal distribution and adopting a spatial correlation function derived from radar measurements. The EXCELL model has also been used for the derivation of joint statistics of rain attenuation in Bosisio and Riva (1998). Matricciani (1994) model is an extension of the two-layer model. NTUA physical models (Kanellopoulos et al., 1990, 1994; Panagopoulos and Kanellopoulos, 2003a,b; Kourogiorgas et al., 2012) compute the unbalanced joint CCDF of rain attenuation in which single-site rain attenuation is assumed lognormal, gamma, and 4, respectively, and the joint statistics of rain attenuation follow the joint bivariate lognormal, gamma, and inverse Gaussian distributions. The spatial correlation adopted for the specific rain attenuation is an extension into two paths of the one proposed by Lin (1975). The objective of the physical statistical models is to calculate the joint exceedance probability of a dual-unbalanced SD scheme that is defined as

$$P_{out} = P[A_1 \geq A_{th1}, A_2 \geq A_{th2}]$$ (2.40)

That means the double integral below has to be calculated:

$$P_{out} = \int\limits_{A_{th1}}^{\infty} \int\limits_{A_{th2}}^{\infty} f_{A_1A_2}(A_1, A_2) \cdot dA_1 \cdot dA_2$$ (2.41)

In Livieratos et al. (2014), a new method for calculating the joint first-order statistics of rain attenuation for spatially separated satellite links has been developed based on copula functions. The advantage of a copula method is that it does not make any assumption on the kind of distribution of rain attenuation and it uses a dependence index that takes into account the dependence of two variables even if these are linked through a nonlinear expression.

The models that have been proposed for prediction of the performance of SD systems are compared using (ITU-R, P. 311-15, 2015) error criterion for the CCDF. Here, we present some comparison results that have been published in Panagopoulos et al. (2005) comparing the NTUA-simplified model performance with some well-accepted models on ITU-R databank (ITU-R SG 3 Databank) (Tables 2.4 and 2.5).

Finally, using the physical statistical model presented in Kourogiorgas et al. (2012), that is based on inverse Gaussian distribution, we have evaluated the performance of a hypothetical SD system located in Athens, Greece, operating at 40 GHz (Q band) for separation distances of 10, 20, and 50 km, respectively, shown in Figure 2.11. The diversity gain as the distance increases is obvious.

TABLE 2.4

Results of a Comparative Test between the Model in Equation 2.36 and Other Models (41 Experiments $D < 15$ km)

Prediction Model	Mean Error (%)	Rms Error (%)
NTUA-simplified model	−3.6	12.7
Hodge model	−8.7	16.1
EXCELL model	4.9	13.8
Matricciani model	8.0	19.2

TABLE 2.5

Results of a Comparative Test between the Model in Equation 2.36 and Other Models (35 Experiments $D > 15$ km)

Prediction Model	Mean Error (%)	Rms Error (%)
NTUA-simplified model	−0.3	16.9
Hodge model	−6.7	16.4
Matricciani model	5.1	15.7

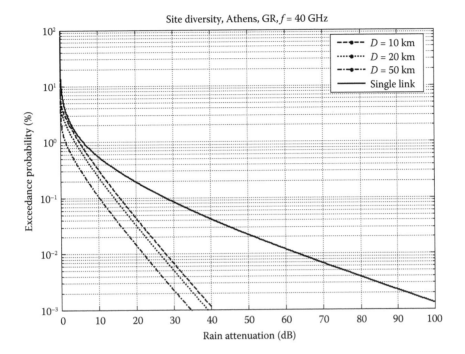

FIGURE 2.11   SD system performance.

Moreover, one can observe that the single-link operation without SD at 99.99% at 40 GHz would be very difficult to be achieved due to the very high-required fade margin.

A different system version of SD technique is the Smart Gateway diversity (Kourogiorgas et al., 2013) that is used as an uplink diversity scheme in multibeam satellite networks for the feeder links to implement the Terabit satellite concept. Its performance can be calculated with similar propagation models. In Panagopoulos et al. (2007), a model for the performance of SD in stratospheric systems using the total attenuation modeling has been presented.

2.3.4.2   Orbital Diversity

SD technique can be considered as the most efficient diversity scheme; from a technical perspective, its cost-effectiveness is under investigation, given that SD requires at least two Earth station installations along with a terrestrial connection (either wireless or through a fiber optic). Another diversity technique is the OD (or SatD) that allows Earth stations to choose between multiple satellites (see Figure 2.12).

Similar to SD, OD also adopts a reroute strategy for the network and, therefore, can be applied only for FSS (Matricciani, 1987). OD is implemented more economically compared to SD, since switching and all the other diversity operations can be carried out in a single Earth station, with backup satellites already considered in an orbit. A possible very good application of OD technique may be its use in medium Earth orbit (MEO) satellite systems in the modern O3b system. As expected, the SD technique in terms of diversity gain is much superior since the separation of the alternative slant paths in SD schemes is greater than the separation found in OD systems; consequently, the decorrelation of the propagation medium is much greater, and the correlation coefficient is much smaller. This point is investigated in Panagopoulos et al. (2004b) that compares the two basic geometrical parameters of SD and OD to obtain the optimum technical solution between the two diversity alternatives. Experimental data coming from OD experiments are far less than relevant data concerning SD systems. In this context, the model in Panagopoulos et al. (2004b) can be used to this direction and converts experimental data of SD systems into OD systems and leads

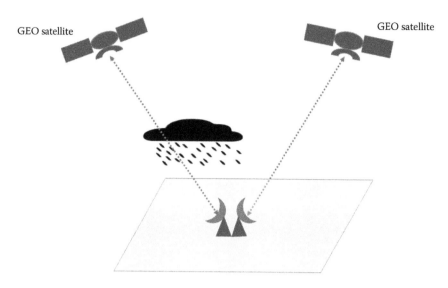

FIGURE 2.12 Configuration of an OD system.

to a reliable design of a satellite system. In Kanellopoulos and Reppas (1991), the dual-OD system is studied and an analytical model based on the bivariate lognormal distribution is given, while in Panagopoulos and Kanellopoulos (2002b), a triple-OD model is presented using trivariate lognormal distribution. Both models have been compared with experimental data with very encouraging results. Finally, among the prediction methods concerning the performance of OD systems, the simple model proposed by Panagopoulos and Kanellopoulos (2003b) and the analytical model proposed by Matricciani (1997) should be stated. The simple pocket calculator formula for the prediction of the OD gain that has been presented in Panagopoulos and Kanellopoulos (2003b) is based on a regression-fitting analysis implemented on an analytical propagation model. The resulting model characterizes OD gain as a function of single-link attenuation depth, angular separation, link frequency, path elevation angles, and the local statistics for the point rainfall rate. The simple model reproduces the analytical method with root mean square (rms) error less than 0.5 dB and moreover, it has been tested with experimental data with quite encouraging results.

Finally, using the physical statistical model presented in Panagopoulos and Kanellopoulos (2002b), that is based on lognormal distribution, we have evaluated the performance of a hypothetical OD system located in Athens, Greece, operating at 20 GHz (Ka-band) with angular separations of 45° and 70°, respectively. As the separation angle increases, the diversity gain is also increased (Figure 2.13).

2.3.4.3 Time Diversity

TD is a less-studied diversity alternative for satellite communication systems. In principle, it is based on retransmitting the corrupted information at times when the channel is expected to have better conditions, that is, at time spacings exceeding the channel coherence time. Actually, it resembles the ARQ FMT implemented in media access control (MAC) layer and falls under the general category of error correction. The difference between the two techniques is that ARQ is characterized by a fixed or random retransmission period, while in the TD system, the information is retransmitted after having the duration of the unfavorable propagation phenomenon estimated. Obviously, the performance of the TD technique is related to the time period selected for retransmission, which ranges from a few seconds to several hours. One of its main advantages when compared to other diversity techniques is that TD does not require additional RF equipment or complicated synchronization procedures, since it involves only a single-satellite link and a single-reception unit. This leads to

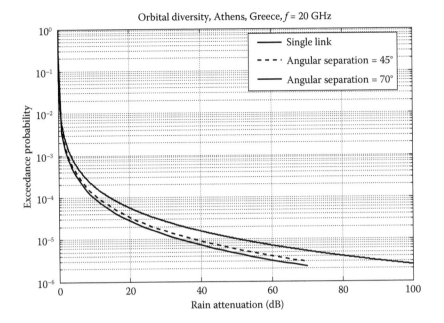

FIGURE 2.13 OD system performance.

the fact that TD option is much more cost efficient. Nevertheless, buffer and registers are required to implement TD in practical systems. However, the application of the TD technique is restricted to delay-tolerant applications (nonreal-time ones), such as video on demand or multimedia and data transfer, and seems particularly attractive for broadcasting services.

In TD technique, an advantage is taken that rainfall rate and rain attenuation are decorrelated in the time domain and therefore, when the time instances of repetition are distanced in the time domain, the probability to have high values of rain attenuation in both time instances is lower than the probability to have high values in one of the two instances. The outage prediction of a TD system is given as

$$P_{TD} = \Pr(A(t) > A_{thr}, \quad A(t + \Delta t) > A_{thr}) \tag{2.42}$$

Here, we have to refer to the analytical models in the literature (Arapoglou et al., 2008; Fabbro et al., 2009; Kourogiorgas et al., 2013b). An empirical correlation coefficient has been presented in Fabbro et al. (2009) that has been obtained after processing the databases. The second approach is based on the exploitation of the transition probability expressions of M–B model and it has been used in Arapoglou et al. (2008) and Kourogiorgas et al. (2013b). It has been shown that the two approaches have similar results.

Finally, using the physical statistical model presented in Kourogiorgas et al. (2013a), that is based on M–B model, we have evaluated the performance of a hypothetical TD system located in Athens, Greece, operating at 30 GHz (Ka-band uplink) and the exceedance curves have been drawn for time lags (delays) of 30, 60, and 120 s, respectively (see Figure 2.14).

Moreover, the concept of hitting-time statistics may be employed in TD systems, to define the time needed to retransmit the data, supposing that the time of the first transmission is experiencing high attenuation, the statistics of the hitting time shall be used, as the latter gives you an estimation of the period between the initial point and the threshold value that the received data are guaranteed. An optimum retransmission protocol based on hitting-time statistics can be designed (Kourogiorgas et al., 2012).

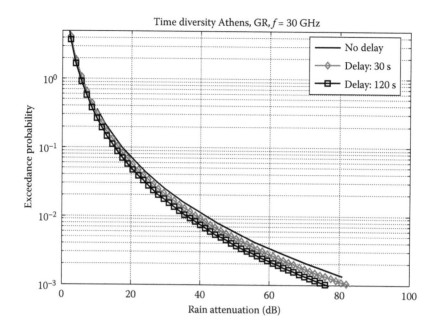

FIGURE 2.14   TD system performance.

2.3.4.4 Frequency Diversity

The satellite link suffers more from precipitation factors (fading, scintillation, depolarization, etc.), as analytically discussed previously. For this reason, since most of the satellite transponders have available onboard repeaters operating at various frequency bands (e.g., Ku, Ka-bands), the lower bands may be exploited when atmospheric phenomena occur. This simple technique is well known as FD, employs the use of high-frequency bands during clear sky conditions, and switches over to spare channels at lower-frequency bands when the attenuation due to rain exceeds a certain threshold. The diversity gain is very significant when the difference in the frequency band increases. There are many drawbacks in this technique since additional equipment-specific RF hardware and an extra antenna must be provided for every Earth station. Nevertheless, with the evolution of software-defined radios, the cost of the implementation of an FD system may be dramatically reduced. Moreover, when we employ the lower-frequency band, the capacity available is much limited; consequently, services that demand a large bandwidth may be excluded.

A general comment is that almost all the models described above mostly refer to rain attenuation and not to the rest of the atmospheric components, that means that the FMTs have been mostly designed to compensate for rain attenuation. This has been done since at lower frequencies, the impact of them is less important, but now, the modern satellite communication networks are moving to higher-frequency bands, and the models and the performance of FMTs should be revisited.

2.3.5 Combined FMTs

In this section, there will be a description of the employment of combined FMTs to combat the tropospheric propagation phenomena. As clearly stated, the above-described FMTs are either cost ineffective, such as FD and SD (even the cost has been reduced to the last decade) or yield to inadequate gains (fade margin reductions) to have DTU applications and services (Panagopoulos et al., 2004a). Consequently, the advanced satellite communication systems use some of these techniques in a combined way (combined FMTs) to design a more sophisticated and powerful fade compensation scheme and increase the time availability. At this point, we will present some general ideas without referring to satellite standards. An important technique for the implementation-combined FMTs is the employment-adaptive multifrequency TDMA (MF-TDMA) schemes. In principle, adaptive MF-TDMA systems may employ a resource-shared approach by reserving a pool of time slots within the frame shared among all Earth terminals during periods of high signal tropospheric attenuation (Panagopoulos et al., 2004a). These time slots are exploited to provide the systems with the capability of lower coding/data rates and modulation schemes of a lower level, to employ ACM and DRR FMTs and thus achieve the necessary additional fade margin to properly operate. Consequently, as a result, when a satellite ground terminal is subjected to fading, an appropriate portion of the shared resource via a change in the time planning is allocated. The gain achieved by employing such a flexible TDMA system is the aggregate gain of the AM, AC, and DRR FMTs. A possible problem may arise if a link outage occurs in case the shared resource is already allocated to other satellite terminals of the network. This relative outage time is influenced by the correlation of the attenuation events, that is, the exceedance of the fade margin in multiple sites simultaneously. For this reason, a concurrent time series of the total tropospheric attenuation in various locations of the world are important for the design of satellite communication networks.

Moreover, except from the ACM techniques, other diversity schemers may be used to increase the availability and further reduce the fade margins. For this reason, to achieve an availability of 99.999%, space diversity techniques are used. Very recently, it has been shown that even for a very small separation distance of the two Earth stations, the diversity gain may be significant. For this, the pico-scale diversity may be used, for example, by a company within its premises and along with ACM achieves low fade margins even for high-frequency bands (above 30 GHz).

For the implementation of radio resources management schemes in multibeam satellite networks, numerous papers have been presented showing that it is important to consider the channel state and the propagation models before proceeding to the allocation of the radio resources (bandwidth and power).

Another adaptive FMT recently proposed is the adaptive reconfigurable antenna front end (PoliMi-Space Engineering). The technique may counteract the attenuation effects due to the propagation in Ka-band by means of a satellite system that is able to allocate the available power in an optimized manner. The meteorological phenomena are generally localized and do not occur simultaneously across the whole area. Instead of a fixed front end, the reconfigurable antenna approach may reduce the power required by the system, modifying the spatial distribution of the transmitted power to counteract the time-variant atmospheric attenuation. The reconfigurable system control is obtained by taking as input the weather conditions all over the served region. Such antennas can be used in multibeam satellites such as in the Terabit/s concept or in broadcasting services. The reconfiguration scheme is composed of a set of multiport amplifiers (MPAs) that is used to feed a beam-forming network of N elements (Paraboni et al., 2009a). An optimization algorithm has been implemented to control the excitation coefficients of the MPA. The NNS parameter (number of nonserved users) has been used as a figure of merit to be minimized.

To sum up, since individual FMTs successfully compensate only a fraction of the total attenuation and, furthermore, correspond to a specific range of availability, they can be simultaneously applied in the form of combined FMTs.

2.3.6 SATELLITE STANDARDS WITH FMTs

The digital video broadcasting (DVB) via a satellite standard has been initially designed to offer digital video services but has been proved to be a very attractive and successful protocol to provide multimedia applications via a satellite. The return protocol known as DVB-return channel via satellite (RCS) employs smaller spot beams while the forward link employs global beams. All the required procedures are controlled by the network control center (NCC) that is installed in the satellite gateway and one of its main task is when an atmospheric event is detecting, a reconfiguration of the burst time plan.

DVB-S2 is the second-generation specification for satellite broadcasting—developed by the DVB Project in 2003 and supported by ESA. It benefits from more recent developments in channel coding (LDPC codes) combined with a variety of modulation formats (QPSK, 8PSK, 16APSK, and 32APSK). When used for interactive applications, such as Internet surfing, it may implement ACM, thus optimizing the transmission parameters for each individual user, depending on the satellite link conditions. The modes that are available are backward compatible allowing the existing operation of DVB-S set-top boxes to continue working in the satellite users premises.

The DVB-S2 system has been designed for several satellite broadband applications: (i) broadcast services for standard-definition television and high-definition television; (ii) interactive services for consumer applications including access to the Internet; (iii) professional applications; and (iv) data content distribution and Internet trunking.

DVB-S2X is a very recent extension of DVB-S2 satellite digital-broadcasting standard. It has been standardized in March 2014 as an optional extension of DVB-S2 standard. It will also become an ETSI (European Telecommunications Standards Institute) standard. Efficiency gains up to 51% can be achieved with DVB-S2X, compared to DVB-S2. The most important transmission capability improvements are higher-modulation schemes (64/128/256APSK), smaller roll-off factors, and general improved filtering making it possible to have a smaller carrier spacing. In principle, DVB-S2X has been designed for the very low SNR regions.

Another standard for the return link is DVB-RCS2 that was approved in 2011 and 2012 with its mobility extensions (DVB-RCS2 + M) to support mobile/nomadic terminals and direct terminal-to-terminal (mesh) connectivity. Its features include handovers between satellite spot beams, spread-spectrum features to meet regulatory constraints for mobile terminals, and continuous-carrier transmission for terminals with high traffic aggregation. It also includes link-layer FEC, used as a countermeasure against shadowing and blocking of the satellite link. The following modulation schemes are eligible in DVB-RCS2 BPSK, QPSK, 8PSK, 16QAM, and constant envelope—CPM and regarding channel coding 16-state PCCC turbo code (linear modulation) SCCC (CPM) exists.

There are numerous papers in the literature that are related to the performance of FMTs in next-generation satellite standards and at this point, we state some also considering the performance of intersystem interference evaluation: (Gremont et al., 1999; Malygin et al., 2002; Noussi et al., 2009; Panagopoulos et al., 2013, 2014; Enserink et al., 2014).

2.4 TELECOMMUNICATION AND PROPAGATION EXPERIMENTS

In the history of the satellite communications propagation experiments, the most significant experiments that have taken place in the last four decades are (a) advanced communications technology satellite (ACTS) experiment, (b) OLYMPUS experiment, and (c) ITALSAT experiment. Other experiments were with the SIRIO satellite in the 1970s.

In the ACTS experiment that was led by National Aeronautics and Space Administration (NASA) Ka-band, satellite propagation measurements were conducted (Gargione, 2002). Beacons signal have been installed on the ACTS satellite: downlink frequencies at 20.185 and 20.195 GHz and uplink frequencies at 27.505 GHz. The sites that have participated in the experiment were in the United States, Canada, Mexico, and generally, they have been chosen as areas with varying climatic characteristics and weather conditions to be able to develop global precipitation prediction models. During this experiment, many raw data have been collected to produce copolar attenuation exceedance curves; fade duration; fade slope; interfade durations; and SD (double and triple). Unfortunately, unwanted effects of water on the antenna reflector surface (wet-antenna effects) were noted, and so the measurement has been considered problematic. There have been developed models (Crane, 2002) to correct the experimental data and some of them have been included in the ITU-R databanks.

ESA has set an experimental telecommunication satellite OLYMPUS in an orbit from 1989 to 1993. The payloads that were carried were including communications payload, broadcast payload, and propagation beacons at Ku (downlink 12.5 GHz) and Ka (downlink 19.77 and 29.66 GHz uplink). From propagation-modeling view, OLYMPUS propagation experiment (OPEX) is of great interest. About 50 beacon receives have been installed across Europe, Ottawa, Blacksburg, and Lewis in North America. From the raw data time series rain attenuation statistics, SD statistics, frequency and polarization scaling, spectra, and the variance of the scintillation have been computed. The depolarization data and studies that have been measured and conducted during OLYMPUS experiment are of great importance. During 4 years of its operation, the satellite was not fully available and there are periods without raw data. Another problem with OLYMPUS satellite was that it was not so stable; so, preprocessing techniques have been developed to derive reliable-attenuation time series by removing the movement of OLYMPUS satellite.

ITALSAT satellite was launched in January 1991 carrying a telecommunications payload and for conducting propagation measurements. Three frequencies were used for the performance 77 of the experiments at 18.7, 39.6, and 49.5 GHz. The beacon at 18.7 GHz covered the Italian territory while the other two beacons covered the whole European continent. The life of the satellite lasted for 9 years and the measuring sites were located at Italy, Germany, Netherlands, and the United Kingdom. However, in most regions, the data are time limited as these were recorded for a small amount of time and cannot be useful to produce the long-term statistics. Significant studies from ITALSAT have been conducted in the United Kingdom (Ventouras et al., 2006).

Since July 2013, ALPHASAT has been launched (25° E), with its telecommunication payload conceived, financed, and realized in Italy; it is actually composed of two separated experimental payloads: a 40/50-GHz telecommunication section that performs a three-spot transponder and a Ka/Q propagation section providing a geographical beacon centered on Europe. The ALPHASAT payload is called ALDO dedicated to the memory of Professor Aldo Paraboni of Politecnico di Milano, who was the principal investigator of this project but passed away in 2011. The main objectives of the ALPHASAT propagation experiment are to provide representative propagation data of the coverage area across Europe of current Ka-band multimedia systems. Moreover, ALPHASAT experiments' objective is to demonstrate the effectiveness of FMTs in improving the achievable data throughput in a real Q/V band satellite link. All the FMTs described in the previous section will be tested and redesigned not only taking into account rain attenuation but all the propagation effects leading to total attenuation.

2.5   OPEN-RESEARCH ISSUES AND CONCLUDING REMARKS

Q/V and W bands are promising solutions for the operation of the future satellite communication systems, either for satellite feeder links or for DTU satellite links. To design reliable satellite systems at these high-frequency bands, new propagation models are required to be developed with a view to capturing better tropospheric effects.

The existing propagation model in the literature and in ITU-R has been tested using previous experiments (the last four decades), that is, they have not been validated with measurements especially above 30 GHz. The empirical models in the literature have been derived taking into account experiments at Ka-band and consequently, their validity at higher frequencies has to be reexamined and tested.

Rain attenuation modeling and all the models for the rest of the tropospheric components in total should be carefully reexamined considering the experimental data from the new experiments. A further research direction on this subject is the development of new statistical radio propagation models tackling the problem of space–time variations of the satellite channel more generally in amplitude and phase that is a very critical point in current multiantenna satellite networks.

Another open-research direction is finally to develop statistical propagation-based models for the radio interference between the adjacent satellite and terrestrial including the space–time variations of the tropospheric propagation phenomena and the assumptions that the radio systems employ FMTs or not. In addition, in the signal-processing techniques that are developed in the framework of cognitive hybrid satellite and terrestrial networks, the propagation phenomena should be incorporated to make them more realistic and accurate.

Summing up, this chapter presents the tropospheric propagation phenomena that deteriorate the radio signal transmission in FSS and services are presented along with well-accepted models for their accurate prediction. Finally, FMTs that are employed in the modern fixed-satellite communication standards are rigorously discussed and evaluated.

ACKNOWLEDGMENT

This chapter has been funded under the framework of THALES-NTUA MIMOSA and si-Cluster ACRITAS.

REFERENCES

Ajayi, G.O. and R.L. Olsen, Modeling of a tropical raindrop size distribution for microwave and millimeter wave applications, *Radio Science*, 20(2), 193–202, 1985.

Allnutt, J.E. and D.V. Rogers, Novel method for predicting site diversity gain on satellite-to-ground radio paths, *Electronic Letters*, 18(5), 233–235, 1982.

Arapoglou, P-D.M., A.D. Panagopoulos, and P.G. Cottis, An analytical prediction model of time diversity performance for earth–space fade mitigation, *HINDAWI International Journal of Antennas and Propagation*, 2008, 5, 2008, doi: 10.1155/2008/142497.

Bertorelli, S. and A. Paraboni, Simulation of joint statistics of rain attenuation in multiple sites across wide areas using ITALSAT data, *IEEE Transactions on Antennas and Propagation*, 53(8), 2611–2622, 2005.

Bosisio, A.V. and C. Riva, A novel method for the statistical prediction of rain attenuation in site diversity systems: Theory and comparative testing against experimental data, *International Journal of Satellite Communications*, 16, 47–52, 1998.

Capsoni, C. et al., Data and theory for a new model of the horizontal structure of rain cells for propagation applications, *Radio Science*, 22(3), 395–494, 1987.

Castanet, L., M. Bousquet, and D. Mertens, Simulation of the performance of a Ka-band VSAT videoconferencing system with uplink power control and data rate reduction to mitigate atmospheric propagation effects, *International Journal of Satellite Communications*, 20(4), 231–249, 2002.

Castanet, L., J. Lemorton, and M. Bousquet, Fade mitigation techniques for new SatCom services at Ku-band and above: A review, in *4th Ka-Band Utilization Conference*, Venice, Italy, November 2–4, 1998.

Castanet, L. et al., SatNEx Ebook, *Influence of the Variability of the Propagation Channel on Mobile*, Fixed Multimedia and Optical Satellite Communications, Shaker-Verlag, Aachen, 2007.

Codispoti, G., C. Riva, M. Ruggieri, T. Rossi, A. Martellucci, J. Rivera-Castro, O. Koudelka, and M. Schoenhuber, The propagation and telecom experiments of the Alphasat Aldo payload (TDP5 Q/V band experiment), in *6th EUCAP*, Prague, Czech, March 26–30, 2012.

COST 255, Radiowave propagation modelling for new Satcom services at Ku-band and above, COST 255 Final Report, ESA Publications Division, SP-1252, 2002.

Crane, R.K., Analysis of the effects of water on the ACTS propagation terminal antenna, *IEEE Transactions on Antennas and Propagation*, 50(7), 954–965, 2002.

Crane, R.K., *Propagation Handbook for Wireless Communication System Design*, CRC Press LLC, Boca Raton, FL, 2003.

Dissanayake, A. and K.T. Lin, Ka-band site diversity measurements and modeling, in *6th Ka-Band Utilization Conference*, Cleveland, Ohio, 2000.

Enserink, S., A.D. Panagopoulos, and M. Fitz, On the calculation of constrained capacity and outage probability of broadband satellite communication links, *IEEE Wireless Communication Letters*, 3(5), 453–456, 2014.

European Standard (Telecommunications Series), *Digital Video Broadcasting (DVB) User Guidelines for the Second Generation System for Broadcasting*, ETSI, Technical Report ETSI TR 102 376, 2005-02.

European Standard (Telecommunications Series), *Digital Video Broadcasting (DVB); Second Generation Framing Structure, Channel Coding and Modulation Systems for Broadcasting*, ETSI, Standard ETSI EN 320 307 v1.1.2, 2006.

Fabbro, V., L. Castanet, S. Croce, and C. Riva, Characterization and modelling of time diversity statistics for satellite communications from 12 to 50 GHz, *International Journal of Satellite Communications*, 27, 87–101, 2009.

Feral, L., H. Sauvageot, L. Castanet, and J. Lemorton, HYCELL—A new hybrid model of the rain horizontal distribution for propagation studies: 1. Modeling of the rain cell, *Radio Science*, 38, 1056, 2003a.

Feral, L., H. Sauvageot, L. Castanet, and J. Lemorton, HYCELL—A new hybrid model of the rain horizontal distribution for propagation studies: 2, *Statistical Modeling of the Rain Rate field*, 38, 1057, 2003b.

Filip, M. and E. Vilar, Adaptive modulation as a fade countermeasure, an OLYMPUS experiment, *International Journal of Satellite Communications*, 8, 31–41, 1990.

Gargione, F., NASA's advanced communications technology satellite (ACTS): Historical development—ACTS program formulation, *Online Journal of Space Communication*, (2), Fall 2002.

Georgiadou, E.M., A.D. Panagopoulos, and J.D. Kanellopoulos, On the accurate modeling of millimeter fixed wireless access channels exceedance probability, *International Journal of Infrared and Millimeter Waves*, 27, 1027–1039, 2006.

Goldhirsh, J., Space diversity performance prediction for earth–satellite paths using radar modelling techniques, *Radio Science*, 17(6), 1400–1410, 1982.

Gremont, B., M. Filip, P. Gallois, and S. Bate, Comparative analysis and performance of two predictive fade detection schemes for Ka-band fade countermeasures, *IEEE Journal on Selected Areas on Communications*, 17(2), 180–192, 1999.

Hodge, D.B., An improved model for diversity gain on earth–space propagation paths, *Radio Science*, 17(6), 1393–1399, 1982.

International Telecommunication Union, http://www.itu.int/ITU-R/Software/study-groups/rsg3/databanks/index.html (Online).

ITU-R. P.311-15, Acquisition, presentation and analysis of data in studies of tropospheric propagation, in *International Telecommunication Union*, Geneva, 2015.

ITU-R. P. 618-11, Propagation data and prediction methods required for the design of earth–space telecommunication systems, in *International Telecommunication Union*, Geneva, 2013.

ITU-R P. 619-1, Propagation data required for the evaluation of interference between stations in space and those on the surface of the Earth, in *International Telecommunication Union*, 1992.

ITU-R. P. 676-10, Attenuation by atmospheric gases, in *International Telecommunication Union*, Geneva, 2013.

ITU-R. P. 835-5, Reference standard atmospheres, in *International Telecommunication Union*, Geneva, 2012.

ITU-R. P. 836-4, Water vapour: Surface density and total columnar content, in *International Telecommunication Union*, 2009.

ITU-R. P. 837-6, Characteristics of precipitation for propagation modeling, in *International Telecommunication Union*, Geneva, 2012.

ITU-R. P. 838-3, Specific attenuation model for use in prediction models, in *International Telecommunication Union*, Geneva, 2005.

ITU-R. P. 839-4, Rain height for prediction methods, in *International Telecommunication Union*, Geneva, 2013.

ITU-R. P. 840-6, Attenuation due to clouds and fog, in *International Telecommunication Union*, Geneva, 2013.

ITU-R. P. 1815, Differential rain attenuation, in *International Telecommunication Union*, Geneva, 2007.

Kanellopoulos, J.D. and S.G. Koukoulas, Analysis of the rain outage performance of route diversity system, *Radio Science*, 22(4), 549–565, 1987.

Kanellopoulos, J.D. S.G. Koukoulas, N.J. Kolliopoulos, C.N. Capsalis, and S.G. Ventouras, Rain attenuation problems affecting the performance of microwave communication systems, *Annals of Telecommunication*, 45(7–8), 437–451, 1990.

Kanellopoulos, J.D. and A.D. Panagopoulos, Ice crystals and raindrop canting angle affecting the performance of a satellite system suffering from differential rain attenuation and cross-polarization, *Radio Science*, 36(5), 927–940, 2001.

Kanellopoulos, J.D., A.D. Panagopoulos, and S.N. Livieratos, A comparison of co-polar and co-channel satellite interference prediction models with experimental results at 11.6 GHz and 20 GHz, *International Journal of Satellite Communications*, 18(2), 107–120, 2000a.

Kanellopoulos, J.D., A.D. Panagopoulos, and S.N. Livieratos, Differential rain attenuation statistics including an accurate estimation of the effective slant path lengths, *Journal of Electromagnetic Waves and Applications*, 14, 663–664, 2000b (Abstract), *Progress in Electromagnetic Research* 28, 101–124, 2000.

Kanellopoulos, J.D. and A. Reppas, A prediction of outage performance of an orbital diversity earth–space system, *European Transactions on Telecommunications*, 2(6), 729–735, 1991.

Kanellopoulos, J.D., S.G. Ventouras, and S. Koukoulas, A model for the prediction of the differential rain attenuation between a satellite path and an adjacent terrestrial microwave system based on the two-dimensional gamma distribution, *Journal of Electromagnetic Waves and Applications*, 8(5), 557–574, 1994.

Kanellopoulos, S.A., A.D. Panagopoulos, and J.D. Kanellopoulos, Calculation of electromagnetic scattering from a Pruppacher–Pitter raindrop using MAS and slant path rain attenuation prediction, *International Journal of Infrared and Millimeter Waves*, 26(12), 1783–1802, 2005.

Kanellopoulos, S.A., A.D. Panagopoulos, and J.D. Kanellopoulos, Calculation of the dynamic input parameter for a stochastic model simulating rain attenuation: A novel mathematical approach, *IEEE Transactions on Antennas and Propagation*, 55(11), 3257–3264, 2007.

Karagiannis, G., A.D. Panagopoulos, and J.D. Kanellopoulos, Multi-dimensional rain attenuation stochastic dynamic modeling: Application to earth–space diversity systems, *IEEE Transactions on Antennas and Propagation*, 60(11), 5400–5411, 2012.

Karagiannis, G., A.D. Panagopoulos, and J.D. Kanellopoulos, Short-term rain attenuation frequency scaling for satellite up-link power control applications, *IEEE Transactions on Antennas and Propagation*, 61(5), 2829–2837, 2013.

Karatzas, I. and S.E. Shreve, *Brownian Motion and Stochastic Calculus*, New York: Springer-Verlag, 1991.

Karlin, S. and H. Taylor, *A First Course in Stochastic Processes*, 2nd ed., New York: Academic Press, 1975.

Karlin, S. and H. Taylor, *A Second Course in Stochastic Processes*, New York: Academic Press, 1981.

Koudelka, O., Q/V-band communications and propagation experiments using ALPHASAT, *Acta Astronautica*, 69(11–12): 1029–1037, 2011. doi: 10.1016/j.actaastro.2011.07.008.Alphasat.

Kourogiorgas, C., G.A. Karagiannis, and A.D. Panagopoulos, Smart gateway diversity outage performance using multi-dimensional rain attenuation synthesizer, in *First CNES-ONERA Workshop on Earth–Space Propagation*, Toulouse, France, January 21–23, 2013.

Kourogiorgas, C. and A.D. Panagopoulos, Interference statistical distribution for cognitive satellite communication systems operating above 10 GHz, in *ASMS 2014*, Livorno, Italy, September 2014.

Kourogiorgas, C., A.D. Panagopoulos, and J.D. Kanellopoulos, On the earth–space site diversity modeling: A novel physical mathematical model, *IEEE Transactions on Antennas and Propagation*, 60(9), 4391–4397, 2012.

Kourogiorgas, C.I., P-D.M. Arapoglou, and A.D. Panagopoulos, Statistical characterization of adjacent satellite interference for earth stations on mobile platforms operating at Ku and Ka bands, *IEEE Wireless Communication Letters*, 4(1), 82–85, 2015a.

Kourogiorgas, C.I. and A.D. Panagopoulos, A new physical–mathematical model for predicting slant-path rain attenuation statistics based on inverse Gaussian distribution, *IET Microwaves, Antennas and Propagation*, 7(12), 970–975, 2013a.

Kourogiorgas, C.I. and A.D. Panagopoulos, A tropospheric scintillation time series synthesizer based on stochastic differential equations, in *19th Ka and Broadband Communications, Navigation and Earth Observation Conference*, Florence, Italy, October 14–17, 2013b.

Kourogiorgas, C.I., A.D. Panagopoulos, P-D.M. Arapoglou, and S. Stavrou, MIMO dual polarized fixed satellite systems above 10 GHz above: Channel modeling and outage capacity evaluation, in *9th EUCAP*, Lisboa, Portugal, 2015b.

Kourogiorgas, C.I., A.D. Panagopoulos, and J.D. Kanellopoulos, A new method for the prediction of outage probability of LOS terrestrial links operating above 10 GHz, *IEEE Antennas and Wireless Propagation Letters*, 12, 516–519, 2013a.

Kourogiorgas, C.I. et al., Rain attenuation hitting time statistical distribution: Application to fade mitigation techniques of future satellite communication systems, in *1st International IEEE-AESS Conference in Europe on Space and Satellite Telecommunications (ESTEL)*, Rome, Italy, October 2012.

Kourogiorgas, C.I. et al., On the outage probability prediction of time diversity scheme in broadband satellite communication systems, *Progress in Electromagnetics Research C*, 44, 175–184, 2013b.

Leitao, M.J. and P.A. Watson, Method for prediction of attenuation on earth–space links based on radar measurements of the physical structure of rainfall, *IEE Proceedings on Microwaves Antennas and Propagation*, 133(4), 429–440, 1986.

Liebe, H., MPM—An atmospheric millimeter-wave propagation model, *International Journal of Infrared and Millimeter Waves*, 10(6), 631–650, 1989.

Lin, S.H., A method for calculating rain attenuation distributions on microwave paths, *Bell System Technical Journal*, 54(6), 1051–1083, 1975.

Livieratos, S.N., C.I. Kourogiorgas, A.D. Panagopoulos, and G. Chatzarakis, On the prediction of joint rain attenuation statistics in earth–space diversity systems using copulas, *IEEE Transactions on Antennas and Propagations*, 62(44), 2250–2257, 2014.

Livieratos, S.N., V. Katsambas, and J.D. Kanellopoulos, A global method for the prediction of the slant path rain attenuation statistics, *Journal of Electromagnetic Waves and Applications*, 14, 713–724, 2000.

Luini, L. and C. Capsoni, MultiEXCELL: A new rain field model for propagation applications, *IEEE Transactions on Antennas and Propagation*, 59(11), 4286–4300, 2011.

Luini, L. and C. Capsoni, Efficient calculation of cloud attenuation for earth–space applications, *IEEE Antennas and Wireless Propagation Letters*, 13(1), 1136–1139, 2014a.

Luini, L. and C. Capsoni, Modeling high resolution 3-D cloud fields for earth–space communication systems, *IEEE Transactions on Antennas and Propagation*, 62(10), 5190–5199, 2014b.

Lyras, N.K., C.I. Kourogiorgas, and A.D. Panagopoulos, Cloud attenuation time series synthesizer for earth–space links operating at optical frequencies, in *ICEAA-AWPC 2015*, Torino, Italy, pp. 638–641, September 2015.

Malygin, A., M. Filip, and E. Vilar, Neural network implementation of a fade countermeasure controller for a VSAT link, *International Journal of Satellite Communications*, 20, 79–95, 2002.

Marshall, J.S. and W.Mc.K. Palmer, The distribution of raindrops with size, *Journal of Meteorology*, 5, 165–166, 1948.

Maseng, T. and P.M. Bakken, A stochastic dynamic model of rain attenuation, *IEEE Transactions on Communications*, COM-29(5), 660–669, 1981.

Matricciani, E., Orbital diversity in resource-shared satellite communication systems above 10 GHz, *IEEE Journal on Selected Areas in Communication*, 5(4), 714–723, 1987.

Matricciani, E., Prediction of site diversity performance in satellite communications systems affected by rain attenuation: Extension of the two layer rain model, *European Transactions on Telecommunications*, 5(3), 27–36, 1994.

Matricciani, E., Prediction of orbital diversity performance in satellite communication systems affected by rain attenuation, *International Journal of Satellite Communications*, 15, 45–50, 1997.

Morita, K. and I. Hihgutti, Prediction methods for rain attenuation distributions of micro and millimeter waves, *Review of the Electrical Communication Laboratory, NTT*, 24(7–8), 651, 1976.

Mousley, T. and E. Vilar, Experimental and theoretical statistics of microwave amplitude scintillation on satellite down-links, *IEEE Transactions on Antennas and Propagation*, AP-30(6), 1099–1106, 1982.

Noussi, E., B. Gremont, and M. Filip, Integration of fade mitigation within centrally managed MF-TDMA/ DVB-RCS networks, *Space Communications*, 22(1), 13–29, 2009.

Olsen, R., D. Rogers, and D.B. Hodge, The aRb relation in the calculation of rain attenuation, *IEEE Transactions on Antennas and Propagation*, 26(2), 318–329, 1978.

Panagopoulos, A.D., *Reliable Simulation Framework for Mobile Computing Systems Using Stochastic Differential Equations, Book Chapter 8 in Simulation Technologies in Networking and Communications: Selecting the Best Tool for the Test*, pp. 213–228, CRC Press, Boca Raton, FL, 2014.

Panagopoulos, A.D., P-D.M. Arapoglou, and P.G. Cottis, Satellite communications at Ku, Ka and V bands, propagation impairments and mitigation techniques, *IEEE Communication Surveys and Tutorials*, 3rd Quarter, 6(3), 1–13, 2004a.

Panagopoulos, A.D., P-D.M. Arapoglou, and P.G. Cottis, Site vs. orbital diversity: Performance comparison based on propagation characteristics at Ku band and above, *IEEE Antennas and Wireless Propagation Letters*, 23(3), 26–29, 2004b.

Panagopoulos, A.D., P-D.M. Arapoglou, J.D. Kanellopoulos, and P.G. Cottis, Long term rain attenuation probability and site diversity gain prediction formulas, *IEEE Transactions on Antennas and Propagation*, 53(7), 2005.

Panagopoulos, A.D., E.M. Georgiadou, and J.D. Kanellopoulos, Selection combining site diversity performance in high altitude platform networks, *IEEE Communication Letters*, 11(10), 787–789, 2007.

Panagopoulos, A.D. and J.D. Kanellopoulos, Adjacent satellite interference effects as applied to the outage performance of an earth–space system located in a heavy rain climatic region, *Annals of Telecommunications*, (9–10), 925–942, 2002a.

Panagopoulos, A.D. and J.D. Kanellopoulos, Prediction of triple-orbital diversity performance in earth–space communication, *International Journal of Satellite Communications*, (20), 187–200, 2002b.

Panagopoulos, A.D. and J.D. Kanellopoulos, On the rain attenuation dynamics: Spatial–temporal analysis of rainfall rate and fade duration statistics, *International Journal of Satellite Communications and Networking*, 21, 595–611, 2003a.

Panagopoulos, A.D. and J.D. Kanellopoulos, A simple model for orbital diversity gain on earth–space propagation paths, *IEEE Transactions on Antennas and Propagation*, 51(6), 1403–1405, 2003b.

Panagopoulos, A.D., T.D. Kritikos, and J.D. Kanellopoulos, Aggravation of radio interference effects on a dual polarized earth–space link by two adjacent interfering satellites under rain fades, *Radio Science*, 40, RS5005, 2005. doi: 10.1029/2004RS003137.

Panagopoulos, A.D., T.D. Kritikos, and J.D. Kanellopoulos, Acceptable intersystem interference probability distribution between adjacent terrestrial and satellite networks operating above 10 GHz, *Space Communications*, 22(2–4), 205–212, 2013.

Panagopoulos, A.D., S.N. Livieratos, and J.D. Kanellopoulos, Interference analysis applied to a double-site diversity earth–space system: Rain height effects and regression-derived formulas, *Radio Science*, 37(6), 1103, 2002.

Panagopoulos, A.D. et al., Interference studies between adjacent satellite communications systems operating above 10 GHz and using power control as fade mitigation technique, *Wireless Personal Communications*, 77(2), 1311–1327, 2014.

Papoulis, A. and S.U. Pillai, *Probability, Random Variables and Stochastic Processes with Errata Sheet*, 4th edn., McGraw-Hill, Boca Raton, FL, 2002.

Paraboni, A. and F. Barbaliscia, Multiple site attenuation prediction models based on the rainfall structures (meso- or synoptic scales) for advanced TLC or broadcasting systems, in *XXVIIth URSI General Assembly*, Maastricht, 2002.

Paraboni, A., M. Buti, C. Capsoni, D. Ferraro, C. Riva, A. Martellucci, and P. Gabellini, Meteorology-driven optimum control of a multibeam antenna in satellite telecommunications, *IEEE Transactions on Antennas and Propagation*, 57(2), 508–519, 2009a.

Paraboni, A., M. Buti, C. Capsoni, P. Gabellini, and A. Martellucci, Long-period statistics of the power distribution of a multi-beam reconfigurable antenna for satellite broadcasting over the European area, in *Proceedings of EuCAP 2009*, Berlin (Germany), April 2009b.

Politecnico di Milano and Space Engineering, Reconfigurable Ka-band antenna front-end for active rain fade compensation: Propagation modeling, ESA/ESTEC contract N° 17877/04/NL/JA.

Pruppacher, H.R. and R.L. Pitter, A semi-empirical of the shape of cloud and rain drops, *Journal of Atmospheric Sciences*, 28, 86–94, 1970.

Resteghini, L., C. Capsoni, L. Luini, and R. Nebuloni, An attempt to classify the types of clouds by a dual frequency microwave radiometer, in *Microwave Radiometry and Remote Sensing of the Environment (MicroRad), 2014 13th Specialist Meeting on*, Pacadena, CA, USA, pp. 90–93, March 24–27, 2014.

Rosello, J. et al., 26-GHz data downlink for LEO satellites, in *Proceedings of the 6th EUCAP*, pp. 111–115, Prague, Czech Republic, March 26–30, 2012.

Salonen, E. and S. Uppala, New prediction method of cloud attenuation, *Electronic Letters*, 27, 1106–1108, 1991.

Sharma, S.K., S. Chatzinotas, and B. Ottersten, Satellite cognitive communications: Interference modeling and techniques selections, in *6th ASMS and 12th SPSC*, Baiona, Spain, September 5–7, 2013a.

Sharma, S.K., S. Chatzinotas, and B. Ottersten, Cognitive radio techniques for satellite communication systems, in *78th VTC Fall*, Las Vegas, USA, September 2–5, 2013b.

Toptsidis, N., P-D.M. Arapoglou, and M. Bertinelli, Link adaptation for Ka band low earth orbit earth observation systems: A realistic performance assessment, *International Journal of Satellite Communications and Networking*, 30(3), 131–146, 2012.

Vassaki, S., M. Poulakis, and A.D. Panagopoulos, Optimal iSINR-based power control for cognitive satellite terrestrial networks, *Transactions on Emerging Telecommunications Technologies*, 2015. doi: 10.1002/ett.2945.

Vassaki, S., M. Poulakis, A.D. Panagopoulos, and Ph. Constantinou, Power allocation in cognitive satellite terrestrial networks with QoS constraints, *IEEE Communication Letters*, 17, 1344–1347, 2013.

Vasseur, H., Prediction of tropospheric scintillation on satellite links from radiosonde data, *IEEE Transactions on Antennas and Propagation*, 47(2), 293–301, 1999.

Ventouras, S. and C. L. Wrench, Long-term statistics of tropospheric attenuation from the Ka/U band ITALSAT satellite experiment in the United Kingdom, *Radio Science*, 41, RS2007, 2006. doi: 10.1029/2005RS003252.

Zhang, W. et al., Prediction of radio waves attenuation due to melting layer of precipitation, *IEEE Transactions on Antennas Propagation*, 42, 492–500, 1994.

3 Mobile Satellite Channel Characterization

Athanasios G. Kanatas

CONTENTS

3.1 INTRODUCTION

The growing demand for comprehensive broadband and broadcast/multicast high-speed wireless communication services, global coverage, and ubiquitous access have prompted the rapid deployment of satellite networks. These networks can strongly support terrestrial backhaul networks by providing uninterrupted radio coverage to stationary, portable, and mobile receivers, bringing communications to sparsely populated or underdeveloped areas, and still maintaining exclusive status in traditional maritime and aeronautical markets from the wide area perspective due to their unique coverage feature. Nevertheless, the development of next-generation systems envisages the synergetic integration of heterogeneous terrestrial and satellite networks (Evans et al., 2005; Giuliano et al., 2008; Kota et al., 2011) with different capabilities, providing voice, text, and multimedia services at frequencies ranging from 100 MHz to 100 GHz as well as at optical frequencies, which gives rise to new services, architectures, and challenges.

As the mobile communications sector has definitely been the fastest-growing area of the telecommunications industry due to its ability to connect people on the move, systems that provide mobile satellite services (MSS) are seen to be critical to the development of current and future networks (Ohmori and Wakana, 1998; Richharia, 2001; Swan and Devieux, 2003; Ilcev, 2005).

According to the Radio Regulations (RR No. S1.25) (ITU, 2012), MSS is a radio communication service between mobile Earth stations (MES) and one or more space stations, or between MES by means of one or more space stations. The MSS includes the land (road or railway) mobile satellite service (LMSS), the maritime mobile satellite service (MMSS), and the aeronautical mobile satellite service (AMSS), depending on the physical locale of the mobile terminals (ITU, 2002). The terminals supported vary from vehicle mounted to handheld devices. Moreover, flexible small-sized Wi-Fi routers have entered the market such as IsatHub by Inmarsat (Inmarsat, 2015), SatSleeve by Thuraya (Thuraya, 2015), and IridiumGO!™ by Iridium (Iridium, 2015), enabling the typical mobile communications smartphones to be used as satellite phones. New broadband networks are designed and are in the process of launching, for example, Global Xpress by Inmarsat (Inmarsat, 2015), providing high-speed services in diverse terminals and vehicles.

An LMSS was first proposed by the Communications Research Center (CRC) of Canada (DOC Report to MOSST, 1981; ADGA/Touch-Ross and Partners, 1982) and later by the Federal Communications Commission (FCC) and National Aeronautics and Space Administration (NASA) in the United States (Anderson et al., 1982; Castruccio et al., 1982; LeRoy, 1982; Abbot, 1984; Cocks et al., 1984). In LMSS, the MES is located on different types of cars, trucks, buses, trains, and other civil or military vehicles, providing commercial, logistics, and business communications. Historically, MMSS was the first mobile satellite service offered in the 1970s by COMSAT and in the 1980s by Inmarsat. In MMSS, the MES is located on-board merchant or military ships, other floating objects, rigs, or offshore constructions, hovercrafts, and/or survival craft stations, providing commercial, logistic, tactical, defense, and safety communications. In AMSS, different types of terminals in airplanes, helicopters, and other civil or military aircraft can be used to provide logistics, air traffic control, safety, telemedicine, in-flight office, business, infotainment, and private high-speed broadband communications, primarily along national or international civil air routes. Figure 3.1 illustrates different types of mobile terminals.

The main characteristics of the new-generation geostationary satellite orbit (GSO) and non-GSO satellite systems are the frequency bands used, the fixed and steerable multiple narrow spot beams, the on-board processing for some of them, and the ability to generate kilowatts of power. Most current MSS operate in the L- and S-bands, although the operation of MSS in higher frequencies up to Ka-band will satisfy the increasing demand for bandwidth (Inmarsat, 2015; Thuraya, 2015; Iridium, 2015). Therefore, a high-speed broadband is expected to be provided by the 20–30 GHz band, using a large number of transponders.

To understand the practical technical difficulties and limitations of MSS, a detailed knowledge and characterization of the underlying radio channel, under various propagation conditions and scenarios, is crucial. Then, efficient and reliable MSS can be designed and accurately tested, before their implementation. Using this knowledge, MSS can be designed to obtain optimal or near-optimal performance. This philosophy has been the driving force behind the research activity on all types of wireless communications systems. Although measurement campaigns are expensive, time-consuming, and difficult to carry out, conducting measurements and collecting measured channel data is a precondition for the successful validation of the results of preliminary theoretical efforts. Hence, the characterization of the satellite radio channel through real-world channel measurements has received the attention of many researchers.

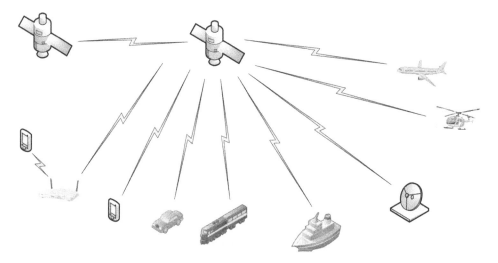

FIGURE 3.1   Different types of mobile terminals of MSS.

Motivated by this observation, this chapter attempts to shed light on a variety of research activities with reference to the characterization of the mobile satellite radio channel through measurement campaigns, when single- or multiantennas are employed. A presentation of the measurement setup and a critical description of the main results obtained from these measurements are included. Apart from considering the vehicular terminal case, the nomadic and pedestrian/handheld terminal cases are also examined in LMSS, which invoke limited mobility and different antenna characteristics. Since scarce information is available in the literature on the experimental characterization of MMSS and AMSS, the spotlight is on the LMSS and especially the multiantenna systems, which have the potential to be at the head of future LMSS due to the possible enhancement in terms of channel capacity and link reliability that can be achieved compared to conventional single-antenna systems (Paulraj et al., 2004; Arapoglou et al., 2011b). Future research directions on the characterization of MSS systems are also underlined.

3.2 PROPAGATION PHENOMENA FOR MSS

The use of MSS involves propagation environments for the transmission of radio signals different from that in conventional mobile terrestrial systems (MTS). The most basic losses that occur in an MSS are losses due to free-space attenuation. Although these losses have a major impact on the signal strength of the line-of-sight (LoS) component, due to the large distance the signals travel from the transmitter to the receiver, the situation is even worse. The transmission to/from a satellite from/to an MES takes place in a complex propagation environment with a moving transmitter/receiver and a multiplicity of obstacles in the path affecting the mobile satellite channel. As shown in Figure 3.2, the radio waves traveling between a satellite and an MES experience several kinds of propagation impairments—the effects of the ionosphere, the effects of the troposphere, and the terrain and local environment fading and shadowing effects (Sheriff and Fun Hu, 2001; Ippolito, 2008).

The combined influence of these impairments on a satellite–Earth link can cause random fluctuations in amplitude, frequency, phase, angles of arrivals, depolarization of electromagnetic waves, and shadowing, which result in degradation of the signal quality, increase in the error rates of the radio links, and degradation of system reliability. Unfortunately, even with state-of-the-art high-power satellites with narrow spot beams or multiple satellite constellations, link availability is not always possible when the signal is blocked or suffers from fading. Therefore, signal degradation caused by attenuation and multipath should be accurately calculated in order to allow for the necessary link margins. Moreover, the received signal is disturbed by radio noise from the receiver antenna, atmospheric gases, rain, clouds, surface emissions, cosmic background noise, and interference from other subscribers or other services. Although the frequency is the major factor affecting the propagation characteristics, the type of service, that is, land, maritime, or aeronautical, is crucial for the determination of the intensity of the phenomena.

For frequencies greater than about 3 GHz, the ionosphere is transparent to radio waves and the main propagation phenomena are caused by the troposphere, the terrain, and the local to the MES environment on Earth. The tropospheric effects include atmospheric gaseous attenuation and precipitation attenuation and depolarization. At frequencies lower than 10 GHz, these effects are not important, but at higher frequencies, one should consider the reduction in signal amplitude, the increase in noise temperature, and the depolarization due to rain and ice particles. The principal interaction mechanism involving the gaseous constituents in the atmosphere (oxygen, nitrogen, argon, etc.) and a radio wave is molecular absorption. The absorption occurs at a specific resonant frequency or narrow band of frequencies. Therefore, a radio wave propagating through a satellite–Earth communication link will experience reduction in the received signal's amplitude level due to attenuation by different gaseous constituents in the atmosphere. The amount of fading due to gases is characterized mainly by the altitude above sea level, the frequency, the temperature, the pressure, and the water vapor concentration (ITU-R P.676-10). The precipitation causes attenuation to electromagnetic waves and can take many forms in the atmosphere. Hydrometeor is the general term

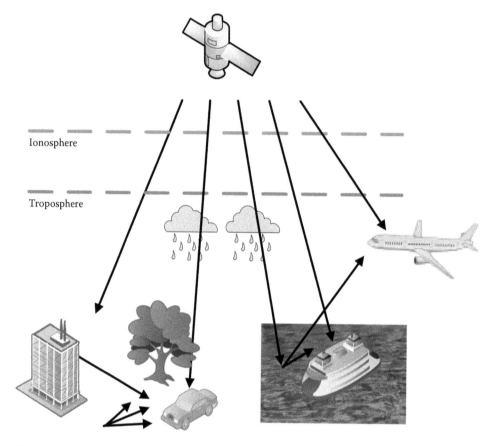

FIGURE 3.2 An MSS suffering from the effects of the ionosphere, the troposphere, and the local environment on the terrain.

referring to the products of condensed water vapor in the atmosphere, observed as rain, cloud, fog, hail, ice, or snow. The latter (hail, ice, and snow) play a minor role in the attenuation. Rain drops absorb and scatter the electromagnetic waves affecting the availability of the communication links (ITU-R P.838-3, 2005; ITU-R P.840-6, 2013; ITU-R P.618-12, 2015).

Another tropospheric effect is refraction and scintillation, that is, the changes of the angle of arrival and rapid signal amplitude variations, caused by corresponding refractive index changes. These two phenomena are important at frequencies greater than 10 GHz and at links with elevation angles lower than 10°. Amplitude scintillations increase with frequency and with the path length, and decrease with the antenna beamwidth. There are two more effects caused by the changes of the refractive index in the troposphere, the decrease in antenna gain due to wavefront incoherence and the beam spreading loss, but both are negligible compared to other propagation phenomena (ITU-R P.618-12, 2015).

The local to the MES environment is responsible for multipath fading, Doppler fading, and shadowing. The multipath fading is the small-scale amplitude and phase variation, caused by the reflections, diffractions, and diffuse scattering by terrain obstacles, man-made structures, and surface roughness. This fading usually degrades the bit error rate (BER) performance for digital links. The same multipath components are responsible for a possible wave depolarization and the corresponding power transfer from the desired polarization state to the orthogonal one, causing an impairment of cross-polarization discrimination.

Shadowing effects are due to the presence of obstacles in the LoS path from the satellite to the MES and result in severe signal degradation. In LMSS, the obstacles and the scatterers may be roadside trees, buildings of varying heights, dimensions, and densities, and other artificial structures, utility poles, mountains and hills, the user's body, etc. (ITU-R P.681-8, 2015). The received signal experiences losses due to shadowing or blocking of the LoS component. The appearance of the foliage medium in the path of the communication link of LMSS has been found to play a significant role on the quality of service (QoS) (Goldhirsh and Vogel, 1987; Mousselon et al., 2003; Sofos and Constantinou, 2004). Generally, the foliage effects are due to a tree, a line or multiple lines of trees, or a forest. Hence, designers of LMSS require information regarding signal degradation effects of foliage at different frequencies and foliage types for various geographic locations. The development of foliage loss prediction methods is also desirable. Discrete scatterers such as the randomly distributed leaves, twigs, branches, and tree trunks can cause attenuation, scattering, diffraction, and absorption of the radiated waves. Figure 3.3 demonstrates the effect of shadowing from the canopies of the trees. In MMSS, multipath components result from reflections to the sea surface and possibly the ship (ITU-R P.680-3, 1999). In AMSS, the phenomenon is limited and is present at low aircraft altitudes (ITU-R P.682-3, 2012). Doppler shift takes place when there is relative movement between the transmitter and the receiver and becomes important for high relative velocity. There are techniques to correct a constant Doppler shift; nevertheless, the problem is bigger when multipath components suffer from different Doppler shifts.

For frequencies greater than 30 MHz and below 3 GHz, the radio waves propagate through the ionosphere but additional propagation phenomena affect the signal transmission. In this frequency region, ionospheric scintillation may occur, that is, rapid fluctuations of the amplitude and phase of the electromagnetic waves due to electron density irregularities in the ionospheric region from 200 to 600 km. These electron density inhomogeneities cause refractive focusing and defocusing of radio waves and lead to amplitude and phase fluctuations. Ionospheric scintillation is important at frequencies less than 1 GHz, at regions with low latitude and at auroral zones. Solar activity is a factor that worsens the scintillation effect. According to ITU-R P.531-12 (2013), at a frequency of

FIGURE 3.3 An LMSS suffering from the effect of shadowing from tree canopies.

1 GHz, the scintillation may take up values peak-to-peak of 20 dB or higher. At 3 GHz, the peak-to-peak value drops to 10 dB. Another phenomenon that is important in this frequency region and for linearly polarized waves is the Faraday rotation, that is, the progressive rotation of the polarization plane of an electromagnetic wave due to the presence of free electrons in the signal path from the satellite to the MES. The rotation is inversely proportional to the frequency squared ($\propto 1/f^2$) and directly proportional to the integrated product of the electron density and the component of Earth's magnetic field along the propagation path. It is negligible for high frequencies. It can be overcome using circularly polarized waves. It has been observed that at L-band, the rotation may reach the value of 50° or greater, depending on the electron density (ITU-R P.531-12, 2013). The free electrons in the ionosphere are also responsible for the group or propagation delay, that is, the reduction of the propagation velocity of the waves. Group delay is inversely proportional to the frequency squared ($\propto 1/f^2$) and is negligible at L- and S-bands. A value of 0.25 µs is given in ITU-R P.531-12 (2013) at 1 GHz, whereas at 3 GHz, this value decreases to 0.028 µs.

3.3   CHANNEL CHARACTERIZATION OF SINGLE-ANTENNA MSS

This section presents MSS channel characterization activities performed using single-antenna systems, that is, targeting single-input single-output (SISO) systems. The majority of the measurement campaigns cover the LMSS channel and the propagation phenomena of shadowing and multipath fading caused by obstacles in the propagation path and scatterers in the local environment. Nevertheless, there are few campaigns for MMSS and AMSS presented herein. LMSS systems provide services in environments such as urban, suburban, or rural areas and often operate under mixed propagation conditions, that is, clear LoS, slight shadowing, and occasional full blockage. This behavior has led the research community to the development of statistical models based on semi-Markov chains. The great importance of the terrain effects makes LMSS vulnerable to outage and decreases the system availability in the range from 80% to 99%. On the contrary, MMSS indicate a very different propagation scenario, where the modeling of the specular ground reflection is largely ignored. In the case of MMSS, reflections from the surface of the sea provide the major propagation impairment, which is especially severe when using antennas of wide beamwidth and operating at a low elevation angle. In particular, terrain and local environment effects for MMSS are blockage from ship structures along the Earth–satellite path, multipath from ship structures, and multipath from the ocean at low grazing angles when low-gain antennas are used. The AMSS radio channels need to be carefully specified due to safety regulations, in order to obtain a high degree of reliability. Note that a land mobile channel-type environment could be envisaged, as long as the aircraft is in the airport. Then, the channel is subject to sporadic shadowing due to buildings, other aircrafts, and obstacles. The aeronautical channel is further complicated by the maneuvers performed by an aircraft during the course of a flight, which could result in the aircraft's structure blocking the LoS to the satellite. The body of the aircraft is also a source of multipath reflections, which need to be considered, whereas the speed of an aircraft introduces large Doppler spreads. Specific broadband multipath models are needed for the AMSS during the aircraft final approach to land when communication availability and reliability and navigation accuracy and integrity are important. Therefore, for landing procedure, the ground reflections and the reflection from the fuselage are significant. The latter is short-delayed but strong multipath affecting broadband communication signals.

3.3.1   DESCRIPTION OF MEASUREMENT CAMPAIGNS AND RESULTS

An extended insight into the typical narrowband single-antenna channels of LMSS, MMSS, and AMSS at several frequency bands, for example, VHF, UHF, L-, S-, Ku-, K-, and Ka-band, and different propagation environments was acquired through experimental research efforts over the last few decades.

3.3.1.1 Measurements for LMSS

The first channel measurements for LMSS were sponsored by NASA in 1977–1978 and were carried out by Hess using the ATS-6 satellite at 860 MHz (UHF-band) and 1550 MHz (L-band), for elevation angles ranging from 19° to 43° (Hess, 1980). The measurements took place in different environments using a mobile test van where the measurement equipment was located. A quarter-wave whip over ground plane antenna was used with right-hand circular polarization. Excess path loss measurements were obtained and the effect of the local environment on the signal propagation was investigated. The experimental results indicated that the excess path loss varies from 25 dB in urban areas to 10 dB to suburban/rural areas. The high values of excess path loss were due to the shadowing created by buildings and the low elevation angles. There was no noticeable difference in excess path loss between 860 MHz and 1550 MHz links. A statistical analysis showed that local environment and vehicle heading are the most important parameters. Moreover, second-order statistics were examined providing decreased level crossing rate (LCR) and increased average fade duration (AFD) compared to typical Rayleigh fading conditions. Therefore, shadowing was considered more important than multipath fading giving a reduced spatial (antenna) diversity gain of 4 dB. Similar communications and position-fixing experiments are reported in Anderson et al. (1981).

In 1981–1983, a satellite channel measurement campaign was conducted at the Communications Research Centre (CRC) through the Canadian Mobile Satellite (MSAT) program (Anderson and Roscoe, 1997). Measured data for the path loss of a satellite system were collected in a suburban area, that is, a residential area of Ottawa, on streets bordered by one- and two-story houses, and rural areas, that is, roads passing through predominantly forested, hilly terrain, at the ultra-high frequency (UHF) band (800 MHz and 870 MHz) and the L-band (1542 MHz) (Butterworth and Matt, 1983; Huck et al., 1983; Butterworth, 1984a,b). A helicopter emulated the satellite transmitter equipped with an inverted conical log-spiral antenna for the UHF-band, whereas for the L-band, Inmarsat's MARECS A satellite was used. The signal was received by a conical log-spiral antenna and a drooping crossed dipole for the UHF-band and the L-band, respectively. Various scenarios were simulated with the helicopter for different elevation angles from 15° to 20°. However, the corresponding elevation angle for the L-band was 19°. Data were recorded as a function of distance traveled with a sampling distance of 5 cm at UHF and 2.5 cm at L-band. The results obtained using a helicopter and the GEO MARECS A satellite in a suburban area at UHF- and L-band revealed that good signal levels appear when the satellite is in LoS (Butterworth and Matt, 1983; Huck et al., 1983; Butterworth, 1984a,b). Besides, severe losses due to shadowing were also visible. This was confirmed by the results in Vogel and Hong (1988), where a difference of 1–1.3 dB was observed. The signal envelope and signal phase data were extracted by transforming the rectangular coordinates, that is, in-phase (I) and quadrature (Q) components, to polar coordinates, that is, envelope and phase. The measurement results of CRC also demonstrated that an excess loss margin of 11.5 dB would be required for a 90% area coverage in West Carleton at an elevation angle of 15° (Butterworth, 1984a). The results also show that the signal level decreases when the elevation angle decreases. Moreover, signal-level spread for various elevation angles is much larger for heavy shadowing than it is for light shadowing. Indeed, there is less shadowing (blockage) by trees for higher elevation angles than there is for lower elevation angles. In addition, light shadowing is generally experienced when traveling through sparsely wooded areas, whereas heavy shadowing is experienced when traveling through densely wooded areas. The results also show that the autocorrelation function almost linearly decreases with time lag due to the strong LoS component (Loo et al., 1986).

In 1983 and 1984, Vogel and Smith carried out preliminary experiments using stratospheric balloons at 869 MHz (Vogel and Smith, 1985). The transmitter was carried for 12 h by the balloon and the measurements were made in a van at elevation angles from 10° to near 35° while traveling on rural U.S. highways. The antennas utilized both at the transmitter and receiver are identical circularly polarized drooping crossed dipoles and the sampling interval was λ/8. Link margins of the order of 2–8 dB were measured and second-order statistics were derived.

Vogel and Goldhirsh carried out extensive channel measurement campaigns mainly sponsored by NASA for LMSS at UHF-, L-, S-, and K-band. The attenuation and fading statistics for shadowing and multipath from roadside trees was investigated using different platforms, that is, helicopter (Goldhirsh and Vogel, 1987, 1989; Vogel and Goldhirsh, 1988), remotely piloted aircraft (Vogel and Goldhirsh, 1986), stratospheric balloon (Vogel and Hong, 1988), tower-mounted transmitter (Vogel and Goldhirsh, 1993; Vogel et al., 1995), or a geostationary satellite (Vogel and Goldhirsh, 1990; Hase et al., 1991; Vogel et al., 1992; Vogel and Goldhirsh, 1995). Goldhirsh and Vogel (1992) presented a review of these campaigns. The elevation angle range covered was from 20° to 60°.

Several different trees were measured and the average single tree attenuation under full foliage conditions was found to range from 6.3 to 15.5 dB in UHF-band. The corresponding largest attenuation measured was from 7.7 to 19.9 dB. The comparison with L-band measurements resulted in a scaling factor of 1.31–1.38 for the attenuation in dB. The range of values reported for the average attenuation at L-band was from 3.5 to 20.1 dB. The results of measurements on fade distributions over mountainous terrain (canyons) implied that at the 1% and 5% levels, 4.8 and 2.4 dB fades were on average exceeded at UHF-band and 5.5 and 2.6 dB at L-band, respectively. This may be compared to roadside tree fading of 15 and 9 dB. Percent level is defined as the percentage of time that a faded signal is exceeded. Results on the effect of trees in full bloom and the dependence of fade distribution as a function of elevation angles are also given in Goldhirsh and Vogel (1989). At K-band, the corresponding 1% fade for different trees ranges from 25 to 43 dB. Moreover, the median attenuation for trees without foliage was found to be 3 dB smaller than the median attenuation at L-band (Vogel and Goldhirsh, 1993). Hase et al. (1991) presented the results from a measurement campaign conducted in Australia where the transmitter was the Japanese Experimental Test Satellite-V (ETS-V). It was shown that the cumulative distributions of fade and nonfade durations follow a lognormal and power law, respectively. At 1% probability, fades last 2–8 m and nonfades 10–100 m.

Narrowband channel measurements at L-band (1.54 GHz), and for elevation angles ranging from 13° to 43° were conducted by the German Aerospace Research Center (Deutsche Forschungsanstalt für Luft- und Raumfahrt [DFVLR], now DLR) in several European cities using MARECS satellite (Lutz et al., 1991). The results of Lutz et al. showed that for an elevation of 24° and at a 9% level, the faded signal is more than 10 dB below an "unfaded" level for a highway environment. For urban areas (Munich, Germany) and for a 10-dB fade as before, the probability level is at 60%, that is, at a 40% level. More narrowband and wideband measurements were conducted by the DLR at 1.82 GHz using a light aircraft carrying a test transmitter (Jahn and Lutz, 1994; Lutz et al., 1995; Jahn et al., 1996) at elevation angles from 15° to 75°. Shadowing of 15–30 dB was recorded whereas a link margin of 10–30 dB was calculated for 95% and 98% service availability in urban and suburban environments, respectively. For the wideband measurements, a signal bandwidth of 30 MHz was used and the delay spread was calculated for urban, suburban, and highway environments from 15° to 75°. The delay spread values range from 500 ns to 2 μs.

A propagation experiment at 1.5 GHz was performed in Europe in the framework of the European Space Agency's (ESA's) PROSAT program using MARECS-B2 in rural, suburban, densely wooded, railroad, and urban environments (Benarroch and Mercader, 1994). The results show that the diffuse multipath signal always was at least 11 dB below the direct signal. The fades for a 10% level and an elevation of 39° were 10.6, 2.0, and 2.0 dB for urban, suburban, and rural areas, respectively, whereas for a 10% level and an elevation of 13°, the fades were 13, 11, and 9 dB for urban, suburban, and rural areas, respectively. The unfaded signal availability was in the range of 80%–90% in rural and suburban environments of Europe with high and intermediate elevation angles. The availability is smaller (60%) at 13° and 30% in urban environments or thick woods at 26°.

Using a light aircraft as the transmitter platform, Renduchintala et al. (1990) conducted measurements at several elevation angles and under diverse terrain conditions. For a 10% level and at an elevation of 40°, the fades are 5.6, 6.0, and 16.4 dB for tree-shadowed, suburban, and dense urban areas, respectively. However, at an elevation of 60° and for a 10% level, the fades are 4, 1, and 14.2 dB for tree-shadowed, suburban, and dense urban areas, respectively.

A light aircraft carrying a continuous-wave (CW) beacon was also used in Smith et al. (1991, 1993) in open, urban, suburban, and tree-shadowed environments, where channel measurements were conducted for an LMSS at L- and S-bands and at elevation angles of 40°, 60°, and 80°. The results of Smith et al. (1993) depicted that the fades were 3.5, 5.5, and 5.9 dB for suburban, urban, and tree-shadowed areas, respectively, for a 10% level.

Butt et al. (1992) used a helicopter to mount a transmitter and simulate an elevation of 60–80° over suburban, wooded, and open areas. The frequencies used in this campaign were 1297.8 MHz (L-band), 2450 MHz and 2320 MHz (S-band), and 10.368 GHz (Ku-band). The main results of this campaign were that the high elevation angles tend to reduce the fade levels due to shadowing and multipath, while attenuation increases with frequency. Numerical results depicted that the fades at L-band and at 60° were 5.0 and 7.5 dB for suburban and wooded areas, respectively, for a 10% level, whereas at 80°, the values decrease to 2.5, 3, and 0.9 dB for suburban, wooded, and open areas, respectively. At S-band, the corresponding values at 60° were 6.0 and 8.8 dB for suburban and wooded areas, whereas at Ku-band the corresponding values were 13 and 18.5 dB, respectively.

A helicopter was also used as a transmitter in Bundrock and Harvey (1988), where the authors conducted measurements for an Australian LMSS. Results are given for two different tree densities, for elevation angles of 30°, 45°, and 60° and for frequencies of 893, 1550, and 2660 MHz. These results show that at 1550 MHz and a 45° elevation angle, attenuation values of 5.0 and 8.6 dB were exceeded 10% of the time for roadside tree densities of 35% and 85%, respectively.

Wakana et al. (1996) and Obara et al. (1993) conducted L-band measurements with ETS-V at expressways for various elevation angles (40–48°) in Japanese cities, whereas Yoshikawa and Kagohara used ETS-V on open areas and main roads in Japanese cities at elevation angles 34° and 47° (Yoshikawa and Kagohara, 1989). Other measurements concerning the ETS-V are reported in Matsumoto et al. (1992), in Kyoto city, and in Ikegami et al. (1993). The results in Wakana et al. (1996) and Obara et al. (1993) showed Rician factor values from 15.5 dB to 24.7 dB and fade levels in the range from 5.3 to 32.3 dB for a 1% level and from 1 to 23.4 dB for a probability of 10%. The results of Yoshikawa and Kagohara showed that the received power distribution in open areas is well fitted by the Nakagami–Rice distribution. The attenuation measured due to utility poles ranges from 6 to 8 dB, whereas the attenuation due to trees ranges from 10 to 20 dB. The received signal level is 3 dB lower than the unfaded LoS signal at a 10% level. The results of Matsumoto et al. (1992) underlined that the received power distribution in roads with light shadowing is well fitted by Rice distribution. Moreover, the received power level fluctuation is dominated by heavy shadowing, whereas both fade and unfade duration distributions for 0 dB threshold are lognormal. In unshadowed areas, the LCR peak value is 1 and 3 m at 0 dB threshold for directional and omni antennas, respectively. The results in Ikegami et al. (1993) showed that the maximum excess delay is within 1 µs and the maximum delay spread is 0.2 µs. The coherence bandwidth was approximately 1 MHz.

Kanatas and Constantinou (1998) conducted measurements in an urban environment, that is, in typical narrow urban streets with heavy traffic and large building blocks, at 1800 MHz, using a helicopter to mount the transmitter and to simulate high elevation angles (60–80°). The transmitter consisted of a signal generator supplying a tone at 1800 MHz, a 10-W power amplifier, a watt meter, and a crossed-drooping dipoles antenna. The receiver was located on a land vehicle moving with constant speed along the streets and consisted of an inverted-V crossed-drooping dipoles antenna, band-pass filter (BPF), low-noise amplifier (LNA), ICOM R7000 receiver, and data-acquisition system. The results in Kanatas and Constantinou (1998) showed that the fades range from 0.9 to 25.8 dB depending on the elevation angle for a 10% level and for an urban environment with narrow streets and high buildings. For a 1% level, the corresponding fade depth ranges from 2.4 to 33.8 dB. Overall, the analysis in Kanatas and Constantinou (1998) indicates a considerable increase of the fade depth compared to suburban and rural environments and that the fade depth decreases with the increasing elevation angle. Moreover, the signal attenuation depends on the width of the streets, average building height, and vegetation at the edges of the streets.

The roadside attenuation due to vegetation or man-made structures on the digital audio radio satellite (DARS) services at 2.33 GHz (S-band) was investigated in Mousselon et al. (2003). The XM Radio DARS system was employed, which consists of two geostationary satellites to cover the North American continent and the measurements were realized in southwest Virginia, which offers terrain that ranges from mountainous rural to open suburban. The measurements considered either unshadowed or vegetative shadowed cases. Unshadowed propagation was assumed to follow a Ricean distribution with an average Ricean factor of 15.75 dB, whereas the shadowed propagation was assumed to follow a vegetative shadowed distribution, which is a combination of a lognormal distribution and a Rayleigh distribution Mousselon et al. (2003).

Propagation measurements at L-band (1600 MHz) and UHF-band (800 MHz) were also performed in Trabzon, Turkey, in 1993 and 1996, respectively, for 14 different tree types and an elevation angle of 30° (Cavdar, 2003). The transmitter was on the top of a high building, whereas the receiver was located in a van. Dipole antennas with vertical polarizations were used at the transmitter side for both bands. The L-band receiver antenna was a drooping dipole and the UHF-band receiver antenna was a quarter-wave dipole antenna. The variations of the tree attenuation were examined during several months, with and without foliage, and the average values of the attenuation were found to be 8.60 and 11.00 dB for UHF- and L-band, respectively. The scaling factor between L- and UHF-band attenuations in dB was determined to be 1.32. Fade depths of 27 dB at L-band and 23.5 dB at UHF-band were calculated at 1% level.

Basari et al. (2010) used the geostationary satellite Engineering Test Satellite VIII (ETS-VIII) of the Japan Aerospace Exploration Agency (JAXA) and a vehicle-mounted antenna system for channel measurement purposes in various environments, that is, open field areas and blockage areas, in Japan at an elevation angle of 48° and at the S-band (2.5 GHz). A global positioning system (GPS) module was utilized to provide accurate information of the vehicle's position and bearing. The received signal power and the average BER were simultaneously retrieved. The results showed that different environments give different degrees of attenuation, which affects the BER performance in terms of fade depth. For LoS measurements, the fade depth at 10% level is 4 dB, whereas the fade depth caused by utility pole and trees is 7 dB at 10% level for single pole and sparse foliage blockage and 11 dB for dense foliage.

Numerous measurements were carried out using Olympus (Pike et al., 1989; Pike, 1993; Loo, 1994, 1996; Murr et al., 1995), Italsat (Damosso et al., 1994; Borghino et al., 1996; D'Amato et al., 1996; Buonomo et al., 1997; Kubista et al., 1998, 2000), EUTELSAT (Dutronc and Colcy, 1990), (NASA's) Advanced Communications Technology Satellite (ACTS) (Pinck and Rice, 1995; Rice et al., 1996), the geostationary satellite Astra (Scalise et al., 2008) and the COMETS satellite (Wakana et al., 2000), in the Ku-, Ka-, and mm-wave frequency bands. Although the mm-wave frequencies cover the range from 30 to 300 GHz, sometimes frequencies above 10 GHz are also called mm-wave frequencies. The use of Ku- and Ka-band for satellite communications is traditionally limited to stationary services and imposes several tropospheric impairments, such as rain attenuation, gaseous absorption, cloud attenuation, scintillation, depolarization, and atmospheric noise. Nevertheless, a good knowledge of the effects due to the environment in the vicinity of the MES is necessary, in order to plan future MSS. Specifically, similar to lower frequencies, shadowing/blockage and multipath are observed.

Loo (1994, 1996) described the propagation measurements conducted at CRC using the Olympus satellite. An experimental 4.2-m Ka-band terminal (Pike et al., 1989; Pike, 1993) was used to access the Olympus satellite at Ottawa, Canada at an elevation angle of 14.2°. A CW signal at 28.072 GHz frequency was transmitted from the 4.2-m terminal and was down-converted at 18.925 GHz and received by a mobile terminal traveling at a speed of 10–20 km/h in the direction of the satellite. The mobile terminal was equipped with a standard gain horn antenna with a beamwidth of 10° at a frequency of 18.925 GHz. The received CW signal was further down-converted to an intermediate frequency (IF) of 4.8 kHz for storing proposes. The main result of the campaign was that the signal envelope for MSS at Ka-band can be modeled as shadowed Rician, and the phase can be modeled as Gaussian. However, weather conditions should be considered and the main assumption was that the two fading processes are independent, yielding a multiplicative model.

Borghino et al. (1996) used Italsat to conduct their measurements in Turin at an elevation angle of 37.8°. Details on the design of the experiment are given in Damosso et al. (1994). Their results showed attenuation of 15–20 and 5–9 dB for trees with and without leaves, respectively. Similar results were presented in D'Amato et al. (1996). Using Italsat, Buonomo et al. (1997) conducted measurements over a range of elevation angles, 30–35°. Their results showed that for a 10% level, the fades are 0.9, 7.2, 9.8, 10.1, and 21.8 dB for open, suburban, mixed, tree-shadowed, and urban areas, respectively, and for a 1% level, the fades are 9.1, 16.3, 15.6, 17, and 26.6 dB, respectively. Kubista et al. (1998, 2000) presented the effects of shadowing/blockage at Ka-band using the 18.7 GHz beacon of the Italsat satellite. The co-polar signal component (horizontal polarization) was measured in a number of representative environment types, that is, urban, suburban, and tree-shadowed, in several European countries and for elevation angles between 28° and 36°. The MES was a van provided by ESA allowing mobile measurements to be carried out at constant velocity of 20 km/h. The receiving antenna was of the Cassegrain type with a gain of 38 dB and a 3-dB beamwidth of 2.4°. The results in Kubista et al. (1998, 2000), underlined that large link margins should be implemented and mitigation techniques, for example, multisatellite diversity, are required to provide reasonable availabilities and reliable MSS at Ka-band both in tree-shadowed and built-up environments. The authors provide availability levels for tree shadowing margins in built-up areas for different MES route orientations with respect to the satellite (0°, 45°, and 90°).

A measurement campaign using EUTELSAT I-F1 satellite and the OmniTRACS land mobile communication system were performed in 1989 at the Ku-band in a total of seven countries of Europe: the United Kingdom, France, Spain, Italy, Germany, the Netherlands, and Switzerland (Dutronc and Colcy, 1990). The measured data collected corresponded to 150 h of operation for the forward link and 150 h of operation for the return link, during which time around 10,000 alpha-numerical messages were sent in each direction. The results suggested that there is about a 5 dB margin in the forward link transmission when reception is on the −2.5 dB contour of the Eurobeam, that is, a zone covering almost all of the EUTELSAT Signatories. The system is very resistant to TV interference and short signal interruptions and ensures service availability. However, although the performance is good in open areas, that is, highways, the message throughput reduces in urban areas due to the increase in the required number of message retransmissions.

Measurements using ACTS at an elevation of 46° were presented in Pinck and Rice (1995) and Rice et al. (1996). The measurements were performed at 19.914 GHz using a high-gain reflector receiver antenna mechanically steered with a pointing error less than 2°. The results of Pinck and Rice showed that for a 10% level, the fades are 0.8, 3.5, and 29.5 dB for lightly, moderately, and heavily shadowed suburban areas, respectively. As a concluding remark, the authors mention that a complete channel characterization should include the fade exceedance levels, the average fade durations, and the percentage of signal outage.

Scalise et al. (2008) describe a measurement campaign performed in Munich at an elevation of about 34° for a Ku-band LMSS system, which was commissioned by the ESA. A test signal using horizontal polarization was transmitted by the geostationary satellite Astra and received by means of a low-gain 10 × 10-cm flat antenna with 19 dBi gain. The antenna and the low-noise block (LNB) were mounted on a mechanically steerable platform, which was placed on top of a measurement van provided by the DLR. Four propagation environments were considered; a rural environment (consisting mainly of relatively open areas, tree alleys, and forests), a suburban environment (composed of small obstacles, such as family houses or villas with gardens), an urban environment (characterized by large and high buildings that produce severe blockage effects), and a highway environment (representing an open area with bridges and tunnels). The measurement results revealed that for the highway and urban environments, the increase of the link margin does not provide significant improvement and alternative techniques should be used, for example, techniques based on long interleavers or time diversity techniques. Moreover, the authors provided a three-state statistical channel model with the corresponding probability density functions (PDFs) describing the fast fading processes.

Wakana et al. (1999, 2000) present results from urban measurements at K-band (20.986 GHz) in Tokyo city using COMETS satellite. This satellite was launched in 1998 and was designed to achieve a geostationary orbit. However, owing to an engine failure, the achieved orbit was highly elliptical and the corresponding elevation angles ranged from 37° to 80°. The authors provide Rician distribution fitting for the received signal envelope with factors from 15 to 20 dB. The propagation channel can be classified into two states in urban environments: LoS or blocked due to building and trees. Moreover, a fade margin of 5 dB was measured for 98% of the total distance at high elevation angles.

In 1998, the DLR performed an extremely high frequency (EHF)-band experiment using a transmitter in an airplane (CESSNA 207) and a receiver in a vehicle (Jahn and Holzbock, 1998). The carrier frequency was 40.175 GHz, so the receiver consisted of two-stage downconverters transferring the signal at 1.8 GHz where channel sounders are available. Narrowband and wideband measurements took place with a bandwidth of 30 MHz for the latter case. Moreover, two types of antennas were used, one omni and one highly directive antenna. The measurements took place in a rural environment with an elevation angle of 25°, 35°, and 45°, and in an urban environment with an elevation angle of 15°, 25°, 35°, and 45°. The omnidirectional antenna picks up more multipath. The Rice factor in nonshadowed conditions for the steered high-gain antenna is 21.5 dB, whereas the omnidirectional antenna yields a Rice factor of 17 dB. The wideband measurement was taken in a tree-shadowed rural environment at 25° elevation using the omnidirectional antenna. There are only few (2–4) echoes with short delays (30–100 ns), and a couple of echoes with delays of 350, 600, and 850 ns. The echo attenuation is in the range of 20–30 dB.

Table 3.1 summarizes the aforementioned measurement campaigns for single-antenna LMSS systems.

3.3.1.2 Measurements for MMSS

For MMSS systems, the multipath fading from the ocean occurs when low-gain antennas are used for low elevation-angle scenarios. An experimental investigation on the multipath effects for MMSS scenarios at 1.5 GHz (L-band) and at elevation angles ranged from 15° to 0° was reported in Fang et al. (1982a,b). In this measurement campaign, a terminal on a ship equipped with an antenna of 1.2 m diameter transmitted/received to/from the MARISAT F-1 satellite over the Atlantic Ocean. Time-division multiplexing (TDM) carriers for Teletype and voice carriers for telephone and data transmissions were monitored and analyzed. The results underlined that the mean carrier reduction and peak-to-peak fluctuations were severe at elevation angles below 2°, where, for example, the peak-to-peak fluctuations of the carrier-to-noise ratio were smaller than 4 dB. Cumulative signal distributions relative to the mean values demonstrated that peak-to-peak fluctuations exceeded 10 dB with a probability of 42% in the angular interval of 0.5–2°. Besides, a mean carrier-to-noise ratio drop of less than 2 dB is observed when passing from 10° to 5° elevation angle. The maximum fade level at 5° caused by signal fluctuations was less than 6.5 dB for 99% of the time, whereas the fading was less than 4.4 dB for 99% of the time, at an elevation angle of 10°.

The characteristics of multipath fading and the fade durations due to sea surface reflections were extensively investigated by Karasawa and Shiokawa (1984, 1987, 1988) and Karasawa et al. (1986). They developed simplified prediction models for the fade depths as a function of elevation angle, wave height, and antenna gain, based on measurements at L-band. The models were adopted by the International Telecommunication Union-Radio Sector (ITU-R) in ITU-R P.680-3 (1999). The fade depth was defined as the dB difference between the signal level of the direct incident wave and a threshold level that the resultant signal level exceeds with a probability of a specific percentage of the time. Karasawa et al. (1990) developed an analytical fading model for what is called "wind-wave" and "swell" and they compared their results with those in Matsudo et al. (1987) and Ohmori et al. (1985).

TABLE 3.1

Summary of Measurement Campaigns for Single-Antenna LMSS

Reference	Year/Location	Environment	Satellite or Emulation	Frequency Band	Elevation Angle
Hess (1980)	1977–1978/USA	Urban/semiurban/suburban/rural	ATS-6	UHF- and L-band	19–43°
Anderson et al. (1981)	1979/USA	Rural (open/mountainous/forested)/highways	ATS-6	L-band	9–23°
Butterworth and Matt (1983), Butterworth (1984a,b), Huck et al. (1983)	1982/Canada	Suburban/rural (forested hilly)	MARECS A/helicopter	UHF- and L-band	15–20°
Vogel and Smith (1985)	1983–1984/USA	Rural	Balloon	UHF	10–35°
Goldhirsh and Vogel (1987, 1989), Vogel and Goldhirsh (1986)	1985–1986/USA	Single trees/roadside trees/mountainous	Helicopter	UHF- and L-band	20–60°
Vogel and Goldhirsh (1986)	1985/USA	Trees	Aircraft	UHF-band	10–40°
Vogel and Hong (1988)	1986/USA	Open farm prairie land/roadside trees	Balloon	UHF- and L-band	25–45°
Vogel et al. (1995), Vogel and Goldhirsh (1993)	1990, 1991, 1993/USA	Trees	Tower	L-, S-, and K-band	20–60°
Vogel and Goldhirsh (1990, 1995), Vogel et al. (1992), Hase et al. (1991)	1987, 1992/USA and Australia	Suburban/rural/hilly/highways	MARECS B2 and ETS-V	L-band	7–14°, 21°, 51°, and 56°
Lutz et al. (1991)	1984–1987/Europe	Urban/suburban/rural/highways	MARECS	L-band	13–43°
Jahn and Lutz (1994), Jahn et al. (1996), Lutz et al. (1995)	1994/Europe	Urban/suburban (open and tree shadowed)/rural/indoor	Aircraft	L-band	15–75°
Benarroch and Mercader (1994)	1991/Europe	Urban/suburban/rural/railways	MARECS	L-band	13°, 26°, and 39°
Renduchintala et al. (1990)	1989/England	Urban/suburban/rural	Aircraft	L-band	40°, 60°, and 80°
Smith et al. (1991, 1993)	1989–1990/England	Urban/suburban/rural	Aircraft	L- and S-band	40°, 60°, and 80°
Butt (1992)	1991–1992/England	Suburban/rural (wooded and open)	Helicopter	L-, S-, and Ku-band	60°, 70°, and 80°
Bundrock and Harvey (1988)	1987/Australia	Trees	Helicopter	UHF-, L-, and S-band	30°, 45°, and 60°
Wakana et al. (1996), Obara et al. (1993)	1992–1995/Japan	Expressways (urban/suburban/rural)	ETS-V	L-band	40–48°
Yoshikawa and Kagohara (1989)	1988/Japan	Roads/building/tree	ETS-V	L-band	34° and 47°
Matsumoto et al. (1992)	1990/Japan	Urban	ETS-V	L-band	47°
Ikegami (1993)	1992/Japan	Urban	ETS-V	L-band	40–47°
Kanatas and Constantinou (1998)	1996/Greece	Urban	Helicopter	L-band	60°, 70°, and 80°
Mousselon et al. (2003)	2002/USA	Urban	DARS	S-band	N.A.
Cavdar (2003)	1993–1996/Turkey	Trees	Tower	UHF- and L-band	30°

(Continued)

TABLE 3.1 (Continued)

Summary of Measurement Campaigns for Single-Antenna LMSS

Reference	Year/Location	Environment	Satellite or Emulation	Frequency Band	Elevation Angle
Basari et al. (2010)	2009/Japan	Open areas/rural with blockage	ETS-VIII	S-band	48°
Loo (1994, 1996)	1993/Canada	Open areas/wooded areas	Olympus	Ka-band	14°
D'Amato, Ossola and Buonomo (1996), Borghino et al. (1996)	1995/Italy	Urban/suburban	Italsat	Ka-band	38°
Buonomo et al. (1997)	1996/Italy	Suburban/open/tree shadowed	Italsat	Ka-band	30–35°
Kubista et al. (1998), Kubista et al. (2000)	1997/Europe	Urban/suburban/tree shadowed	Italsat	Ka-band	28–36°
Dutronc and Colcy (1990)	1989/Europe	Urban/open highways/mountainous (forested)	EUTELSAT I-F1	Ku-band	N.A.
Pinck and Rice (1995), Rice (1996)	1994/USA	Suburban	ACTS	K- and Ka-band	46°
Scalise et al. (2008)	2002/Germany	Urban/suburban/rural/highways	Astra	Ku-band	34°
Wakana et al. (1999, 2000)	1998/Japan	Urban	COMETS	K-band	37–80°
Jahn and Holzbock (1998)	1997/Germany	Urban/rural	Aircraft	EHF-band	15–45°

Note: N.A. stands for not available.

It was shown that the fading depth tends to peak for wave heights between 1 and 2 m, and the multipath fading reaches 8 dB, 5 dB, and 4 dB for antenna gains 15 dBi, 21 dBi, and 15 dBi at elevation angles of 5°, 5°, and 10°, respectively. In the same paper, Karasawa et al. provided a relationship between frequency, wave height, and fading depth.

Higuchi and Shinohara (1988) describe the results of the basic propagation characteristics of Standard-C system obtained from the field trial using Inmarsat's satellite of Indian Ocean Region at L-band. During the experiment, the ship Earth station was installed on a small vessel, which sailed around the Osaka bay in the western part of Japan where the Indian Ocean Region (IOR) satellite is seen at 6° of elevation angle. The mean wave height during the experiment was less than 1 m. The ship Earth station antenna was a nonstabilized omnidirectional antenna called quadrifilar helix. The signal fading depth for 99% of time was calculated at around 7 dB.

Several narrowband field trials were undertaken by DLR in 1979–1980 and 1983 to characterize the MMSS at L-band using the in-phase and quadrature components of the received signal with different antennas, at different elevation angles and sea conditions (Hagenauer et al., 1987). The field trials were categorized in three tests using different satellites, namely, the Indian Ocean MARISAT at 73°E, the Atlantic Ocean MARISAT at 15°W, and the Atlantic Ocean MARECS at 26°W. The routes followed by the ships carrying the receiver stations included the Mediterranean Sea, with elevation angles from 0° to 15°, the North Atlantic, with elevation angles from 0° to 15°, and the North Sea and Atlantic Ocean, with elevation angles from 4° to 30°. It was found that the AMSS channel can be modeled by Rice distribution with a Ricean factor, or carrier over multipath power ratio, taking its lowest value (8–9 dB) at the edge of satellite coverage at 5° elevation angle. The Ricean factor increases with elevation angle. The influence of the sea condition is insignificant compared to the influence of elevation. Shadowing effects due to the ship's superstructure cause a signal degradation of 8 dB. The 99% fading range at 5° is 12 dB. The average duration of fades at −3 to −5 dB below average signal power is approximately 0.1 s. Finally, the Doppler offsets are of the order of ±10 Hz with a rate of 1 Hz/s.

Perrins and Rice (1997) describe shipboard measurements at Ka-band using the ACTS. The receive antenna of the mobile terminal preserved at 20 GHz, a gain greater than 18.8 dBi over a 12° elevation beamwidth for elevation angles between 30° and 60°. During the measurement campaign, six significant fades above 10 dB were noted.

Table 3.2 summarizes the aforementioned measurement campaigns for single-antenna MMSS systems.

TABLE 3.2
Summary of Measurement Campaigns for Single-Antenna MMSS

Reference	Location	Ship/Platform	Satellite or Emulation	Frequency Band	Elevation Angle
Fang et al. (1982a,b)	1978/USA	Mobil Aero Oil Tanker	MARISAT	L-band	0–15°
Karasawa and Shiokawa (1984, 1987a,b), Karasawa (1986)	1983/Japan	Ferry boat "Saroma"	MARISAT	L-band	5–15°
Ohmori et al. (1985)	1980 and 1983/ Japan	Coast tower	Inmarsat IOR	VHF- and L-band	5°
Higuchi and Shinohara (1988)	1987/Japan	Small vessel	Inmarsat IOR	L-band	6°
Hagenauer et al. (1987)	1979, 1980, and 1983/Europe	Research vessels "Bannock," "Meteor," and "Gauss"	MARISAT IOR, MARISAT AOR, MARECS AOR	L-band	0–15° 0–15° 4–30°
Perrins and Rice (1997)	1996/USA	U.S. Navy USS Princeton	ACTS	Ka-band	30–60°

3.3.1.3 Measurements for AMSS

The first successful experiments for AMSS were carried out in 1964 (Ilcev, 2005). Pan Am airlines and NASA program succeeded in achieving aeronautical satellite links at the very high frequency (VHF) band using the Syncom III GEO spacecraft.

The first extended measurement campaign to demonstrate the feasibility of satellite communications to mobile platforms, especially aircrafts, was undertaken by the MIT Lincoln Laboratory in 1965–1966 (Lebow et al., 1971). The LES-3 satellite was used as a transmitter and various U.S. Air Force aircrafts were equipped with receiving facility in the 225–400 MHz (VHF/UHF) band. The measurements were taken under a variety of satellite elevation angles. Most of them over the ocean, although some were made over land, ice, and snow. The AMSS channel was measured and parameters such as airborne noise, radio-frequency interference in the aircraft and the satellite, ionospheric scintillation, multipath over ocean, and fading conditions were examined. The received signals were processed and measures such as correlation of fading across the frequency band, correlation of fading with time, and the fading probability distribution were derived. Especially for the fading, the four alternative solutions are discussed in Lebow et al. (1971) to minimize the degrading effects of multipath propagation, including vertical or circular polarization, antenna pattern, and frequency diversity. The roughness of the ocean surface was examined and a Rician distribution was observed for very rough sea surfaces where diffusion components dominate.

Sutton et al. (1973) present results from an extensive measurement campaign involving a KC-135 jet airplane, the ATS-5 satellite, and a ground station at L-band. The measurements were performed over the ocean between August 1971 and May 1972 using an experimental airborne antenna system of a 15 dBi quad-helix and a 13 dBi downward-looking crossed-dipole antenna placed to receive the multipath components. The Doppler spectrum of the scattered signal was Gaussian shaped with bandwidth that increased with the grazing angle. This agrees with the model provided by Bello (1973). Moreover, it was observed that upon reflection of an incident circularly polarized wave, a predominate polarization sense reversal occurred. The observed frequency selective fading characteristics were in agreement with theoretical time dispersion predictions. Typical values from the measured data are given in Tables 1 and 2 of ITU-R P.682-3 (2012).

Channel measurements for an AMSS at the L-band were also presented in Neul et al. (1987). The 17 flights were performed with a twin jet "Falcon" along routes with a constant elevation angle with respect to the Atlantic-Ocean-Region MARECS satellite. Four different antenna units were installed at the aircraft; a conformal, electronically steerable, phased array mounted on the upper left side for reception of the direct signal, a mechanically steerable helix antenna installed near the tail, pointing toward the specular reflection point, a printed patch element, and a crossed dipole both fixed behind one of the windows for the reception of the combined direct plus reflected signal. The L-band signal was downconverted to an IF frequency of 70 MHz in the front end, fed through an IF amplifier, mixed down to baseband in the channel prober, filtered, and finally recorded on analog tape. The received signal at the aircraft was mixed down to baseband to acquire the in-phase (I) and quadrature (Q) components in order to estimate the amplitude and phase variations. Based on the measured data, results regarding the PDF of the received signal were obtained. These results showed that the reflected signal followed a Rayleigh distribution, when the direct and received signals were received separately. However, a Rician distribution is suitable, when the composite direct and reflected signals are simultaneously received. The root mean square (RMS) Doppler spread was evaluated to be in the range of 10 Hz at 5° elevation to 330 Hz at 44° elevation angle. The carrier to multipath power ratio was found to range from 5 dB at 5° elevation to 15 dB at 55° elevation angle. The mean fade duration with respect to a threshold of 5 dB below mean power was 4 ms at 5° elevation and decreases to 0.3 ms at 44° elevation angle.

In 1989, a NASA/FAA/COMSAT/Inmarsat collaborative experiment at L-band was performed to demonstrate and evaluate a digital mobile satellite terminal developed by Jet Propulsion Laboratory (JPL) (Jedrey et al., 1991). The experiment included the establishment of a satellite link between COMSAT coast Earth station in Southbury and a Boeing 727 B100 aircraft flying along the east

TABLE 3.3
Summary of Measurement Campaigns for Single-Antenna AMSS

Reference	Location	Aircraft/Platform	Satellite or Emulation	Frequency Band	Elevation Angle
Lebow et al. (1971)	1965–1966/USA	U.S. Air Force aircrafts	LES-3	VHF/UHF-band	N.A.
Sutton et al. (1973)	1971–1972	KC-135 jet airplane	ATS-5	L-band	9–31°
Neul et al. (1987)	1986/Europe	Twin jet Falcon	MARECS AOR	L-band	4–70°
Jedrey et al. (1991)	1989/USA	Boeing 727 B100	MARECS B2	L-band	N.A.
Hoshinoo (1991)	1990/Japan	Boeing 747	ETS-V	L-band	10–45°
Hozbock et al. (1999)	1998/Germany	Dornier 228 D-CALM	ITALSAT-F1	K/Ka-band	35°

Note: N.A. stands for not available.

coast of the United States, through the Inmarsat MARECS B2 satellite. The worst-case performance was observed in the presence of heavy turbulence and was approximately 1.0 dB worse than that measured in the laboratory for additive white Gaussian noise (AWGN). The fading-induced degradation for clear weather conditions was estimated to be 0.3 dB.

Hoshinoo (1991) presents results from measurement performed using the ETS-V satellite and a Boeing 747 on the North Pacific route at L-band. The range of elevation angle was from 10° to 45°. The main results drawn from the experiment was that the carrier-to-multipath power ratio was over 16 dB, whereas the 1%–99% fading range due to sea reflections was from 1.5 to 3.5 dB. The LCR increases from 20 to 200 as the satellite elevation angle increases.

Field trials for an AMSS operating at the K/Ka-band using the ITALSAT-F1 geostationary Earth orbit (GEO) satellite were presented in Holzbock et al. (1999). The measurements were performed in different flight scenarios and weather conditions, that is, rainy, cloudy, sunny, flights below, throughout, and above clouds, in an area of Munich, Germany. The measurement setup consisted of an antenna rack and a rack holding the control, monitoring, and downconverter units. The receiver was equipped with a steered Cassegrain antenna produced by the DLR with a beamwidth of 4.2° and consisted of two-stage demodulators transferring a 18.685 GHz signal to 2.15 GHz and then to 70 MHz IF band. The antenna platform was top mounted on the aircraft between the wings and tail structure. Rotary joints enabled full-service coverage from −7° to 90° in elevation and several rotations in azimuth. Geographical data, that is, latitude, longitude, altitude, and attitude data, that is, pitch, roll, magnetic heading, and Doppler compensation data were monitored during all flight trials and stored. The results highlighted that during the normal flight cruising with no shadowing, the received power was nearly constant and the variation from the mean value could be described using a Rican probability density function, with a high Rician factor of 34 dB. However, when multipath fading or shadowing occurred due to maneuvers or at low elevation angles, the reception level decreased during the inclination of the body of the airplane and the wing disturbed the signal reception. Reflection due to the tail or nose structure or mean received power variation by clear or rainy weather was also observed. Fade values of up to 15 dB caused by the wings of the airplane were measured during flight maneuvers. Fade depths due to diffraction and shadowing of 2–3 dB were observed when the tail crossed the satellite path. Shadowing or diffraction may also occur on ground by buildings when the airplane is at the terminal or on taxiway. The mean difference between two measured signal levels while flying under and above clouds was 0.78 dB. Table 3.3 summarizes the aforementioned measurement campaigns for single-antenna MSS systems.

3.4 CHANNEL CHARACTERIZATION OF MULTIPLE-ANTENNA MSS

During the last few decades, the multiple-input multiple-output (MIMO) technology (Paulraj et al., 2004) has played an important role in revolutionizing the terrestrial wireless networks and brought

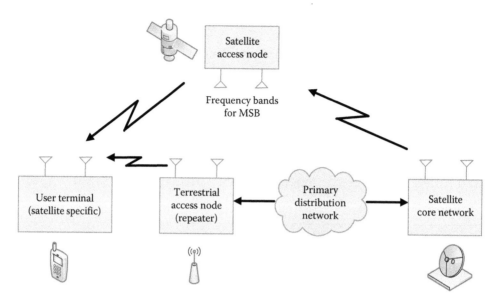

FIGURE 3.4   An MIMO DVB-SH hybrid satellite–terrestrial network.

wireless gigabit vision closer to reality, leading to growing acknowledgment from the research community, industry, and wireless standardization bodies, for example, IEEE 802.11n, 3rd Generation Partnership Project 2 (3GPP2), Ultra Mobile Broadband (UMB), and Digital Video Broadcasting–Second Generation Terrestrial (DVB-T2). Satellite communication systems have not been immune from this wave of innovation and the application of MIMO techniques to satellite systems has gained great interest due to the standardization activities on the finalized DVB-Satellite to Handheld (DVB-SH) standard (ETSI EM 302 583, 2007; Alamanac et al., 2009; Kyröläinen et al., 2014) and the prospective DVB-Next Generation Handheld (DVB-NGH) (Sangchul et al., 2012; Gomez-Barquero et al., 2014) standard. Figure 3.4 illustrates a typical structure of a MIMO DVB-SH hybrid satellite–terrestrial network.

The information theory behind MIMO technology suggests the application of spatial multiplexing and/or space–time coding techniques, where time is complemented with the spatial dimension inherent in the use of multiple spatially distributed antennas. To successfully exploit MIMO advances and retain MIMO performance gain, the existence of sufficient antenna spacing as well as a rich scattering environment, which renders the fading paths between the antenna elements of the transmitter/receiver uncorrelated, is a prerequisite. Then, the fading signal paths between the antenna elements of the transmitter/receiver are typically independent. However, the terrestrial and the satellite channels differ substantially, which makes the applicability of MIMO techniques to satellite systems a challenging subject (Arapoglou et al., 2011a). In particular, the huge distance between the satellite segment and the terrestrial stations degrades the satellite radio link to an effective keyhole channel with only one transmission path. Then, the correlation among the MIMO subchannels caused by a deficient multipath environment leads to a substantial loss in channel capacity with respect to the ideal level (Tulino et al., 2005). According to (King et al., 2005), an antenna elements separation of at least 1.5×10^5 wavelengths is required to achieve low correlation. Therefore, the deployment of multiple antennas at single satellites does not seem beneficial due to spatial limitations. Although the use of multiple satellites overcomes the aforementioned constraint, several drawbacks then appear, such as waste of the limited satellite bandwidth for the transmission of the same signal, lack of synchronization in reception, scheduling issues, intersymbol interference, and high implementation cost.

The ambitious adoption of MIMO technology in the satellite segment has recently begun to be theoretically and experimentally investigated by academia and space agencies by applying

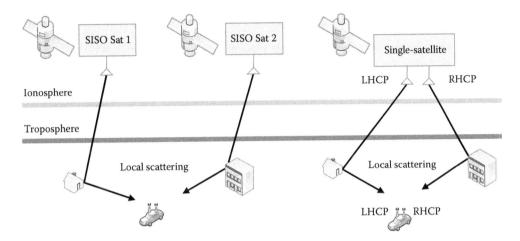

FIGURE 3.5 A simple representation of an LMSS system exploiting dual-satellite or polarization diversity.

multielement antennas on either one or both sides of the satellite radio link forming multiple-input single-output (MISO), single-input multiple-output (SIMO), and/or MIMO satellite channels. Most of the research efforts have been directed toward applying MIMO in MSS rather than in fixed satellite system (FSS), since MSS usually exhibits multipath propagation near the terrestrial end of the satellite link. The overwhelming majority of previous work has focused on exploiting the following aspects of diversity (Liolis et al., 2007) or a combination of them:

1. Site diversity, where multiple cooperating and sufficiently separated terrestrial stations communicate with a single satellite.
2. Satellite diversity, also called angle or orbital diversity, through multiple sufficiently separated satellites and a single terrestrial station equipped with multiple single polarization antennas or diversity through hybrid satellite–terrestrial MIMO systems.
3. Polarization diversity, where a single dual-orthogonal polarized satellite communicates with a single terrestrial station equipped with a dual-orthogonal polarized antenna. Figure 3.5 demonstrates the principles of satellite and polarization diversity.

Most of the research activities are in favor of the polarization domain (Horváth et al., 2007), which stepped up due to the recent advances in MIMO compact antennas (Getu and Andersen, 2005) and intends to overcome possible space limitations of single satellites and counter potential drawbacks of multiple satellite constellations. However, the polarization diversity can only increase the throughput by a factor of two, whereas satellite multiplicity can succeed in m-fold capacity increase, where m is the number of satellites. To further increase channel capacity and exploit space-polarization-time coding, the satellite and polarization diversity could be combined using multiple satellites each utilizing a dual-polarization scheme (Pérez-Neira et al., 2011). In general, the performance of polarization diversity is limited by the on/off signal blockage phenomena and the highly correlated rainfall medium dominating at frequency bands well above 10 GHz (Liolis et al., 2007). Faraday rotation in the ionosphere can also significantly affect polarization diversity. Therefore, satellite systems opt for circular polarization (Saunders and Aragón-Zavala, 2007), whereas legacy terrestrial systems usually employ linear polarization.

3.4.1 DESCRIPTION OF MEASUREMENT CAMPAIGNS

An LMSS SIMO configuration using S-band payload on ESA ARTEMIS GEO satellite was evaluated in the framework of the ORTIGIA research project (Heuberger et al., 2008) co-funded by ESA.

The scope of this project was to verify the techniques introduced by the DVB-SH standard to counteract fading in satellite and terrestrial environments. The experiments were performed in rural, tree-shadowing, highway, suburban, and urban areas in Erlangen, Germany, and comprised satellite, terrestrial, and hybrid satellite–terrestrial architectures. The mobile reception unit included a van equipped with a directive and omnidirectional antennas. The follow-up of the ORTIGIA project was the J-ORTIGIA project (Pulvirenti et al., 2010), which aimed at the performance optimization of the DVB-SH standard in hybrid satellite–terrestrial networks and performed onfield trials using the satellite S-band payload on Eutelsat W2A satellite in urban, suburban, highway, and rural environments in Pisa, Italy. A DVB-SH receiver developed by the Fraunhofer IIS was used and allowed for antenna diversity setup with up to four antennas.

Arndt et al. (2011) present a measurement campaign conducted in 2009 along the east U.S. coast for land mobile satellite (LMS) applications, covering a distance of 1000 km. The campaign targeted the achievable antenna diversity gain for various antenna configurations using a high-power satellite (ICO-G1) acting as a transmitter. The carrier frequency used was 2.185 GHz (S-band) and various environments, that is, highways, rural, suburban, urban areas, and areas with trees were examined. The receiver was positioned in a van equipped with four antennas placed front, back, left, and right on the rooftop of the vehicle. A channel measurement equipment developed by Fraunhofer IIS was used to record in parallel the signals from each antenna. Moreover, a GPS receiver logged details related to the exact measurement time and the vehicle position, whereas a front camera behind the windscreen and a fisheye camera on the van's rooftop captured the environmental characteristics. Along the measurement route, the elevation angle varied between 29° and 46°.

Heuberger (2008) presents a measurement campaign using two broadcasting satellites in geostationary orbit with 30° separation in urban, residential, and rural environments. The results include fade correlation in time and space of the signals from the two satellites. The measurements were performed in the United States using two operational XM satellites (S1 at 85°W and S2 at 115°W) in the 2332.5–2345 MHz band with a bandwidth of 1.5 MHz and an active quadrafilar helix left-hand circularly polarized (LHCP) antenna. The length of the measurement trials was 2300, 4300, and 4600 m for the urban, residential, and rural area, respectively.

Heyn et al. (2011) and Arndt et al. (2012) present two experiments carried out by ESA in the frame of the Mobile satellite channeL with Angle DiversitY (MiLADY) project. The campaigns used satellites from the existing GEO, Sirius highly elliptical orbit (HEO) and medium Earth orbit (MEO) constellations. The first campaign took place along the east coast of the United States using 2 XM GEO satellites and 3 HEO satellites for Satellite Digital Audio Radio Services (SDARS) at S-band. The received signals were sampled with a high sample rate (2.1 kHz) in urban, suburban, tree-shadowed, forest, commercial, and highway/open environments. The second campaign was conducted around Erlangen in Germany using 20 Global Navigation Satellite Systems (GNSS) MEO satellites belonging to GPS and/or GLONASS and transmitting at L-band. The latter campaign covered different propagation environments with a high variability of angle diversity constellations. The permanent availability of at least eight satellites enabled the detailed analysis of slow fading correlation for a wide range of elevation and azimuth angle combinations of multiple-satellite constellations. The GNSS campaign was divided in two parts. In the first part, the GNSS antenna was mounted on a van at a height of 2 m. A measurement round-trip of 38 km was driven 10 times, covering several environments, that is, suburban, forest, open, and commercial, in and around Erlangen. In the second part, the GNSS antenna was mounted at a height of 3.1 m onto two city buses, driving along an identical route for 3 days.

The potential MIMO channel capacity of a land mobile dual-satellite system was experimentally investigated in King and Stavrou (2006a), where a measurement campaign was carried out in Guildford, UK. Two collocated in the same orbital slot, low elevation (mean elevation angle 15°) geostationary satellites operating at 2.45 GHz carrier frequency with 200 MHz bandwidth were emulated using a hilltop. Each emulated satellite was separated by 10 wavelengths corresponding to two GEO satellites residing in a cluster and carrying a right-hand circularly polarized (RHCP) antenna.

Besides, a dual-antenna vehicle containing two RHCP antennas separated by four wavelengths was used in main road, suburban, and urban environments.

Lacoste et al. (2010) provide results on spatial and circular polarization diversity techniques for the LMSS, MISO, and SIMO channel at S-band. The experiment was performed in the small French city called Auch using a channel sounder developed by Centre National d'Etudes Spatiales (CNES) able to perform multilink simultaneous measurements with a bandwidth of 100 kHz. A helicopter equipped with two collocated transmitters was employed to emulate a GEO satellite at a 35° elevation angle, flying at about 2600 m above the ground. The first transmitter at 2172 MHz was connected to an RHCP antenna and the second one at 2172.5 MHz to an LHCP antenna. The receive antennas were two identical vertically polarized (V-polarized) dipoles spatially separated by a distance of 11.5 cm. These antennas were located on a mast at the rear of the measurement van to emulate the configuration for pedestrian use. The measurement data were collected in different propagation environments: a tree-lined road with quite dense vegetation, a suburban area with relatively spaced buildings, a high-density built-up urban area, and a continuous combination of all these environments.

The French Aerospace Lab ONERA and CNES designed and performed one more experiment to investigate dual circular polarization MIMO and SIMO configurations and to characterize both GEO LMSS and nomadic satellite channels at 2.2 and 3.8 GHz (Lacoste et al., 2012). The term "nomadic" refers to the nonstationary characteristics of the environment close to the receiver, for example, a tree swaying due to wind. To emulate the GEO satellite, two terrestrial transmitters were situated on a mountain surrounding Saint Lary village in France, maintaining elevation angles between 20° and 30°. The configuration guaranteed a free first Fresnel ellipsoid. The transmitter was equipped with two RHCP and LHCP patch antennas, whereas the receiver was located on the van rooftop and equipped with two types of antennas, a dual-polarized (RHCP and LHCP) antenna (2.2 and 3.8 GHz) and two V-polarized dipoles.

The measurement campaign presented in King and Stavrou (2006b, 2007) was the first to investigate the wideband characteristics of LMSS MIMO systems equipped with dual-orthogonal polarized antennas. The measurements took place in Guildford, UK, and three different environments were investigated, tree-lined road, suburban, and urban environments. The emulated satellite transmitter provided low elevation angles ranged from 7–18°, 5–10°, and 5–15° in the three environments, respectively. An Elektrobit Propsound wideband MIMO channel sounder configured for a 2.45 GHz carrier frequency and 200 MHz bandwidth was used. This sounder system was based on exceptional fast data acquisition and storage concept providing real-time assessment of virtually any channel data during the ongoing channel measurement in the field. Both the transmitter and receiver were controlled through a personal computer and were capable of controlling radio-frequency (RF) switches, which were synchronized in order to make time-multiplexed MIMO measurements. To emulate a downlink satellite scenario, the transmitter was a terrestrially based artificial platform on a hilltop containing directional RHCP and LHCP antennas, whereas the receiver was a van employing omnidirectional RHCP and LHCP antennas. The scattering created some depolarization from RHCP to LHCP and from LHCP to RHCP, which were represented in a 2×2 polarized MIMO channel matrix. Figure 3.6 shows these four channels, where the subscripts R and L denote the RHCP and LHCP antennas at each end of the link and n_R and n_L represent the AWGN at each antenna. A similar measurement setup was also realized in Mansor et al. (2010). However, the receiver terminal employed two RHCP and LHCP reference antennas and one dual circularly polarized contrawound quadrifilar helix antenna (CQHA). This antenna is a vertical array of two miniaturized quadrifilar helix antennas with opposite winding direction to radiate orthogonal circular polarization that allows the performance investigation of collocated antennas in MIMO systems. Mansor et al. (2010) compare the performance of the CQHA antenna with the one achieved by the spatially separated reference antennas. The measurements included obstructed-LoS (OLoS) and non-line-of-sight (NLoS) areas with different characteristics of the co-polarized and cross-polarized received signals. The satellite transmitter again was emulated by positioning the transmitter platform on top of a hill, maintaining an elevation angle of 10°.

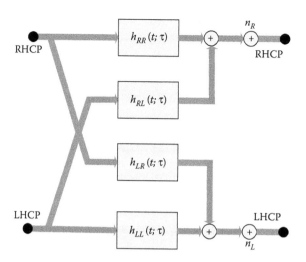

FIGURE 3.6 The outline of the dual-polarized MIMO channel model.

In order to develop a stochastic channel model for LMS MIMO channels, the authors not only utilized existing measurements provided in King and Stavrou (2006b, 2007), Mansor et al. (2010), and King et al. (2012), but also carried out two measurement campaigns in Guildford, UK, to accommodate higher elevation angles (Ekpe et al., 2011). In the first measurement campaign, the satellite was emulated by mast-mounted directional RHCP and LHCP antennas placed on a hill. The carrier frequency used was 2.43 GHz. The vehicular mobile receiver used omnidirectional RHCP and LHCP antennas mounted on the roof of a vehicle driven along preselected routes in a rural environment. The second measurement campaign was conducted in a suburban area for a 2.5 GHz carrier frequency and 15–37° elevation angles. To emulate the satellite, a mast mounting two RHCP and LHCP directional antennas was installed on a tower block.

The performance of the SS-DP MIMO concept for mobile reception using small car-roof antennas was the main target of the project MIMOSA funded by ESA. The measurement campaigns were designed at 2.187 GHz carrier frequency and 5 MHz bandwidth and were conducted to statistically analyze a dual-polarized 2 × 2 MIMO channel from a single EUTELSAT W2A satellite acting as the transmitter (Eberlein et al., 2011). This configuration enables mobile reception and is typical for applications such as DVB-SH. Both the transmitter and the receiver were equipped with LHCP and RHCP antennas and different propagation environments were measured, that is, urban, suburban, rural highway, and forest. To facilitate the separation of the four subchannels, the RHCP and LHCP signals were transmitted on interlaced grids with a small frequency offset of the LHCP with respect to RHCP. The mobile measurement equipment was capable of simultaneously recording six antenna ports. Therefore, multiple antennas were used at the mobile terminal with different characteristics to test SIMO, MISO, and MIMO configurations providing 12 measured subchannels in each run.

Table 3.4 summarizes the aforementioned measurement campaigns for multiple-antenna LMSS systems.

3.4.2 Description of Experimental Results

King and Stavrou (2006a,b) present the channel capacity improvement of a dual-satellite single-polarized (DS-SP) LMSS MIMO system and a single-satellite dual-polarized (SS-DP) LMSS MIMO system over a single-satellite single-polarized (SS-SP) LMSS SISO system. The available channel capacity was estimated for a signal-to-noise ratio (SNR) of 15 dB. The main results are illustrated in Table 3.5. It is obvious that the SS-DP and DS-SP MIMO systems significantly outperform the SS-SP SISO system in terms of outage capacity. These results also depict that the

TABLE 3.4
Summary of Measurement Campaigns for Multiantenna LMSS

Reference	Location	Environment	Satellite Emulation	Multiantenna System	Carrier Frequency/ Bandwidth	Elevation Angle
King and Stavrou (2005)	Guildford, UK	Main road, suburban, and urban	Artificial platform on a hilltop	Dual-satellite MIMO	2.45 GHz/200 MHz	15°
King and Stavrou (2006b, 2007)	Guildford, UK	Urban, suburban, and tree-lined road	Artificial platform on a hilltop	Single-satellite dual-polarized MIMO	2.45 GHz/200 MHz	5–18°
Heuberger et al. (2008)	Erlangen, Germany	Rural, tree-shadowing, highway, suburban, and urban	ARTEMIS	SIMO	1–3 GHz/1.5 MHz	–
Heuberger (2008)	USA	Urban, residential, rural	Two XM-radio satellites	Dual-satellite MISO	2332–2345 MHz/1.5 MHz	28–44°
Lacoste et al. (2010)	Auch, France	Tree-lined road, suburban, high-density built-up urban area	Helicopter equipped with two collocated transmitters	Single-satellite dual-polarized MISO and SIMO	2.2 GHz/100 kHz	35°
Pulvirenti et al. (2010)	Pisa, Italy	Urban, suburban, highway, and rural	EUTELSAT W2A	SIMO	2170–2200 MHz/4.75 MHz	–
Mansor et al. (2010)	Guildford, UK	Hilly rural area with tree-lined roads	Artificial platform on a hilltop	Single-satellite dual-polarized MIMO	2.43 GHz/50 MHz	10°
Eberlein et al. (2011)	Erlangen, Germany	Urban, suburban, rural highway, and forest	EUTELSAT W2A	Single-satellite dual-polarized MIMO	2187 MHz/5 MHz	30–40°
Arndt et al. (2011)	USA, East Coast	Highways, rural, suburban, urban areas, and areas with trees	ICO-G1 satellite	SIMO	2185 MHz/Up to 34.67 MHz (SISO), up to 52 MHz (multiantenna)	29–46°
Ekpe et al. (2011)	Guildford, UK	Rural and suburban	Mast placed on (a) a hill and (b) a tower block	Single-satellite dual-polarized MIMO	2.5 GHz/200 MHz	15–37°
Heyn et al. (2011), Arndt et al. (2012)	U.S. East Coast and Erlangen, Germany	Urban, suburban, tree-shadowed, forest, commercial, and highway (open)	SDARS satellites and GNSS	Dual-satellite MISO	2.3 GHz (U.S. East Coast) 1.575 GHz (Erlangen, Germany)	10–90°
Lacoste et al. (2012)	Saint Lary, France	Rural, built-up areas, tree alleys, and open areas	Two transmitters on a mountain	Single-satellite dual-polarized MIMO and SIMO	2.2 GHz and 3.8 GHz/CW	20–30°

TABLE 3.5

A 50% Outage Capacity of SS-SP SISO, SS-DP MIMO, and DS-SP LMSS MIMO Channels for Different Propagation Environments and 15 dB SNR

| | 50% Outage Capacity (bits/s/Hz) | | | | 10% Outage Capacity (bits/s/Hz) | | | |
| | SS-DP | | DS-SP | | SS-DP | | DS-SP | |
SNR = 15 dB	SISO	MIMO	SISO	MIMO	SISO	MIMO	SISO	MIMO
Road	0.39	0.96	0.62	1.13	0.02	0.14	0.06	0.12
Suburban	0.8	1.35	0.44	0.84	0.09	0.37	0.06	0.12
Urban	0.27	0.67	0.33	0.63	0.03	0.26	0.12	0.24

Source: Adapted from King, P.R. and S. Stavrou, *IEEE Vehicular Technology Conference (VTC-Fall)*, Montreal, Canada, 2006a; King, P.R. and S. Stavrou, *IEEE Antennas Wireless Propagation Letters*, 5(1), 98–100, 2006b.

SS-DP system performs better than the DS-SP system and suggest that using dual polarization is more beneficial than implementing two distinct transmitters. An interesting observation from the SS-DP measurement campaign was that at low received powers, the co-polarized and cross-polarized components carry similar power in all environments. This is not true in urban environments and for higher received power. The channel capacity for several configurations (2.2 GHz, 3.8 GHz, SISO, SIMO, and SS-DP MIMO) was also investigated in Lacoste et al. (2012). The capacity results showed that the MIMO configuration outperforms the SISO, as expected. In addition, for outage probabilities lower than 15%, the SIMO system performs evenly well.

Mansor et al. (2010) evaluated and compared the channel capacity of the CQHA antenna with that of spatially separated reference antennas. The experimental results depicted that the CQHA has nearly the same capacity performance as the reference antennas in NLoS environments, whereas in obstructed LoS (OLoS) areas with Ricean K-factor equal to approximately 5 dB, suitable orientation of the CQHA is required. Besides, the co-located CQHA gives lower correlation than the reference antenna in OLoS areas, while the correlation reduces with decreasing Ricean K-factor in all scenarios. Results for the 50% outage SISO and MIMO channel capacity and for 20 dB SNR are shown in Table 3.6.

The model developed by Ekpe et al. (2011) is a stochastic channel model for small-scale fading characteristics. The effects of channel correlation on the capacity of an SS-DP LMSS MIMO channel were investigated using this model. Specifically, the capacity for the dual circular polarization multiplexing (DCPM) was compared with the theoretical capacity of an equal power allocation MIMO system. The results showed that at high correlation (co-polar 0.95 and cross-polar 0.7) and

TABLE 3.6

A 50% Outage Capacity of LMSS SISO and LMSS MIMO Channels Using CQHA for OLoS and NLoS Scenarios at 20 dB SNR

| | 50% Outage Capacity (bits/s/Hz) | |
SNR = 20 dB	OLoS	NLoS
SISO	5.5–6.5	6
MIMO	8–11	9.5–10.5

Source: Adapted from Mansor, M.F.B., T.W.C. Brown, and B.G. Evans, *IEEE Antennas Wireless Propagation Letters*, 9, 712–715, 2010.

with SNR values less than 10 dB, the DCPM provides slightly better channel capacity than equal power allocation MIMO. However, at SNR values above 12 dB, traditional MIMO gives better performance, while lower channel correlation (co-polar 0.5 and cross-polar 0.3) negatively influences the capacity of DCPM but has an insignificant impact on the capacity of equal power allocation MIMO. Therefore, DCPM does not succeed with reducing channel correlation, whereas low receiver SNRs in highly correlated MIMO channels ensure the satisfactory performance of DCPM. Moreover, it was found that the relationship between the Rice factor and the channel correlation follows an exponential growth pattern.

Two measurement campaigns were designed and carried out during the MIMOSA project. The first campaign targeted the statistical analysis of the MIMO channel characteristics (Eberlein et al., 2011). The overall statistics of the MIMO channel and the large-scale fading channel states (good/ bad states) were experimentally investigated and analyzed. The results showed that the correlation for the dual-polarized antenna pair is higher than the correlation of the two single-polarized antennas. The relationship between the co-polarized signal and the cross-polarized subchannel for different environments was also shown. Specifically, for suburban environments the cross-polarized signal may be even stronger than the co-polarized signal. In addition, the strength of the direct component (LoS) and multipath component (NLoS) was also estimated. The Rice factor varies from 0 to 20 dB for tree shadowing, whereas values between −10 and 5 dB where observed for the cross-polarized signal.

The cross-polarization discrimination (XPD) in the NLoS and OLoS areas was used to study the depolarization of the received signal due to the channel conditions (Mansor et al., 2010). The XPD is defined as the ratio of co-polarized average received power to the cross-polarized average received power. Therefore, two ratios are defined, one for the RHCP and one for the LHCP. The results showed that the increasing density of trees along the measurement road in the NLoS area increases channel depolarization. The results in Lacoste et al. (2012) indicated that the S-band antennas XPD is higher than the XPD measured with C-band antennas. Multipath power statistics to characterize the influence of antenna types (RHCP, LHCP, or V-polarized dipole) and frequency band were also estimated. A difference lower than 1 dB between 2.2 and 3.8 GHz dual-polarized antennas was observed, whereas a difference higher than 1 dB in terms of multipath power was recorded between circularly polarized and dipole antennas.

King et al. (2012) developed a stochastic channel model based on existing measurements in a typical tree-lined road of a suburban environment in Guildford, UK. The model requires the calculation of a correlation matrix of all MIMO subchannels. Therefore, the correlation matrix for large-scale fading was initially constructed based on the calculation of the correlation of the shadowing between all the combinations of receiver branches. For a 2×2 MIMO system, the dimension of the correlation matrix is 4×4. The large-scale fading correlation coefficient between LMSS MIMO narrowband channels in the combined spatial/polarization domain was also estimated in King (2007). The results in Table 3.7 confirmed that strong correlation exists (close to 1) between each pair of these channels for all propagation environments examined. These values were expected, since both antennas were co-located at one satellite and likewise co-located at one vehicle. The results in King (2007) also indicated that the correlation coefficients get similarly high values, in the case of higher elevation angles. Further experimental results or accurate simulations using ray-tracing models are required in other areas with different propagation conditions in order to investigate the dependence of the correlation values on the specific sites. These average large-scale correlation coefficient matrices have already been used for the evaluation of MIMO satellite scenarios by ESA (Yamashita et al., 2005; Liolis et al., 2010b; Arapoglou et al., 2011c; Shankar et al., 2012; Kyröläinen et al., 2014).

King et al. (2012) derived the standard deviation and mean values of shadowing from the measured data. In NLoS areas, the mean value for the co-polar component was −20.5 dB, whereas that for the cross-polar component was −21.5 dB. The standard deviation was evaluated as 6.5 and 6 dB correspondingly. In LoS areas, the mean value for the co-polar component was −1.5 dB, whereas

TABLE 3.7

Average Large-Scale Correlation Coefficients between LMSS MIMO Channels for Different Propagation Environments

Large-Scale Fading		Correlation Coefficients (King, 2007)				Correlation Coefficients (King et al., 2012)			
		R/R	L/L	R/L	L/R	R/R	L/L	R/L	L/R
Urban	R/R	1.00	0.86	0.86	0.92				
	L/L	0.86	1.00	0.89	0.85				
	R/L	0.86	0.89	1.00	0.93				
	L/R	0.92	0.85	0.93	1.00				
Suburban	R/R	1.00	0.76	0.76	0.83				
	L/L	0.76	1.00	0.83	0.75				
	R/L	0.76	0.83	1.00	0.78				
	L/R	0.83	0.75	0.78	1.00				
Main road	R/R	1.00	0.86	0.85	0.90	1.00	0.86	0.85	0.9
	L/L	0.86	1.00	0.91	0.87	0.86	1.00	0.91	0.87
	R/L	0.85	0.91	1.00	0.88	0.85	0.91	1.00	0.88
	L/R	0.90	0.87	0.88	1.00	0.9	0.87	0.88	1.00

Source: Adapted from King, P.R. et al. *IEEE Transactions on Antennas and Propagation*, AP-60(2), 606–614, 2012; King, P.R., Modelling and measurement of the land mobile satellite MIMO radio propagation channel. PhD dissertation, University of Surrey, 2007.

that for the cross-polar component was −4.5 dB. The standard deviation was evaluated as 4.0 and 3.0 dB correspondingly. The measurements also provided data concerning the Rice K-factor, for which values ranging from 0 to 10 for co-polar data were observed. The results depict that a high Rice K-factor corresponds to an inherently high co-polar correlation and high XPD (15 dB), whereas a low Rice K-factor leads to low correlation with greater variance and XPD close to 0 dB.

King and Stavrou (2007) properly normalized the measured raw data extracted from the wideband MIMO channel sounder with respect to the free-space loss (FSL) and delay, at each distance between the emulated satellite and the vehicle. Then, the wideband fading characteristics were studied. Results for the first delay bin for co-/cross-polarized channels are gathered and displayed in Table 3.8. The coherence distance was found to increase with excess delay. In addition, the results concerning the coherence time showed that the median normalized coherence time, τf_d, where f_d

TABLE 3.8

Coherence Distance for the First Delay Bin, RMS Delay Spread, and Correlation Coefficient for Different Propagation Environments

Propagation Environment	Coherence Distance First Delay Bin (M)		RMS Delay Spread (NS)		Correlation Coefficient	
	Co-	Cross-	Co-	Cross-	Co-	Cross-
Tree-lined road	29	31	54	58	0.62	0.72
Suburban	132	160	29	44	0.38	0.42
Urban	204	291	63	58	–	–

Source: Adapted from King, P.R. and S. Stavrou, *IEEE Transactions on Wireless Communications*, 6(7), 2712–2720, 2007.

is the maximum Doppler frequency, for co-polarized tree-lined road channels was 0.86 at 0 ns, decreasing to 0.44 at 10 ns, 0.28 at 20 ns, and 0.18 at 30 ns. Hence, the coherence time decreases with increasing delay. However, higher excess delays had similar coherence times to the 30 ns delay point. The delay domain fading correlation coefficients were also obtained between each pair of bins in the delay domain and it was shown that increased delay separations have a lower correlation coefficient due to the longer and different shadowing path. Moderate correlations were found in the first delay bin for the tree-lined road and suburban environments, whereas no significant correlation was found in the first delay bin for the urban environment. The direct path correlation coefficients for co-/cross-polarized channels and the RMS delay spread of the impulse response are also presented in Table 3.8. Since the small-scale correlation between MIMO channels directly controls the performance of MIMO systems, an investigation of the correlation between MIMO channels in the LMSS dual polarization system was also obtained and showed that the correlation decreases with increasing excess delays. Besides, identically located scatterers, that is, nonisotropic scattering, cause high correlation, while separately located scatterers, that is, isotropic scattering, lead to low correlation.

Heuberger (2008) presented fade correlation results in time and space from the two broadcasting GEO satellites. Three types of environment were analyzed—urban, residential, and rural areas. The diversity gain was calculated by simulation based on measured fading characteristics and for several network configurations, that is, for single-satellite space diversity, two-satellite space diversity, single-satellite time diversity, and two-satellite space and time diversity. Typical values range from 4.1 dB for two-satellite space and time diversity to 2.3 dB for two-satellite space diversity in residential areas and to 0.3 dB for one-satellite time diversity. The gain is below 1 dB in the rural area for all network configurations. The analysis from single-satellite measured data showed that the coherence distance is around 4 m in the rural area, 7 m in the residential area, and up to 18 m in the urban area due to the local blockage variation introduced by the buildings. Moreover, the correlation coefficient of dual satellite fades is sufficiently small and below 0.3 for the rural and residential areas, whereas the correlation reached values of up to 0.7 in the urban area.

Lacoste et al. (2010) provide fading margin results versus percentage coverage for SISO, SIMO, and MISO measurements of the LMSS channel at S-band. The first observation is related to the SISO channels unbalance, which is expected to degrade the performance of diversity combining techniques. Next, it shows that MISO and SIMO diversity techniques reduce the fading margin for a given target coverage (better than 90%) for high availability systems. However, for coverage less than 80%, diversity techniques are not effective due to the high values of the Rice K-factor usually observed in suburban areas. SIMO spatial diversity performs better than polarization diversity for high availability LMSS systems. Specifically, possible maximum ratio combining (MRC) gains from 4 dB up to 8 dB can be achieved depending on the target coverage in a mixed environment. Nevertheless, combining circular polarization diversity with spatial diversity leads to a slight reduction in the fading margin.

Arndt et al. (2011) based the evaluation of antenna diversity gain on the estimation of the Word Error Rate (WER) for physical layer configuration provided in ETSI Satellite Digital Radio (ETSI SDR). The actual channel capacity was primarily estimated from the measured data and the WERs were derived. Then the authors calculated the required carrier-to-noise ratio (C/N) for a target WER of 5% for single- and multiantenna reception using selection combining (SC) and MRC in six different environments. Figure 3.7 depicts typical calculated values of the required C/N for single-antenna reception and four antennas using MRC in several propagation environments. One observes that the required C/N highly depends on the probability of obstruction by several obstacles in the environment. The problem is severe in urban areas where terrestrial repeaters might be required to ensure adequate QoS. It was shown that the corresponding antenna diversity gain of each antenna configuration is independent from the environment. The MRC technique that provides a power combining gain, a fading reduction gain, and a gain due to an improved overall antenna pattern, performs better than the SC technique. The SC would require four antennas to achieve the same performance as a two-antenna MRC.

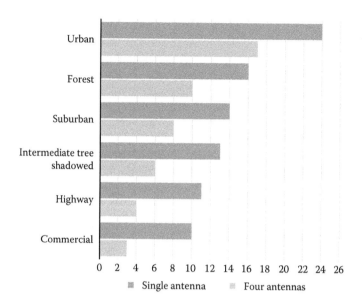

FIGURE 3.7 Required C/N in dB for 5% WER using one or four antennas in several propagation environments.

Arndt et al. (2012) presented the results of the GNSS and SDARS systems in terms of state probabilities and state duration statistics for different environments and orbital positions of the satellites for both single- and dual-satellite reception. Two states were used to model the received signal. The state identification was performed by using a threshold of 5 dB below LoS. The single-satellite analysis showed that the "bad"-state probability (Lutz et al., 1991; Prieto-Cerdeira et al., 2010), which corresponds to heavy shadowing/blockage, increases with the increase of the angle between satellite azimuth and driving direction from 0° to 90°. However, the "bad"-state probability decreases with increasing satellite elevation and for elevation angles larger than 70°, the influence of the driving direction is insignificant. The "bad"-state probability in urban and forest areas is generally higher than in other environments. In addition, the SDARS and GNSS measurements extracted similar results for urban areas, whereas lower signal availability is obtained for SDARS in suburban, forest, commercial, and open areas. The dual-satellite and multisatellite analysis indicated that the reception depends on the type of the environment, the elevation angle of each satellite, and the azimuth of each satellite relative to the driving direction. The same analysis yielded results concerning the correlation coefficient between the states of two satellites as a function of the azimuth separation derived. For small azimuth separations, both state sequences are highly correlated, with typical correlation values of 0.9. However, the correlation is minimized for azimuth separations between 60° and 120° and slightly increases toward 180° azimuth separation. The correlation coefficient also depends on the elevation angle separation between two satellites.

Heuberger et al. (2008) experimentally investigated the service availability of the satellite component of DVB-SH in S-band in several propagation environments. The results showed that the availability is extremely high, almost always 100%, in highway and rural environments with affordable satellite equivalent isotropically radiated power (EIRP). In light tree-shadowed and suburban areas, the service availability is still rather high. However, in urban, dense-urban, and indoor scenarios, the quality of the service is degraded and the deployment of a terrestrial repeater is necessary. Then, the hybrid configuration leads to service availabilities in the order of 90%. The performance enhancement obtained using hybrid reception in areas, where neither the satellite nor the terrestrial segment allows for unexceptionable reception, was also demonstrated in Pulvirenti et al. (2010). The results depicted that the service availability is over 90%.

The measurements in Liolis et al. (2010b), Ekpe et al. (2011), Arndt et al. (2012), and King et al. (2012) resulted in channel models that can be used as a cost-effective and time-saving approach for the test, optimization, and performance evaluation of multiantenna LMSS systems. Specifically, a simple stochastic channel model was generated and validated in Ekpe et al. (2011) by exploiting DCPM, under LoS and OLoS fading conditions. King et al. (2012) implemented and validated an empirical-stochastic dual circularly polarized LMSS MIMO narrowband channel model based on the Markov chain approach to account for the "on/off" nature of the observed data at low elevation angles, while different approaches for dual-satellite state narrowband channel modeling based on experimental data were presented in Arndt et al. (2012). Nevertheless, all the previously presented results are site-, scenario-, channel sounder equipment-, antenna-, and frequency-dependent and therefore, further measurement campaigns with multiantenna elements and new technology equipment should be designed and are required in order to unify them into a MIMO satellite radio channel within ITU-R. Another important issue is to take advantage of the MIMO LMSS real-world measurements in order to develop step-by-step methodologies for the simulation and time-series generation of MIMO LMSS channel models (Liolis et al., 2010b).

3.5   SUMMARY AND FUTURE RESEARCH DIRECTIONS

This chapter presented a review on the progress of channel measurements for MSS obtained from many investigators worldwide and the important milestones that have been already achieved. Various platforms, such as satellites, balloons, helicopters, airplanes, and transmitters placed on the top of a high building or hill as pseudosatellites were used to emulate the satellite component at various elevation angles and propagation environments. Much data was collected in frequencies ranging from UHF-band (800 MHz) to Ka-band (20/30 GHz). Overall, the following conclusions can be drawn from the described channel measurement campaigns for single-antenna MSS:

- The LMSS strongly depend on the propagation environment, for example, urban, suburban, or rural.
- High values of excess path loss are expected due to shadowing created by buildings, especially at low elevation angles.
- The signal attenuation increases with the frequency, the density of vegetation, and the decrease of the elevation angle.
- In urban environments, signal attenuation depends on the width of the streets, the average building height, and the vegetation at the edge of the streets.
- Weather conditions are of major importance at frequencies greater than 10 GHz and large link margins should be foreseen at K- and Ka-band.
- Moreover, mitigation techniques should be implemented at large frequencies in order to provide reasonable availabilities and reliable services in tree-shadowed and built-up areas.
- In MMSS, reflections from the sea surface should be considered especially for antennas with wide beamwidth and at low elevation angles.
- In MMSS, multipath components from ship structures and the ocean are important.
- In AMSS and during the normal flight cruising, multipath from the aircraft body cause signal degradation.
- In AMSS, the maneuvers increase the fade depth due to the interaction of the signal with the aircraft body or local scatterers when the airplane is at the terminal or on taxiway.
- In AMSS, the aircraft speed introduces increased Doppler spread.

Nevertheless, as satellite communication infrastructures have been widely used for the provision of communication services, the idea of implementing and experimenting with multiple antennas on both sides of the link allows for a further expansion of the boundaries of satellite systems performance. It is envisaged that multiantenna satellite systems will be potentially capable of providing and delivering a compelling range of current and next-generation services. In order to exploit the advantages of applying MIMO technology to satellite networks, the satellite operators and system designers will design and construct new types of satellite systems. The experimental results for multiple-antenna MSS described the benefits of polarization and satellite diversity mainly in terms of the achievable data rate and suggested that exploiting dual polarization fairly better succeeds in upgrading the system performance than invoking two distinct satellite transmitters. Nevertheless, several critical issues should be readdressed and revised, before official system implementations take place. Specifically, the results underlined that higher correlation is maintained for the dual-polarized antennas due to the co-located antennas, high availability LMSS give prominence to spatial diversity rather than polarization diversity at the receiver, and the two-satellite diversity prospers, when short interleavers are considered. The results also examined particular scenarios with distinct characteristics and underlined several special circumstances. For instance, although MIMO LMSS are preferred, the performance of SIMO LMSS systems for low outage probability is adequate. Moreover, CQHA can be used instead of spatially separated reference antennas to overcome space limitations and achieve significant capacity improvement, while DCPM can be used to attain the benefits of MIMO technology over highly correlated channels and low SNR. Furthermore, the results highlighted practical issues and reported that the correlation coefficient between the MIMO channels and the XPD are functions of the strength of the LoS component and the density of the scatterers in NLoS environments and increase as the Rice K-factor increases and the multipath becomes sparse. The correlation and the XPD are also directly related to the geometry of the link, that is, the azimuth and elevation angles, the orientation of the antenna arrays, and the excess delay. Besides, the results revealed the demand for hybrid terrestrial–satellite configurations to preserve satisfactory QoS in areas where high C/N is necessary, for example, urban areas. Overall, the most promising MIMO satellite and hybrid satellite scenarios from a system point of view are (Kyröläinen et al., 2014): (a) a GEO satellite with two orthogonally polarized antennas set up as a dual polarization per beam payload and a satellite terminal with the two co-located orthogonally polarized (RHCP/LHCP) receiver antennas and (b) a hybrid satellite/terrestrial system with one satellite and one terrestrial base station or repeater jointly transmitting data to users, employing either single- or dual-polarized antennas in single- and multifrequency networks.

Although the results from current real-world channel measurement campaigns constitute a basis for the characterization of the single- and multiple-antenna satellite channel and are indicative of a promising evolution for MSS, the experiments were performed in specific areas, where the influence of the fluctuating direct and diffuse components of the signal significantly differs. Therefore, future research efforts may be devoted to collecting measured channel data in various areas and improve and/or extend the validity of current results. Then, accurate channel models for satellite systems can be developed for as much as thoroughly validated against experimental data.

Sophisticated precoding (Diaz et al., 2007) and space-polarization-time coding techniques, for example, Golden codes (Belfiore et al., 2005) can be utilized and experimentally tested and then introduced into the DVB-Second Generation (DVB-S2) and DVB-SH standards. Moreover, experimental characterization of MIMO MSS operating at frequencies above 10 GHz is necessary in order to introduce mobility into DVB-S2 systems and study the performance of future broadband applications, such as the provision of high-speed Internet access, audio and video on demand, and file transfer to vehicles, airplanes, trains, and ships. Then, both local environment propagation effects, for example, multipath, shadowing, and blockage due to the local environment in the vicinity of the terrestrial receiver, and tropospheric effects, for example, rainfall, oxygen absorption, water vapor, clouds, and precipitation, are involved as shown by some preliminary studies (Li et al., 2001; Andersen et al., 2006; Morlet et al., 2007; Alamanac et al., 2010; Liolis et al., 2010a; Arapoglou et al., 2012). In this regard, several measurement campaigns at the Ku- and Ka-bands were previously conducted for conventional single-antenna LMSS systems (Kubista et al., 2000; Scalise et al., 2008), while the DVB-Return Channel via Satellite for Mobile Applications (DVB-RCS + M) specification (EN 301 790 V1.5.1., 2009) has been recently approved. Finally, it would be interesting to investigate the feasibility and possible benefits of tri-polarized MIMO (Quitin et al., 2012) in LMSS scenarios with reference to the data rate enhancement and the robustness of the MIMO performance.

REFERENCES

Abbot, T.M., Requirements of a mobile satellite service, *International Conference on Mobile Radio Systems and Techniques*, University of York, UK, 238, 212–215, 1984.

ADGA/Touch-Ross and Partners, *MSAT Commercial Viability Study, DOC-CR-SP-82-028*, 1982.

Alamanac, A.B., P. Burzigotti, R. De Gaudenzi, G. Liva, H.N. Pham, and S. Scalise, In-depth analysis of the satellite component of DVB-SH: Scenarios, system dimensioning, simulations and field trial results, *International Journal of Satellite Communications and Networking*, 7, 215–240, 2009.

Alamanac, A.B., P. Chan, L. Duquerroy et al., DVB-RCS goes mobile: Challenges and technical solutions, *International Journal of Satellite Communications and Networking*, 28(3–4), 137–155, 2010.

Anderson, R.E., R.L. Frey, and J.R. Lewis, Technical feasibility of satellite aided land mobile radio, *ICC Conference Record*, Philadelphia, USA, 7H.2.1–2.5, 1982.

Anderson, R.E., R.L. Frey, J.R. Lewis, and R.T. Milton, Satellite-aided mobile communications: Experiments, applications, and prospects, *IEEE Transactions on Vehicular Technology*, VT-30(2), 54–61, 1981.

Andersen, B.R., O. Gangaas, and J. Andenaes, A DVB/Inmarsat hybrid architecture for asymmetrical broadband mobile satellite services, *International Journal of Satellite Communications and Networking*, 24(2), 119–136, 2006.

Anderson, R.E. and O.S. Roscoe, How MSAT came about, *International Mobile Satellite Conference*, Pasadena, CA, pp. 3–9, 1997.

Arapoglou, P.-D., P. Burzigotti, M. Bertinelli, A.B. Alamanac, and R. De Gaudenzi, To MIMO or not to MIMO in mobile satellite broadcasting systems, *IEEE Transactions on Wireless Communications*, 10(9), 2807–2811, 2011a.

Arapoglou, P.-D., K.P. Liolis, M. Bertinelli, A.D. Panagopoulos, P. Cottis, and R. De Gaudenzi, MIMO over satellite: A review, *IEEE Communications Surveys & Tutorials*, 13(1), 27–51, 2011b.

Arapoglou, P.-D., K.P. Liolis, and A.D. Panagopoulos, Railway satellite channel at Ku band and above: Composite dynamic modeling for the design of fade mitigation techniques, *International Journal of Satellite Communications and Networking*, 30(1), 1–17, 2012.

Arapoglou, P.-D., M. Zamkotsian, and P. Cottis, Dual polarization MIMO in LMS broadcasting systems: Possible benefits and challenges, *International Journal of Satellite Communications and Networking*, 29(11), 349–366, 2011c.

Arndt, D., A. Ihlow, A. Heuberger, and E. Eberlein, Antenna diversity for mobile satellite applications: Performance evaluation based on measurements, *5th European Conference on Antennas and Propagation (EuCAP)*, Rome, Italy, pp. 3729–3733, 2011.

Arndt, D., A. Ihlow, T. Heyn, A. Heuberger, R., Prieto-Cerdeira, and E. Eberlein, State modelling of the land mobile propagation channel for dual-satellite systems, *EURASIP Journal on Wireless Communications and Networking*, 228, 2012.

Basari, K. Saito, M. Takahashi, and K. Ito, Field measurement on simple vehicle-mounted antenna system using a geostationary satellite, *IEEE Transactions on Vehicular Technology*, 59(9), 4248–4255, 2010.

Belfiore, J.-C., G. Rekaya, and E. Viterbo, The golden code: A 2 × 2 full-rate space-time code with nonvanishing determinants, *IEEE Transactions on Information Theory*, 51(4), 1432–1436, 2005.

Bello, P.A. Aeronautical channel characterization, *IEEE Transactions on Communications*, 21(5), 548–563, 1973.

Benarroch, A. and L. Mercader, Signal statistics obtained from a LMSS experiment in Europe with the MARECS satellite, *IEEE Transactions on Communications*, 42, 1264–1269, 1994.

Borghino, L., S. Buonomo, L. D'Amato, and C. Molinari, Measurements and analysis on Ka land mobile satellite channel, *Mobile and Personal Satellite Communications*, 2, 456–473, 1996.

Bundrock, A. and R. Harvey, Propagation measurements for an Australian land mobile satellite system, *International Mobile Satellite Conference*, JPL, Pasadena, USA, pp. 119–124, 1988.

Buonomo, S., L. D'Amato, and B. Arbesser-Rastburg, Mobile propagation measurements at Ka band: statistical results, *3rd Ka Band Utilization Conference*, Sorrento, Italy, pp. 121–125, 1997.

Butt, G., B.G. Evans, and M. Richharia, Narrowband channel statistics from multiband propagation measurements applicable to high elevation angle land-mobile satellite systems, *IEEE Journal on Selected Areas in Communications*, 10(8), 1219–1226, 1992.

Butterworth, J.S. *Propagation Measurements for Land-mobile Satellite Services in the 800 MHz Band*, Ottawa: Communications Research Centre, 1984a.

Butterworth, J.S. *Propagation Measurements for Land-mobile Satellite Systems at 1542 MHz*, Ottawa: Communications Research Centre, 1984b.

Butterworth, J.S. and E.E. Matt, The characterization of propagation effects for land mobile satellite services, *International Conference on Satellite Systems for Mobile Communication & Navigation*, London, UK, pp. 51–55, 1983.

Castruccio, P.A., C.S. Marantz, and J. Freibaum, Need for, and financial feasibility of, satellite aided land mobile communications, *ICC Conference Record*, Philadelphia, USA, pp. 7H.1.1– 7H.1.7, 1982.

Cavdar, I.H., UHF and L band propagation measurements to obtain log-normal shadowing parameters for mobile satellite link design, *IEEE Transactions on Antennas and Propagation*, 51(1), 126–130, 2003.

Cocks, R.J., P.S. Hansell, R. Krawec, and D. Lewin, Personal mobile communications by satellite, *International Conference on Mobile Radio System and Techniques*, University of York, UK, 238, pp. 135–139, 1984.

D'Amato, L., P. Ossola, and S. Buonomo, Statistical analysis of attenuation measurements due to vegetation at 18 GHz, *2nd Ka Band Utilization Conference*, Florence, Italy, pp. 407–414, 1996.

Damosso, E., G. Di Bernardo, A.L. Rallo, M. Sforza, and L. Stola, Propagation models for the land mobile satellite channel: Validation aspects, *Mobile and Personal Satellite Communications*, Ananasso, F. and Vatalaro, F. (Eds), London: Springer, pp. 263–276, 1994.

Diaz, M.A., N. Courville, C. Mosquera, G. Liva, and G.E. Corazza, Non-linear interference mitigation for broadband multimedia satellite systems, *International Workshop on Satellite and Space Communications*, Salzburg, Austria, pp. 61–65, 2007.

DOC Report to MOSST. *Mobile Communication via Satellite*, 1981.

Dutronc, J. and J.N. Colcy, Land mobile communications InKu-band. Results of a test campaign on Eutelsat I-F1., *International Journal of Satellite Communications*, 8, 43–63, 1990.

Eberlein, E., F. Burkhardt, C. Wagner, A. Heuberger, D. Arndt, and R. Prieto-Cerdeira, Statistical evaluation of the MIMO gain for LMS channels, *5th European Conference on Antennas and Propagation (EuCAP)*, Rome, Italy, pp. 2695–2699, 2011.

Ekpe, U.M., T.W.C. Brown, and B.G. Evans, Channel characteristics analysis of the dual circular polarized land mobile satellite MIMO radio channel, *IEEE-APS Topical Conference on Antennas and Propagation in Wireless Communications*, Torino, Italy, pp. 781–784, 2011.

ETSI EN 301 790 V1.5.1. (2009-05), Digital Video Broadcasting (DVB); Interaction channel for satellite distribution systems, 2009.

ETSI EN 302 583. *Digital Video Broadcast (DVB), Framing Structure, Channel Coding and Modulation for Satellite Services to Handheld Devices (SH) below 3 GHz*, 2007.

Evans, B., M. Werner, E. Lutz et al. Integration of satellite and terrestrial systems in future multimedia communications, *IEEE Wireless Communications*, 12(5), 72–80, 2005.

Fang, D., F.-T. Tseng, and T. Calvit, A low elevation angle propagation measurement of 1.5-GHz satellite signals in the Gulf of Mexico, *IEEE Transactions on Antennas and Propagation*, AP-30(1), 10–15, 1982a.

Fang, D., F.-T. Tseng, and T. Calvit, A measurement of the MARISAT L-Band signals at low elevation angles onboard mobil aero, *IEEE Transactions on Communications*, 30(2), 359–365, 1982b.

Getu, B.N. and J.B. Andersen, The MIMO Cube—A compact MIMO antenna, *IEEE Transactions on Wireless Communications*, 4(3), 1136–1141, 2005.

Giuliano, R., M. Luglio, and F. Mazzenga, Interoperability between WiMAX and broadband mobile space networks, *IEEE Communications Magazine*, 46(3), 50–57, 2008.

Goldhirsh, J. and W. Vogel, Roadside tree attenuation measurements at UHF for land mobile satellite systems, *IEEE Transactions on Antennas and Propagation*, AP-35(5), 589–596, 1987.

Goldhirsh, J. and W.J. Vogel, Mobile satellite system fade statistics for shadowing and multipath from roadside trees at UHF and L-band, *IEEE Transactions on Antennas and Propagation*, 37(4), 489–498, 1989.

Goldhirsh, J. and W.J. Vogel, Propagation effects for land mobile satellite systems: Overview of experimental and modeling results, *NASA Reference Publication*, 1274, 1992.

Gomez-Barquero, D., C. Douillard, P. Moss, and V. Mignone, DVB-NGH: The next generation of digital broadcast services to handheld devices, *IEEE Transactions on Broadcasting*, 60(2), 246–257, 2014.

Hagenauer, J., F. Dolainsky, E. Lutz, W. Papke, and R. Schweikert, The maritime satellite communication channel–channel model, performance of modulation and coding, *IEEE Journal on Selected Areas in Communications*, 5(4), 701–713, 1987.

Hase, Y., W.J. Vogel, and J. Goldhirsh, Fade-durations derived from land-mobile-satellite measurements in Australia, *IEEE Transactions on Communications*, 39(2), 664–668, 1991.

Hess, G.C., Land-mobile satellite excess path loss measurements, *IEEE Transactions on Vehicular Technology*, VT-29(2), 290–297, 1980.

Heuberger, A., Fade correlation and diversity effects in satellite broadcasting to mobile users in S-band, *International Journal of Satellite Communications and Networking*, 26(5), 359–379, 2008.

Heuberger, A., H. Stadali, B. Matuz et al. Experimental validation of advanced mobile broadcasting waveform in S-band, *4th Advanced Satellite Mobile Systems*, Bologna, Italy, pp. 140–148, 2008.

Heyn, T., D. Arndt, and E. Eberlein, MiLADY CCN: Mobile satellite channel model for satellite-angle diversity, *ESA Propagation Workshop*, Noordwijk, The Netherlands, 2011.

Higuchi T. and T. Shinohara., Experiment of Inmarsat Standard-C system, *Fourth International Conference on Satellite Systems for Mobile Communications and Navigation*, London, UK, 1988.

Holzbock, M., E. Lutz, and G. Losquadro., Aeronautical channel measurements and multimedia service demonstration at K/Ka band, *4th ACTS Mobile Communications Summit*, Sorrento, Italy, 1999.

Horváth, P., G.K. Karagiannidis, P.R. King, S. Stavrou, and I. Frigyes, Investigations in satellite MIMO channel modeling: Accent on polarization, *EURASIP Journal on Wireless Communications and Networking*, 098942, 2007.

Hoshinoo, K., Aeronautical mobile satellite communication propagation characteristics in flight experiment using ETS-V, *Seventh International Conference on Antennas and Propagation (ICAP)*, York, UK, 2, 690–693, 1991.

Huck, R.W., J.S. Butterworth, and E.E. Matt, Propagation measurements for land mobile satellite services, *33rd IEEE Vehicular Technology Conference*, Toronto, Ontario, pp. 265–268, 1983.

Ikegami, T., Y. Arakaki, H. Wakana, and R. Suzuki, Measurement of multipath delay profile in land mobile satellite channels, *International Mobile Satellite Conference*, Pasadena, USA, pp. 331–336, 1993.

Ilcev, S.D., *Global Mobile Satellite Communication for Maritime, Land, and Aeronautical Applications*, Dordrecht: Springer, 2005.

Inmarsat, http://www.inmarsat.com, Accessed August 21, 2015.

Ippolito, L.J., *Satellite Communications Systems Engineering: Atmospheric Effects, Satellite Link Design, and System Performance*, Chichester, West Sussex, England: Wiley, 2008.

Iridium, Iridium Satellite Phone Communications. https://www.iridium.com, Accessed August 21, 2015.

ITU, *Handbook on Satellite Communications*, 3rd Edition. New York, NY: Wiley-Interscience, 2002.

ITU, *Radio Regulations*, 2012.

ITU-R P.531 12, *Ionospheric propagation data and prediction methods required for the design of satellite services and systems*, 2013.

ITU-R P.618-12, *Propagation data and prediction methods required for the design of earth–space telecommunication systems*, 2015.

ITU-R P.676-10, *Attenuation by atmospheric gases*, 2013.

ITU-R P.680-3, *Propagation data required for the design of earth–space maritime mobile telecommunication system*, 1999.

ITU-R P.681-8, *Propagation data required for the design of earth–space land mobile telecommunication systems*, 2015.

ITU-R P.682-3, *Propagation data required for the design of earth–space aeronautical mobile telecommunication systems*, 2012.

ITU-R P.838-3, *Specific attenuation model for rain for use in prediction methods*, 2005.

ITU-R P.840-6, *Attenuation due to clouds and fog*, 2013.

Jahn, A., H. Bischl, and G. Heiss, Channel characterisation for spread spectrum satellite communications, *International Symposium on Spread Spectrum Techniques and Applications*, 3, 1996.

Jahn, A. and M. Holzbock, EHF-band channel characterisation for mobile multimedia satellite systems, *48th IEEE Vehicular Technology Conference*, Ottawa, Canada, 1998.

Jahn, A. and E. Lutz, DLR channel measurement programme for low Earth orbit satellite systems, *3rd IEEE International Conference on Universal Personal Communications*, San Diego, California, USA, 1994.

Jedrey, T.C., K.I. Dessouky, and N.E. Lay, An aeronautical-mobile satellite experiment, *IEEE Transactions on Vehicular Technology*, 40(4), 741–749, 1991.

Kanatas, A.G. and P. Constantinou, City center high-elevation angle propagation measurements at L band for land mobile satellite systems, *IEEE Transactions on Vehicular Technology*, 47(3), 1002–1011, 1998.

Karasawa, Y. T. Matsudo, T. Shiokawa, and T. Shiokawa, Wave height and frequency dependence of multipath fading due to sea reflection in maritime satellite communications, *Electronics and Communications in Japan Part 1*, 73(1), 95–106, 1990.

Karasawa, Y. and T. Shiokawa, Characteristics of L-band multipath fading due to sea surface reflection, *IEEE Transactions on Antennas and Propagation*, 32(6), 618–623, 1984.

Karasawa, Y. and T. Shiokawa, Fade duration statistics of L-band multipath fading due to sea surface reflection, *IEEE Transactions on Antennas and Propagation*, 35(8), 956–961, 1987.

Karasawa, Y. and T. Shiokawa, A simple prediction method for L-band multipath fading in rough sea conditions, *IEEE Transactions on Communications*, 36(10), 1098–1104, 1988.

Karasawa, Y., M. Yasunaga, S. Nomoto, and T. Shiokawa, On-board experiments on L-band multipath fading and its reduction by use of the polarization shaping method, *Transaction IECE Japan*, E69(2), 124–131, 1986.

King, P.R., Modelling and measurement of the land mobile satellite MIMO radio propagation channel. PhD dissertation, University of Surrey, 2007.

King, P.R. and S. Stavrou, Land mobile-satellite MIMO capacity predictions, *Electronics Letters*, 41(13), 749–751, 2005.

King, P.R. and S. Stavrou, Characteristics of the land mobile satellite MIMO channel, *IEEE Vehicular Technology Conference (VTC-Fall)*, Montreal, Canada, 2006a.

King, P.R. and S. Stavrou, Capacity improvement for a land mobile single satellite MIMO system, *IEEE Antennas and Wireless Propagation Letters*, 5(1), 98–100, 2006b.

King, P.R. and S. Stavrou, Low elevation wideband land mobile satellite MIMO channel characteristics, *IEEE Transactions on Wireless Communications*, 6(7), 2712–2720, 2007.

King, P.R., T.W.C. Brown, A. Kyrgiazos, and B.G. Evans, Empirical-Stochastic LMS-MIMO channel model implementation and validation, *IEEE Transactions on Antennas and Propagation*, AP-60(2), 606–614, 2012.

Kota, S., G. Giambene, and S. Kim, Satellite component of NGN: Integrated and hybrid networks, *International Journal of Satellite Communications and Networking*, 29(3), 191–208, 2011.

Kubista, E., F. Pérez-Fontán, M.A.V. Castro, S. Buonomo, B. Arbesser-Rastburg, and J.P.V.P. Baptista, LMS Ka-band blockage margins in tree-shadowed areas, *IEEE Transactions on Antennas and Propagation*, 46(9), 1397–1399, 1998.

Kubista, E., F.P. Fontan, M.A.V. Castro, S. Buonomo, B.R. Arbesser-Rastburg, and J.P.V.P. Baptista, Ka-band propagation measurements and statistics for land mobile satellite applications, *IEEE Transactions on Vehicular Technology*, 49(3), 973–983, 2000.

Kyröläinen, J., A. Hulkkonen, J. Ylitalo et al., Applicability of MIMO to satellite communications, *International Journal of Satellite Communications and Networking*, 32(4), 343–357, 2014.

Lacoste, F., F. Carvalho, F. Pérez-Fontán, A. Nunez Fernandez, V. Fabbro, and G. Scot, MISO and SIMO measurements of the land mobile satellite propagation channel at S-band, *4th European Conference on Antennas and Propagation (EuCAP)*, Barcelona, Spain, pp. 1–5, 2010.

Lacoste, F., J. Lemorton, L. Casadebaig, and F. Rousseau, Measurements of the land mobile and nomadic satellite channels at 2.2 GHz and 3.8 GHz, *6th European Conference on Antennas and Propagation (EuCAP)*, Prague, Czech Republic, pp. 2422–2426, 2012.

Lebow, I.L., K.L. Jordan, and P.R. Drouilhet, Satellite communications to mobile platforms, *Proceedings of the IEEE*, 59(2), 139–159, 1971.

LeRoy, B.E., Satellite-aided land mobile communications system implementation considerations, *ICC Conference Record*, Philadelphia, USA, pp. 7H.3.1–3.5, 1982.

Li, W, C. Look Law, V.K. Dubey, and J.T. Ong, Ka-band land mobile satellite channel model incorporating weather effects, *IEEE Communications Letters*, 5(5), 194–196, 2001.

Liolis, K.P., A.D. Panagopoulos, and P.G. Cottis, Multi-satellite MIMO communications at Ku-band and above: Investigations on spatial multiplexing for capacity improvement and selection diversity for interference mitigation, *EURASIP Journal on Wireless Communications and Networking*, 059608, 1–11, 2007.

Liolis, K.P., A.D. Panagopoulos, and S. Scalise, On the combination of tropospheric and local environment propagation effects for mobile satellite systems above 10 GHz, *IEEE Transactions on Vehicular Technology*, 59(3), 1109–1120, 2010a.

Liolis, K.P., J.G. Vilardebo, E. Casini, and A. Perez-Neira, Statistical modeling of dual-polarized MIMO land mobile satellite channels, *IEEE Transactions on Communications*, 58(11), 3077–3083, 2010b.

Loo, C. Land mobile satellite channel measurement at Ka band using olympus, *IEEE Vehicular Technology Conference*, Stockholm, Sweden, 1994.

Loo, C. Statistical models for land mobile and fixed satellite communications at Ka band, *IEEE Vehicular Technology Conference*, Atlanta, USA, 1996.

Loo, C., E.E. Matt, J.S. Butterworth, and M. Dufour, Measurements and modelling of land-mobile satellite signal statistics, *36th IEEE Vehicular Technology Conference*, Dallas, USA, pp. 262–267, 1986.

Lutz, E., D. Cygan, M. Dippold, F. Dolainsky, and W. Papke, The land mobile satellite communications channel-recording, statistics, and channel model, *IEEE Transactions on Vehicular Technology*, 40(2), 375–386, 1991.

Lutz, E., A. Jahn, M. Werner, and H. Bischl, DLR activities in the field of personal satellite communications systems, *AIAA/ESA Workshop on International Cooperation in Satellite Communications*, Noordwijk, The Netherlands, 1995.

Mansor, M.F.B., T.W.C. Brown, and B.G. Evans, Satellite MIMO measurement with colocated quadrifilar helix antennas at the receiver terminal, *IEEE Antennas and Wireless Propagation Letters*, 9, 712–715, 2010.

Matsudo, M., Y. Karasawa, M. Yasunaga, and T. Shiokawa, Results of on-board experiment for reduction technique of multipath fading due to sea surface reflection (IV), *Tech. Group on Antennas and Propagation IEICE Japan*, pp. AP87–23, 1987.

Matsumoto, Y., R. Suzuki, K. Kondo, and M.H. Khan, Land mobile satellite propagation experiments in Kyoto city, *IEEE Transactions on Aerospace and Electronic Systems*, 28(3), 718–727, 1992.

Morlet, C., A.B. Alamanac, G. Gallinaro, L. Erup, P. Takats, and A. Ginesi, Introduction of mobility aspects for DVB-S2/RCS broadband systems, *Space Communications*, 21(1–2), 5–17, 2007.

Mousselon, L., R.M. Barts, S. Licul, and G. Joshi, Radio wave propagation measurements for land-mobile satellite systems at 2.33 GHz, *IEEE Antennas and Propagation Society International Symposium. Digest. Held in Conjunction With: USNC/CNC/URSI North American Radio Sci. Meeting (Cat. No.03CH37450)*, Columbus, USA, 2003.

Murr, F., B. Arbesser-Rastburg, and S. Buonomo, Land mobile satellite narrowband propagation campaign at Ka band, *International Mobile Satellite Conference*, Ottawa, Canada, pp. 134–138, 1995.

Neul, A., J. Hagenauer, W. Papke, F. Dolainsky, and F. Edbauer, Propagation measurements for the aeronautical satellite channel, *37th IEEE Vehicular Technology Conference*, Tampa, USA, 37, 90–97, 1987.

Obara, N., K. Tanaka, S. Yamamoto, and H. Wakana, Land mobile satellite propagation measurements in Japan using ETS-V satellite, *International Mobile Satellite Conference*, Pasadena, USA, pp. 313–318, 1993.

Ohmori, S., A. Irimata, H. Morikawa, K. Kondo, Y. Hase, and S. Miura, Characteristics of sea reflection fading in maritime satellite communications, *IEEE Transactions on Antennas and Propagation*, 33(8), 838–845, 1985.

Ohmori, S., H. Wakana, *Mobile Satellite Communications*, Boston: Artech House, 1998.

Paulraj, A.J., D.A. Gore, R.U. Nabar, and H. Bolcskei, An overview of MIMO communications—A key to gigabit wireless, *Proceedings of the IEEE*, 92(2), 198–218, 2004.

Pérez-Neira, A.I., C. Ibars, J. Serra, A. Del Coso et al., MIMO channel modeling and transmission techniques for multi-satellite and hybrid satellite–terrestrial mobile networks, *Physical Communication*, 4(2), 127–139, 2011.

Perrins, E. and M. Rice, Propagation analysis of the ACTS maritime satellite channel, *Fifth International Mobile Satellite Conference*, Pasadena, USA, pp. 201–205, 1997.

Pike, C.J. EHF (28/19 GHz) suitcase satellite terminal, *Olympus Utilization Conference*, Sevilla, pp. 203–207, 1993.

Pike, C.J., D.R. Bradely, and D.J. Hindson, Canadian EHF (28/19 GHz) satellite communication terminals for the Olympus program, *Canadian Satellite User Conference*, Ottawa, Canada, pp. 364–369, 1989.

Pinck, D.S. and M.D. Rice, K/Ka-band channel characterization for mobile satellite systems, *IEEE 45th Vehicular Technology Conference*, Chicago, USA, pp. A3–A10, 1995.

Prieto-Cerdeira, R., F. Pérez-Fontán, P. Burzigotti, A. Bolea-Alamañac, and I. Sanchez-Lago, Versatile two-state land mobile satellite channel model with first application to DVB-SH analysis, *International Journal of Satellite Communications and Networking*, 28(5–6), 291–315, 2010.

Pulvirenti, O., D. Ortiz, A. Del Bianco, S. Sudler, R. Hoppe, and M. Pannozzo, Performance assessment based on field measurements of mobile satellite services over hybrid networks in S-band, *5th Advanced Satellite Multimedia Systems Conference and the 11th Signal Processing for Space Communications Workshop*, Cagliari, Italy, pp. 315–324, 2010.

Quitin, F., C. Oestges, A. Panahandeh, F. Horlin, and P. De Doncker, Tri-polarized MIMO systems in real-world channels: Channel investigation and performance analysis, *Physical Communication*, 5(4), 308–316, 2012.

Renduchintala, V.S.M., H. Smith, J.G. Gardiner, and I. Stromberg, Communications service provision to land mobiles in Northern Europe by satellites in high elevation orbits-propagation aspects, *40th IEEE Conference on Vehicular Technology*, Orlando, USA, 1990.

Rice, M., J. Slack, B. Humpherys, and D.S. Pinck, K-band land-mobile satellite channel characterization using acts, *International Journal of Satellite Communications*, 14, 283–296, 1996.

Richharia, M., *Mobile Satellite Communications: Principles and Trends*, London: Pearson Education Limited, 2001.

Sangchul, M., K. Woo-Suk, D. Vargas, D. Gozalvez Serrano, M.D. Nisar, and V. Pauli, Enhanced spatial multiplexing for rate-2 MIMO of DVB-NGH system, *19th International Conference on Telecommunications (ICT)*, Jounieh, Lebanon, 2012.

Saunders, S.R. and A. Aragón-Zavala. *Antennas and Propagation for Wireless Communications*, 2nd Edition, London: Wiley, 2007.

Scalise, S., H. Ernst, and G. Harles, Measurement and modeling of the land mobile satellite channel at Ku-band, *IEEE Transactions on Vehicular Technology*, 57(2), 693–703, 2008.

Shankar, B., P.-D. Arapoglou, and B. Ottersten, Space-frequency coding for dual polarized hybrid mobile satellite systems, *IEEE Transactions on Wireless Communications*, 11(8), 1–9, 2012.

Sheriff, R.E. and Y. Fun Hu, *Mobile Satellite Communication Networks*, Chichester, West Sussex: Wiley, 2001.

Smith, H., J.G. Gardiner, and S.K. Barton, Measurements on the satellite channel at L and S bands, *International Mobile Satellite Conference*, Pasadena, USA, pp. 319–324, 1993.

Smith, H., M. Sforza, B. Arbesser-Rastburg, J.P.V. Baptista, and S.K. Barton, Propagation measurements for S-band land mobile satellite systems using highly elliptical orbits, *Second European Conference on Satellite Communications*, Liège, Belgium, pp. 517–520, 1991.

Sofos, T. and P. Constantinou, Propagation model for vegetation effects in terrestrial and satellite mobile systems, *IEEE Transactions on Antennas and Propagation*, 52(7), 1917–1920, 2004.

Sutton, R., E. Schroeder, A. Thompson, and S. Wilson, Satellite-aircraft multipath and ranging experiment results at L band, *IEEE Transactions on Communications*, 21(5), 639–647, 1973.

Swan, P.A. and C.L. Devieux Jr., *Global Mobile Satellite Systems: A Systems Overview*, Boston: Kluwer Academic Publishers, 2003.

Thuraya, http://www.thuraya.com, Accessed August 21, 2015.

Tulino, A.M., A. Lozano, and S. Verdu, Impact of antenna correlation on the capacity of multiantenna channels, *IEEE Transactions on Information Theory*, 51(7), 2491–2509, 2005.

Vogel, W.J. and J. Goldhirsh, Tree attenuation at 869 MHz derived from remotely piloted aircraft measurements, *IEEE Transactions on Antennas and Propagation*, AP-34(12), 1460–1464, 1986.

Vogel, W.J. and J. Goldhirsh, Fade measurements at L-band and UHF in mountainous terrain for land mobile satellite systems, *IEEE Transactions on Antennas and Propagation*, 36(1), 104–113, 1988.

Vogel, W.J. and J. Goldhirsh, Mobile satellite system propagation measurements at L-band using MARECS-B2, *IEEE Transactions on Antennas and Propagation*, 38(2), 259–264, 1990.

Vogel, W.J., J. Goldhirsh, and Y. Hase, Land-mobile-satellite fade measurements in Australia, *Journal of Spacecraft and Rockets*, 29, 123–128, 1992.

Vogel, W.J. and J. Goldhirsh, Earth satellite tree attenuation at 20 GHz: Foliage effects, *Electronics Letters*, 29(18), 1640–1641, 1993.

Vogel, W. J. and J. Goldhirsh, Multipath fading at L band for low elevation angle, land mobile satellite scenarios, *IEEE Journal on Selected Areas in Communications*, 13(2), 197–204, 1995.

Vogel, W.J. and U.-S. Hong, Measurement and modeling of land mobile satellite propagation at UHF and L-band, *IEEE Transactions on Antennas and Propagation*, 36(5), 707–719, 1988.

Vogel, W.J. and E.K. Smith, Theory and measurements of propagation for satellite to land mobile communication at UHF, *35th IEEE Vehicular Technology Conference*, Boulder, USA, pp. 218–226, 1985.

Vogel, W.J., G.W. Torrence, and H.-P. Lin, Simultaneous measurements of L- and S-band tree shadowing for space–Earth communications, *IEEE Transactions on Antennas and Propagation*, 43(7), 713–719, 1995.

Wakana, H., N. Obara, K. Tanaka, S. Yamamoto, and N. Yoshimura, Fade statistics measured by ETS-V in Japan for L-band land-mobile satellite communication systems, *Electronics Letters*, 32(6), 518, 1996.

Wakana, H., H. Saito, S. Yamamoto et al., COMETS experiments for advanced mobile satellite communications and advanced satellite broadcasting, *International Journal of Satellite Communications*, 18, 63–85, 2000.

Wakana, H., S. Yamamoto, N. Obara, A. Miura, and M. Ikeda, Fade characteristics for K-band land-mobile satellite channels in Tokyo measured using COMETS, *Electronics Letters*, 35(22), 1912, 1999.

Yamashita, F., K. Kobayashi, M. Ueba, and M. Umehira, Broadband multiple satellite MIMO system, *IEEE 62nd Vehicular Technology Conference (VTC-2005-Fall)*, Washington, DC, USA, pp. 2632–2636, 2005.

Yoshikawa, M. and M. Kagohara, Propagation characteristics in land mobile satellite systems, *IEEE 39th Vehicular Technology Conference*, San Fransisco, USA, pp. 550–556, 1989.

4 Land Mobile Satellite Channel Models

Fernando Pérez-Fontán

CONTENTS

4.1 INTRODUCTION

The land mobile satellite (LMS) channel has received increasing attention in the past few years for a number of very interesting applications, including bidirectional communications, broadcast, and global navigation positioning. The kind of terminals involved includes mobile, handheld, and nomadic. Moreover, the attention toward this channel was initially centered on frequencies in the L- and S-bands with some research also in C-band. However, the interest has been extended to higher-frequency bands such as Ku and Ka, where there is a greater availability of larger bandwidths. Moreover, this channel covers various satellite orbit options, including geostationary (GEO), low/medium Earth orbits (LEO, MEO), highly elliptical orbits (HEO), etc. LMS systems also involve the possibility of using constellations of satellites and may be aided with an ancillary terrestrial component such as a network of urban gap fillers.

To plan new systems with adequate error rate performances, a deep knowledge of the propagation channel is needed. This has triggered the development of a number of channel models. In this chapter, an overview of the various propagation mechanisms intervening will be presented followed by a survey of the existing modeling approaches developed in the past few years, including purely statistical through to physical–statistical. The models presented deal with the various issues of interest starting with purely narrowband models on to models dealing with multiple satellite configurations trying to improve the coverage quality. Then, the discussion goes on to look into wideband models.

Special attention is paid to the requirements imposed on the modeling by global navigation systems, where continuity has to be preserved along the timeline for all multipath components, in contrast to the "drop" or snapshot approach used in communications. Other issues to be dealt with are scenarios such as satellite-to-indoor links. Topics such as the use of MIMO configurations, including those relying on the use of the polarization dimension or satellite aeronautical cases, are dealt with in other chapters of this book.

Throughout, the attention will be focused on modeling alternatives whereas model parameters are mostly left out but can be consulted in the abundant references provided.

4.2 COMPONENTS AND PROPAGATION IMPAIRMENTS

The LMS link traverses three distinct layers: the ionosphere, the troposphere, and the local environment in the vicinity of the terminal. Among the effects caused by the ionosphere, which is populated by free electrons forced to resonate by the passage of the transmitted waves, are scintillation, Faraday rotation, delays, and dispersion. For communications applications, all such effects are already very small at L-band and negligible at higher frequencies. Faraday rotation can be countered by the use of circular polarization. However, in navigation applications, the above effects are very important when estimating the distances to each satellite, and very precise models of the ionosphere must be used. In some terminals, the use of a second frequency helps estimate the current state of the ionosphere. More detailed information on ionospheric effects can be found in ITU-R Rec. P.531-12 (2013).

In the case of the troposphere, several effects can also be considered affecting the received signal. Again there are attenuations, dispersion, delay, scintillation, etc. The main source of attenuation at low frequencies like L-band is rain but this is practically negligible. As we go higher in frequency to Ku- and Ka-bands, rain can significantly degrade link performance. The presence of variable amounts of humidity in the atmosphere is the source of changing refraction conditions, which affects the delay in navigation system distance measurements. The presence of gases such as oxygen or water vapor also presents peaks of attenuation at given frequency bands. More information on tropospheric effects can be found in ITU-R documents such as ITU-R Rec. P.676-10 (2013), ITU-R Rec. P.618-11 (2013), and ITU-R Rec. P.681-8 (2015).

In this chapter, we pay attention only to the modeling of local effects. This third layer affecting the link is made up of natural and man-made features, such as trees, bridges, buildings, and smaller elements such as passing cars or pedestrians and small urban elements such as lamp posts and traffic signs. Again, the signal is subjected to attenuation, spreading, depolarization, delay, etc. (Karaliopoulos and Pavlidou, 1999). This chapter concentrates on this third layer where issues such as slow and fast signal variations, specular and diffuse scattering, etc. will be discussed, which all together give rise to a very challenging channel to model.

As to the types of users being addressed by LMS services, one possible classification is as a function of their mobility. In this respect, most available models have addressed the vehicular user or the pedestrian, handheld terminal user with hemispheric, nondirective antennas picking up the multipath contributions from all directions. More and more, especially with the use of higher-frequency bands, directive antennas are possible, for example, in public transport systems capable of providing higher data rates. Special attention should be given to the nomadic user wanting to perform high rate transmissions with the special characteristic that he/she is willing to select the best possible spot for performing the communication in the vicinity of the point where the decision to communicate took place.

Normally, the modeling of the LMS channel is based on the supposition that there are three main components making up the channel: the direct, the specular, and the diffuse scattered contributions. In the low-pass equivalent representation, they can be represented by phasors (Davarian, 1987; Karaliopoulos and Pavlidou, 1999), that is,

$$\tilde{r}(t) = \tilde{H}(f_c, t) = a_{Dir} \exp(j\phi_{Dir}) + \sum a_{Spec} \exp(j\phi_{Spec}) + \sum a_{Diff} \exp(j\phi_{Diff}) \tag{4.1}$$

where each of the three terms contains amplitudes and phases. Equation 4.1 presents a narrowband view of the channel, that is, at a single frequency, for example, the carrier f_c, where delay is considered negligible. The coherent, amplitude and phase combination of all the above contributions gives rise to time variations, especially due to changes in the phase term as the terminal moves. Also the static user experiences fade due to moving scatterers.

The direct, wanted signal may be subjected to blockage or shadowing as the terminal moves and goes behind an obstacle or when a moving obstacle such as a truck intercepts the link. This is

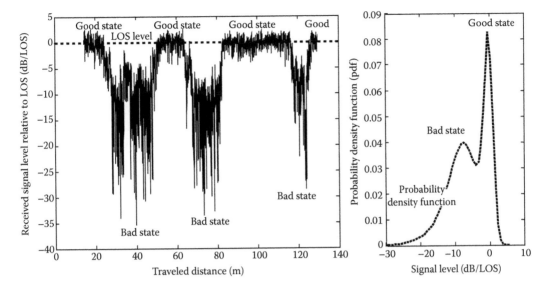

FIGURE 4.1 Variability of the received signal and need for defining "states."

a significant effect given that the link margin is quite limited in many applications and, especially, in the uplink in bidirectional communication applications. The received signal conditions change drastically from the clear to the shadowed or blocked conditions and thus, the so-called states are normally defined to describe these markedly different conditions (Figure 4.1).

Moreover, the modeling of the direct signal and the other contributors is carried out as a function of the type of environment where the terminal is at or traveling through, thus classifications and model parameterizations in terms of scenarios such as urban, suburban, open, highway, etc. are very common.

The presence of obstacles and features in the traversed scenario also gives rise to scattering phenomena where rays are generated off various illuminated surfaces. These rays are not very strong compared to the direct signal but they still can contribute significant energy levels, especially in the case the direct signal is partially or totally blocked. When combined at the receiver, they give rise to fading effects. Moreover, they arrive delayed with respect to the maybe much attenuated direct signal, giving rise to delay dispersion effects, which in turn give rise to intersymbol interference and frequency-selective effects.

If the obstacles reflecting the signal in the direction of the terminal are small or large but rough, they tend to reradiate in all directions, thus the name diffuse scattering. Not so common is to find large features, for example, a large smooth building face giving rise to strong specular reflections in the direction of the terminal. In these cases, the reflected, specular component could be as strong as the clear direct signal.

Specular rays will normally undergo a significant attenuation when the pseudo-Brewster angle is exceeded and the sense of rotation of the incoming circular polarized component is switched. We will consider that specular rays are rarely observed and will pay more attention to the diffuse contributions.

It is important to find out whether it is necessary to carry out a wideband analysis or whether a simpler narrowband model will suffice. This is not an absolute decision but depends on the bandwidth of the signals to be transmitted. The criterion is to compare the channel's delay spread or its coherence bandwidth with the characteristics of the transmitted signal: its symbol rate or its bandwidth. With respect to existing measurements (Arndt, 2015), Parks et al. (1997) report delay spreads for L- and S-bands to be in the order of 30–90 ns. Jahn et al. (1995) reported that the measured delay spreads were of 500 ns to 2 µs at L-band; this is equivalent to coherence bandwidths from 0.5 to 2 MHz. Prieto-Cerdeira et al. (2010) reported that narrowband conditions are supposed for DBV-SH signals with a bandwidth of 5 MHz.

Arndt (2015) reviewed a number of current systems citing their respective bandwidth, that is, Iridium operating at L-band uses a bandwidth of 31.5 kHz (Evans, 1997), XM radio operating at S-band has a bandwidth of 1.84 MHz (Michalski, 2002), Sirius radio also at S-band has a bandwidth of approximately 4 MHz (Akturan, 2008), DVB-SH at S-band BW = 1.7, 5, 6, 7, and 8 MHz (DVB BlueBook A120, 2008-05), S-UMTS at L/S-band BW = 5 MHz (ESTEC, 2000), and Inmarsat's BGAN system L-band BW = 200 kHz (Franchi et al., 2000). However, satellite navigation systems such as GPS or Galileo, especially their enhanced versions, spread over bandwidths greater than 10 MHz. This means that most communication systems at low frequencies L-, S-bands will normally be narrowband while navigation systems will require wideband channel modeling.

In addition to time spreading, the various contributions pertaining to the three categories discussed are subjected to Doppler. Each component depending on its direction of arrival with respect to the direction of travel of the terminal undergoes a Doppler shift. As the received signal is composed of various components each with different Dopplers, there is an effect known as Doppler spread or frequency spreading. When producing synthetic time series for testing new physical layers, it is important to define adequate Doppler spectra, which affect the so-called second-order statistics such as the coherence time of the channel, the fade duration, or level crossing rate.

There is another Doppler-related effect, which is due to the movement of non-GEO satellites. For instance, low Earth orbit (LEO) satellites stay overhead for only a few minutes and thus travel at important speeds. This gives rise to a significant Doppler shift that has to be compensated for in both the transmit and receive directions.

4.3   EMPIRICAL/STATISTICAL CHANNEL MODELS

Here, empirical and statistical models are dealt with in the same section since their main difference is that statistical models describe in more detail the characteristics of the channel by making physical assumptions that are built into the selected distributions. But, in any event, the parameters of empirical and statistical models are drawn from experimental data.

A model classification would include empirical, statistical, and deterministic/physical models. Hybrid approaches are also common as well as practical and will be discussed in detail in another section. Empirical models try to describe the evolution of one or several parameters, for example, the attenuation, as a function of the elevation angle, using experimental data as a starting point. Normally, these models are developed using single or multiple regression techniques depending on whether the model depends on a single or multiple inputs. A discussion on empirical models is provided in Kanatas and Constantinou (1996).

Physical or deterministic models require detailed inputs describing the propagation scenario: terrain (Braten, 2000), building, land usage databases, and the use of theoretical models. The ultimate case would be a direct application of Maxwell's equations. Normally, ray-tracing techniques (Dottling et al., 2001) combined with the uniform theory of diffraction (UTD), or physical optics (PO) techniques provide very detailed results with any time resolution as required, for each contribution given that their trajectories are tracked throughout the propagation scenario, for example, double-reflected and reflected–diffracted rays. Moreover, theoretical propagation models are also able to provide polarimetric information, that is, for the co- and cross-polar received contributions. The disadvantage is that the time required for computing simple results is quite long (Aguado Agelet et al., 2000). The applicability of these techniques is important, for example, in analyzing very specific situations such as the propagation environment around a GPS reference station. In addition, these are necessary to gain more insight into the propagation phenomena present at a particular location and its comparison with detailed analyses of measurements, for example, through high-resolution and ray-tracing techniques (Fleury et al., 1999; Jost et al., 2012). Such efforts are required for making new progress in our knowledge of propagation phenomena.

Statistical models make physical assumptions as to how the channel is arranged: the presence of a direct and a number of scattered contributions, for example. However, their parameterization has to be carried out from experimental data. Normally, such models are produced for a given operational scenario such as urban, suburban, and so on, and their parameters are produced for a number of satellite elevation ranges, for example, from 20° to 30°, from 30° to 40°, and so on. The selected statistical distributions should be flexible enough so as to be valid through the whole range of elevations, that is, the same phenomena are assumed to be present with different intensities, reflected in the different parameters for the same distribution.

However, the received signal's dynamics are normally much more involved, requiring the description, at the same time, of the direct signal and the multipath, with possible shadowing effects superposed varying at different rates. This has led to the use of combined distributions, and finally, the use of states when the signal cannot be described by a single distribution or combination of distributions. Normally, single distributions are valid for short sections of traveled route; for longer stretches, mixed distributions are necessary; and finally for complete routes of tens to hundreds of kilometers, the use of states is mandatory.

To overcome the shortcomings of the above models, hybrid techniques combining the advantages of the various basic approaches have been defined under the general category of *physical–statistical* or *virtual city* models, which will be discussed after the statistical modeling approach has been reviewed.

As mentioned earlier, if the inverse of the delay spread induced by the local environment, the coherence bandwidth, is small compared with the system's bandwidth, we can safely follow a narrowband modeling approach. As discussed in Arndt (2015), current satellite communication and broadcasting systems with bandwidths of about 5 MHz can be modeled using a narrowband approach.

We analyze next the statistical distributions used for LMS channel modeling. Normally, for coverage predictions only, the voltage or power distribution is required. For more detailed analyses and for generating time series for simulation of the physical layer, second-order statistics are required where the rate of variation is accounted, for example, in the form of Doppler spectra or autocorrelation functions for the fast multipath-induced variations and autocorrelation functions for the slow, shadowing-induced variations. Finally, Markov chains are used for state transitions, for example, for larger-scale modeling.

Statistical distributions have built-in assumptions on the nature of the propagation phenomenon and its components. Typical distributions used in radio wave propagation modeling are the Rayleigh, Rice, lognormal, and Nakagami-m (Abdi et al., 2003).

However, simple, common distributions, in some instances, cannot fully describe the behavior of the received amplitude variations. This is why combinations of distributions are also usually employed. When there is a mixture of fast and slow variations, it is common to use distributions requiring two or three parameters. Still stationary conditions may be assumed from the point of view of the distribution parameters, which should remain constant.

Typical distributions used for this are the Loo (1984) (Figure 4.2), Suzuki (1977), or Corazza-Vatalaro (1994) models. Loo's and Corazza-Vatalaro's are three parameter distributions while Suzuki's has only two parameters. There have been a number of other, more sophisticated approaches where up to four parameters are needed (Table 4.1) (Arndt, 2015). Other combinations of distributions can be found in Patzold et al. (1998), Hwang et al. (1997), Vatalaro (1995), Xie and Fang (2000), and Simunek et al. (2013). A discussion and comparison of the various models can be found in Abdi et al. (2003). In Figure 4.2, a phasor representation of Loo model's physical assumptions is illustrated: a direct signal phasor is subjected to lognormal variations while the diffuse scattering phasor is the result of the coherent addition of many individual contributions. The diagram presents an instantaneous snapshot. In time, each phasor is rotating due to changes in the terminal, satellite, or scatterer positions. The Loo probability density function (pdf) is as follows:

$$f(r) = \frac{8.686r}{\sigma^2 \Sigma \sqrt{2\pi}} \int_0^\infty \frac{1}{a} \exp\left(-\frac{r^2 + a^2}{2\sigma^2}\right) \exp\left[-\frac{(20 \log a - M)^2}{2\Sigma^2}\right] I_0\left(\frac{ra}{\sigma^2}\right) da \quad r \geq 0 \qquad (4.2)$$

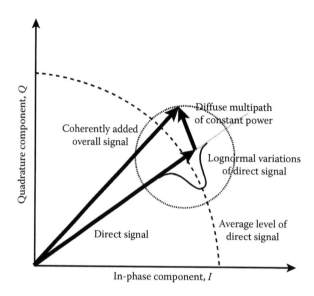

FIGURE 4.2 Loo model assumptions. (After Loo, C., *ICC'84—Links for the Future: Science, Systems and Services for Communications, Proceedings of the International Conference on Communications*, 2, 588–594, May 14–17, 1984.)

TABLE 4.1
Combined Distribution Models for the Land Mobile Satellite Channel

Model	Complex Channel $\tilde{r} = r_I + jr_Q$	Distributions
Suzuki (1977)	$\tilde{r} = RS\exp(j\phi)$	R: Rayleigh, S: lognormal, ϕ: uniform
Loo (1984)	$\tilde{r} = R\exp(j\phi_M) + S\exp(j\phi_0)$	R: Rayleigh, S: lognormal, ϕ_M, ϕ_0: uniform
Corazza and Vatalaro (1994)	$\tilde{r} = RS\exp(j\theta)$	R: Rice, S: lognormal, θ: uniform
Vatalaro (1995)	$\tilde{r} = RS\exp(j\theta) + x_1 + jy_1$	R: Rice, S: lognormal, x_1, y_1: Gaussian
Hwang et al. (1997)	$\tilde{r} = AS_1\exp(j\phi) + RS_2\exp[j(\theta+\phi)]$	R: Rayleigh, A: constant, S_1, S_2: lognormal
Patzold et al. (1998)	$\tilde{r} = S\exp(j\theta) + x_1 + jy_1$	S: lognormal, x_1, y_1: Gaussian
Xie and Fang (2000)	$\tilde{r} = RS\exp(j\theta)$	R: Beckmann, S: lognormal
Abdi et al. (2003)	$\tilde{r} = R\exp(j\phi_M) + S\exp(j\phi_0)$	R: Rayleigh, S: Nakagami

Source: Adapted from Arndt, D., On channel modelling for land mobile satellite reception, PhD thesis, Technische Universität, Ilmenau, 2015.

Note: S corresponds to the direct signal and slow-fading processes and R corresponds to the multipath component.

where r is the received signal envelope normalized with respect to the line-of-sight (LOS) level, I_0 the zeroth-order modified Bessel function, and the model parameters M and Σ, both in dB, are the mean and the standard deviation of the direct signal's amplitude, a, and $MP = 10\log(2\sigma^2)$, also in dB, is the normalized diffuse multipath power.

To model the nonstationary behavior of the channel for longer sections of the terminal route and include the overall peak-to-peak variations in the received signals, we need to define states, where the above combined distributions are applicable. Moreover, state transitions and their probabilities have to be defined. Arndt (2015) collated a list of such model, which is reproduced in Table 4.2.

TABLE 4.2

Available Multistate Models for the Narrowband LMS Channel

Model	Model Structure	Comment
Lutz et al. (1991)	Rice + Suzuki	
Wakana (1991)	Rice + Rice/Rice	Different attenuations for non-LOS states are assumed
Barts and Stutzman (1992)	Rice + Loo	
Karasawa et al. (1995)	Rice + Loo + Rayleigh	
Akturan and Vogel (1995)	Rice + Loo + Loo or Rayleigh	Fish-eye photos for state extraction
Pérez-Fontán et al. (1998)	Loo + Loo + Loo	Reference model for DVB-SH (DVB BlueBook A120, 2008-05)
Rice and Humpherys (1997)	Rice/Rice + Suzuki	Extension of two-state Lutz model for K-band with two "line-of-sight" states regarding antenna pointing errors
Mehrnia and Hashemi (1999)	Rice + Nakagami	
Pérez-Fontán et al. (2007a)	Loo + Loo	Enhanced version of three-state model
Ming et al. (2008)	Rice/Rice + Loo + Rayleigh/Rayleigh	

Source: Adapted from Arndt, D., On channel modelling for land mobile satellite reception, PhD thesis, Technische Universität, Ilmenau, 2015.

The model that first introduced the concept of state in LMS channel modeling was due to Lutz et al. (1991). A two-state model, "good" and "bad," was proposed for LOS and NLOS (non-LOS) conditions. In the good state, the variations were fitted to a Rice distribution while the bad state was characterized by a Suzuki (1977) distribution, which assumes Rayleigh fading for short distances whose local average power is subjected to slower lognormal variations. State transitions were modeled by a two-state Markov chain. Figure 4.3 illustrates the state model and Figure 4.4 its associated time series generator.

Other models were later proposed where more states, mainly three, were proposed (Karasawa et al., 1995; Akturan and Vogel, 1997; Pérez-Fontán et al., 1997) corresponding to three conditions: LOS, moderate shadowing, and heavy shadowing/blockage conditions. We discuss briefly with reference to Figure 4.5 the model proposed in Pérez-Fontán et al. (1997). The aim was to use the same statistical framework for all states. To cover the wide dynamic range encountered, three states were selected all following the same Loo (1984) distribution, that is, the combination of local Rice variations where the multipath power remained constant within each state but the direct signal was subjected to slow lognormal variations (Figure 4.2). The very slow variations were left for the states, which absorbed most of the possible dynamic range in the signal.

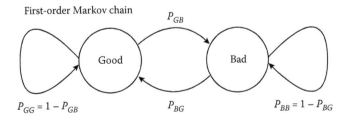

FIGURE 4.3 First-order Markov chain model with two states proposed by Lutz. (After Lutz, E. et al. *IEEE Transactions on Vehicular Technology*, 40, 375–386, 1991.)

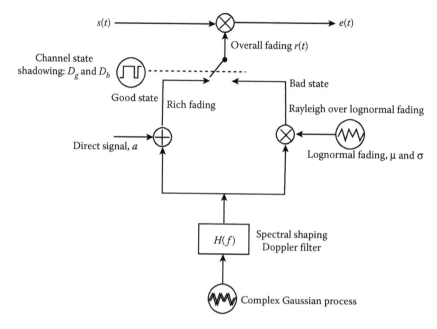

FIGURE 4.4 Lutz model time series generator. (After Lutz, E. et al. *IEEE Transactions on Vehicular Technology*, 40, 375–386, 1991.)

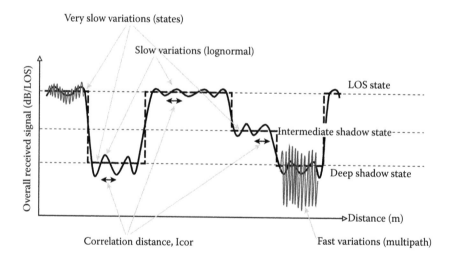

FIGURE 4.5 Three-state model. (After Pérez-Fontán, F. et al. *International Journal of Satellite Communications*, 15, 1–15, 1997.)

A full parameterization based on ESA measurement campaigns was provided in Pérez-Fontán et al. (2001); moreover, details on the time variation structure were also proposed in terms of shadowing correlation lengths and minimum state durations. The assumed Doppler spectrum was based on a uniform distribution of scatterers about the moving terminal. This is equivalent to the so-called Clarke/Jakes spectrum. Instead of using a U-shaped Doppler filter, a generator based on a point scatterer model was used. In fact, this allowed the introduction of an associated wideband model as discussed in another section. The model is used as the reference model for the DVB-SH standard (DVB BlueBook A120, 2008-05).

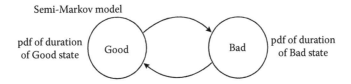

FIGURE 4.6 Semi-Markov model.

State-based models with more than three states (five and six) were proposed in Ming et al. (2008) and Dongya et al. (2005).

A key element in the evaluation of the received signal quality is an adequate model for the state durations. Short outages where the received signal goes below a given E_b/N_0 threshold can be overcome by forward error correction (FEC) coding. Longer fades are overcome by means of interleaving. Sizing up the parameters employed in these techniques very much depends on the availability of adequate state duration models (Braten and Tjelta, 2002).

The modeling of state transitions was first approached using first-order Markov chains. However, the simulated durations produced in this way, once significantly long measured data was available, was found not to match well with the measurements. ITU-R Rec. P.681-8 (2015), Braten and Tjelta (2002), and Milojevic et al. (2009) indicated that this behavior does not fit the experimental observations. Two approaches were proposed to overcome this limitation and lack of flexibility, one was the use of *semi-Markov* chains (Braten and Tjelta, 2002) (Figure 4.6), and the other was using *dynamic Markov* chains (Ming et al., 2008; Milojevic et al., 2009).

In Markov models, the probability of being in a given state and the duration of the states are determined by the transition probabilities, that is,

$$p_i(N = n) = p_{i|i}^{n-1}(1 - p_{i|i}) \quad \text{with } n = 1,2,\ldots \tag{4.3}$$

$$p_i(N \le n) = (1 - p_{i|i}) \sum_{j=1}^{n} p_{i|i}^{j-1} \quad \text{with } n = 1,2,\ldots \tag{4.4}$$

where $p_{i|i}$ means probability of remaining in state i, p_i is the probability of being in state i, n is the time step index, and N is a given number of time steps, that is, the duration of the dwelling in a particular state.

In semi-Markov models, however, state duration distributions are defined: typically, lognormal for the bad state and power law (Braten and Tjelta, 2002) for the good state, (Vogel et al., 1992; ITU-R Rec. P.681-8, 2015), that is,

$$p_{Good}(D \le d) = 1 - \beta d^{-\gamma} \tag{4.5}$$

$$p_{Bad}(D \le d) = 0.5\left[1 + erf\left(\frac{\ln(d) - \ln(\alpha)}{\sigma\sqrt{2}}\right)\right] \tag{4.6}$$

where d is the duration variable and D is a given duration. The other terms in the equations are model parameters.

In the case of dynamic Markov models, the transition probabilities change depending on the current state duration (Milojevic et al., 2009).

An evolution of state models was later proposed based on these ideas in Pérez-Fontán et al. (2007a) and Prieto-Cerdeira et al. (2010). Previously, each state, good, bad, and intermediate, were assumed to follow the same distribution with constant parameters. This is unrealistic giving fairly

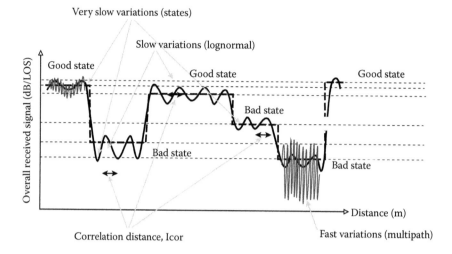

FIGURE 4.7 Two-state model with different parameter sets in each state event. (After Prieto-Cerdeira, R. et al. *International Journal of Satellite Communications and Networking*, 28, 291–315, 2010.)

nonuniform behaviors in each specific state event (Figure 4.5). One simplification, and improvement was to go back to the initial idea of using two states (Lutz et al., 1991) but having with each outcome a distribution with a different parameter set (Figure 4.7).

Each time a state, good or bad, is drawn, the general distribution used in both states, Loo's, is assigned a different parameter set drawn from empirical expressions derived for each state in the measurements. In previous models, a single distribution encompassing all the data belonging to a single state was analyzed together to provide a single parameter set.

Each state was modeled using the Loo distribution with parameters M_A, Σ_A, and MP; however, as mentioned above, these parameters themselves follow specific distributions, that is,

- The mean of the lognormal variations in the direct signal was fitted to a Gaussian distribution

$$f(M_A) \equiv N(\mu_1, \sigma_1) \tag{4.7}$$

- While the standard deviation of the lognormal variations in the direct signal were fitted to another Gaussian distribution

$$f(\Sigma_A \mid M_A) \equiv N(\mu_2, \sigma_2) \tag{4.8}$$

whose parameters depend on M_A through empirically fitted second-order polynomials. For more details, see Prieto-Cerdeira et al. (2010).
- Finally, the multipath power parameter was also fitted to another Gaussian distribution independent of A,

$$f(M_A) \equiv N(\mu_1, \sigma_1) \tag{4.9}$$

It was later observed, using extensive CNES (French Space Agency) data (Pérez-Fontán et al., 2013) that MP could be linked to the direct signal as illustrated in Figure 4.8, where K(dB) is the instantaneous Rice factor or carrier to multipath ratio, defined as

$$K(\text{dB}) = A(\text{dB}) - MP(\text{dB}) \tag{4.10}$$

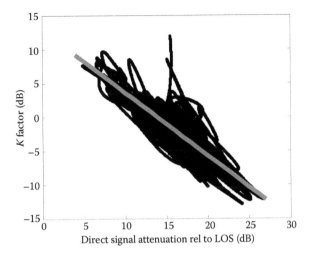

FIGURE 4.8 Experimental linkage between A and MP through the K factor.

This idea can be included in time series generators implementing the Loo's model within each state (Pérez-Fontán et al., 2008a) as illustrated in Figure 4.9.

The model proposed in Prieto-Cerdeira et al. (2010) also adopted the semi-Markov approach for both states "good" and "bad," but fitted lognormal distributions for both states (Pérez-Fontán et al., 2007a). The model was validated with L- and S-band data and further enhanced based on CNES data for S- and C-bands (Pérez-Fontán et al., 2013).

Further enhancements proposed in Pérez-Fontán et al. (2013) include an improved modeling of the transition region between the states in order not to produce unrealistic sharp changes in signal level. This was achieved by introducing such transition regions whenever necessary in order not to exceed certain experimentally derived slope (dB/m) values in the variations of A.

Statistical models in general address specific operational scenarios such as urban, suburban, and so on, and specific elevation angles. Such models fail to take into account a fundamental element of

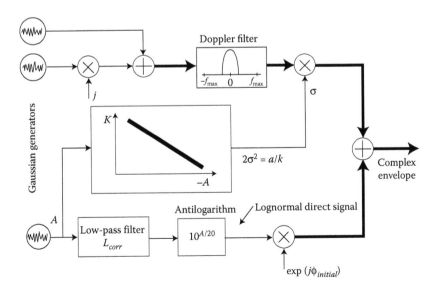

FIGURE 4.9 Time series generator implementing the Loo model within one state.

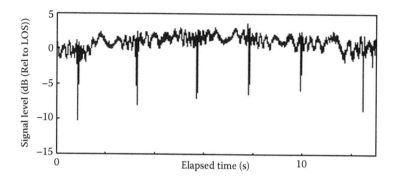

FIGURE 4.10   Tram line effects at Ka-band.

the channel, which is the driving direction with respect to the satellite link. Imagine a user driving in a Manhattan-like scenario. If the link is aligned with the driving direction, the coverage would be good, perfect in fact, regardless of the elevation angle. If we now drive along transversal streets, the coverage would be extremely poor. This fact is lost in traditional statistical models such as the ones discussed previously, since the model parameters would originate from the actual mix of along, across, and intermediate angles traveled during the associated measurement campaign.

More recent statistical models have paid attention to the driving direction relative to the link. To be able to do so, there is the need for a huge measurement campaign encompassing a fair mix of driving directions only achieved during ESA's MiLADY project (MiLADY, 2010). Rieche et al. (2014) presented a model that, in addition to elevation, takes into account the driving direction, still remaining within the frame of the two-state model approach discussed above.

As we go higher in frequency, the channel presents more features that were not so visible or blurred at the lower bands (L-, S-, and C-band). This is the case shown in Figure 4.10 measured at Ka-band (Pérez-Fontán et al., 2011a). In this respect, as the wavelength decreases, smaller obstacles can clearly be seen in the measured data. This means that, in addition to states that reflect macro-blockage/shadowing events, there are small spikes in the time series, which can be modeled using periodic processes as in the case of the poles supporting tramline overhead electric wiring, as shown in Figure 4.10.

4.4   PHYSICAL–STATISTICAL MODELING

An alternative that is more computationally intensive but intrinsically takes into account such things as the driving direction is the so-called physical statistical or virtual city approach. The basic idea is to represent the various operational environments in terms of a small number of canonical shapes: plates, cylinders, boxes, etc., which are easy to analyze from an electromagnetic point of view. These synthetic scenarios have to represent the characteristics of the actual operational scenarios to be encountered by the user, for example, in terms of building height distributions, separation between buildings, and so on. From the electromagnetic modeling side, there is the wide range of options, for example, building effects can initially be treated as blocking elements using a simple on–off model. A step forward is using knife-edge diffraction, and further improvements can be based on Fresnel–Kirchhoff theory, UTD, and others. Similarly, EM models can be initially scalar and then enhanced to take into account the polarization.

Similarly, building scattering effects can be modeled using ray-tracing, UTD, PO, etc. The degree of sophistication in the definition of buildings can go from the assumption of flat, smooth surfaces to including some roughness-related parameters.

In addition, small urban features such as traffic signs and lights or lamp poles, drawn randomly in the synthetic scene according to specific laws derived from observations at different operational scenarios can further enhance their representation and hence, the reproduction of the path loss and its variability, as well as other features such as angles of arrival (AoA), times of arrival, etc.

Vegetation modeling can also be included in this approach. In this case, tree trunks and canopies can be modeled as cylinders. Distributions of positions, separations, sizes, etc. can be obtained from visual observation. This information can also be drawn from fish-eye pictures or movies taken in the chosen representative locations.

A physical statistical model/simulator should be able to synthesize wideband channel parameters with a particular emphasis on realistic multipath modeling, including amplitude, phase, polarization, delay, DoA (direction of arrival), and Doppler as key parameters. To enhance its versatility, the design of the simulator shall not require extensive measurement campaigns so that there is no need to tune the selected physical models. Moreover, it has to handle satellite diversity, receiving space diversity, frequency diversity, and polarization diversity, and MIMO configurations. Models following this approach have been proposed, for example, in Karasawa et al. (1995), Tzaras et al. (1998), Oestges and Vanhoenacker-Janvier (2001), Pérez-Fontán et al. (2007b), and Ait-Ighil et al. (2013).

The direct path's blockage model represents the attenuation due to building shadowing. Several models could be used, depending on the system requirements. For example, as presented later on for analyzing shadowing correlation effects in multiple satellite diversity configurations, a simple ray intersection algorithm from the transmitter to the receiver could be used. In this simple model, if the direct path is intersected, a constant attenuation is applied to the direct path in addition to the path loss. Very efficient in terms of time computation, this model is mainly suitable for generating on/off time series for satellite diversity studies, replacing the generator of correlated Markov chains based on statistical models. An alternative could be based on UTD (Aguado Agelet et al., 2000) to reproduce the diffraction phenomenon taking place on building edges. This alternative is more time consuming.

A simpler, scalar (nonpolarimetric) approach is based on Fresnel–Kirchhoff diffraction modeling (Vauzelle, 1994; Hristov, 2000), similar to knife edge but taking into account more than one building edge, represented by boxes, not just the top edge or a side edge, and combing edge effects coherently. The surfaces of calculation of F–K diffraction are illustrated in Figure 4.11. A simple 2D approach could be improved by using a 3D approach, thus improving the reproduction of the signal in the state transition areas.

For multipath contributions, using UTD (Aguado Agelet et al., 2000) plus ray-tracing on flat smooth cuboids, would just yield the specular contributions. These can exist if real smooth surfaces are encountered and if the pseudo-Brewster angle is not exceeded, otherwise a circular polarization switch to the cross-polar component takes place. Second-, third-, and higher-order specular reflections would complete the otherwise very sparse channel impulse response obtained. A more

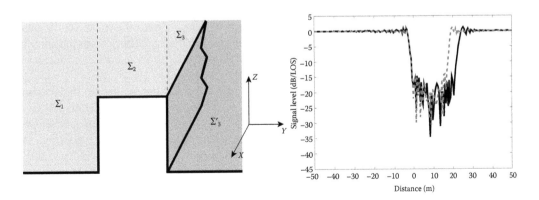

FIGURE 4.11 Scalar Fresnel–Kirchhoff diffraction calculation. 2D and 3D, and improvement in transition area calculations. On the left the integration surfaces in both cases are illustrated.

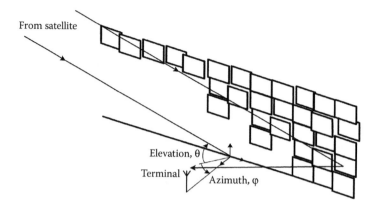

FIGURE 4.12   Description of a rough building face using squared protrusions on a dielectric plate.

realistic approach is to take into account the presence of diffuse scattering happening at rough surfaces and small urban features. These can originate everywhere in the scene and not just where the possible specular reflection points are located.

One possible approach is to assume isotropic scattering (Pérez-Fontán et al., 2007b) originating at small surface elements on the facades, mainly on the side of the street opposite the satellite. Same street side scattering is generated when dual bounce effects occur. This discretization could be produced in small squares with sizes depending on the required resolution. They can be modeled in terms of normalized radar cross section (RCS) values, σ^0. The actual RCS of one tile would be $\sigma = S\sigma^0$ m^2, with S being the area of the tile. Tiles, however, have to be small enough, especially for short distances to the terminal on the street so as not to produce physically excessive scattered power using the bistatic radar equation since the RCS is in fact a far-field concept. The values of the normalized RCS have to be derived, in principle, empirically, which should also provide pattern-related information and elevation angle dependencies.

A theoretically based approach to calculate the above RCS values supposes that the building facades can be described by squared protruding and receding plates with homogeneous dielectric constants (Figure 4.12) (Valtr and Pérez-Fontán, 2011). The plates represent the architectural features that can be usually found on a facade such as windows. Physical optics (PO) (Diaz and Milligan, 1996) is an appropriate method to study such a complex shape of the facade since it allows a discrete meshing of the scatterers; moreover, it can be used in near-field conditions. A great deal of work has been carried out in the modeling of rough building surfaces as discussed in Degli-Esposti et al. (2007).

One further step could be to assume a more complex building face roughness pattern as illustrated in Figure 4.13 (Ait-Ighil et al., 2013). In this case, windows especially make up the more salient feature in a building face; these are usually composed of an area set back from the nominal surface, which produces some dihedral features. Moreover, we can consider in the rest of the facade a general regular roughness characteristic.

The 3CM (three-component model) has been proposed in the context of mobile satellite communications and navigation. It is a simplified EM building scattering model (Ait-Ighil et al., 2013). Its principle is to divide the scattering function of any complex facade into three different mechanisms depending on its architecture: specular reflections, backscattering, and incoherent scattering. The specular reflection takes place on smooth and wide surfaces such as flat walls or windows, and is based on physical optics (PO) techniques. The incoherent scattering component is due to rough surfaces and small elements present on complex facades. In this case, the model developed by Barrick (1968) is used. The backscattering component comes from double bounce reflections taking place on protruding and receding corners such as balconies or windows. In this case, a hybrid geometrical optics (GO)–physical optics (PO) approach is used, as illustrated in Figure 4.13.

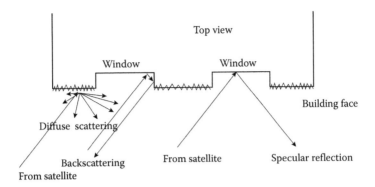

FIGURE 4.13 3CM—Three-component model building scattering contributions. (After Ait-Ighil, M. et al. SCHUN—A hybrid land mobile satellite channel simulator enhanced for multipath modelling applied to satellite navigation systems, in *EuCAP 2013, 7th European Conference on Antennas and Propagation*, Gothenburg, Sweden, April 8–12, 2013.)

In any realistic operational scenario, the presence of trees is common and thus presents a significant influence on channel behavior. One possible approach is the one presented in Figure 4.14 where a tree is assumed to consist of two elements treated separately: the trunk and the canopy. The trunk is assimilated either to a screen with no depth on which the Kirchhoff diffraction scalar theory is applied. Alternatively, PO can be used to model its effects, not only as a contributor to shadowing but also to scattering, both in forward and backward directions. The canopy is modeled as a cylindrical volume V, with a radius R and a height H_c, containing randomly distributed branches. It is assumed to be responsible for both attenuation of the direct signal and diffuse multipath. Pérez-Fontán et al. (2011a) proposed a model for Ku-band where there is a diffraction–absorption-related component on which a stochastic component is superposed.

Abele et al. (2010) proposed the use of the radiative energy transfer (RET) theory (Caldeirinha, 2002; Rogers et al., 2002). The random nature of tree foliage supports the use of the transport theory for modeling propagation through the canopy (Rogers et al., 2002).

Another alternative was proposed in Cheffena and Pérez-Fontán (2010), Cheffena and Pérez-Fontán (2011), Cheffena et al. (2012), and Israel et al. (2014), where multiple scattering theory (MST) (Ishimaru, 1999) was extended based on the ideas in de Jong and Herben (2004) from horizontal

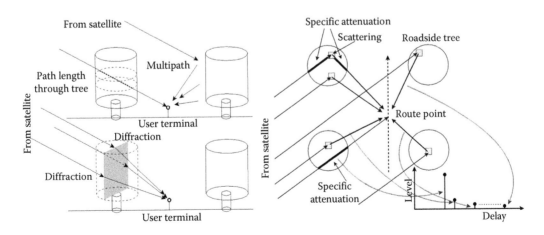

FIGURE 4.14 Simplified model of a tree. Tree scattering and attenuation model based on multiple scattering theory. (After Ishimaru, A., *Wave Propagation and Scattering in Random Media*, Wiley-IEEE Press, Piscataway, NJ, 1999.)

paths to the slant path geometry, typical in satellite channels. This model provides us with three components, a loss through the canopy, and a narrow-angle coherent component about the direct path plus a wider-angle incoherent component. This last component allows the calculation of multipath contributions in the backscattering direction as illustrated in Figure 4.14. This model has also been linked to a wind swaying tree model (Cheffena and Ekman, 2009) describing the signal variations for static terminals in Cheffena and Pérez-Fontán (2011).

4.5　SATELLITE DIVERSITY

LMS is a very harsh channel where outages can be unsurmountable given the limited fade margins available. One possible solution is using satellite diversity, that is, shadowing and blockage can be overcome if the terminal can have an alternative satellite to connect to. A more sophisticated concept is also using a terrestrial component providing an alternative connection, especially in the very hard dense urban environments: gap fillers. A key element to the success of such systems is that the blockage/shadowing effects for the alternative links are uncorrelated. This has led research into shadowing cross-correlation models as well as into simultaneous multiple-link models.

Shadowing cross-correlation tends to decrease as the angular spacing between links increases. Measurements have been reported in Jahn and Lutz (1994) where the multiple satellite link geometry was simulated using circular flight measurements. Additional results are provided in Robert et al. (1992) and Tzaras et al. (1998).

Alternative ways for producing cross-correlation information was using zenith pointing, fish-eye camera pictures conveniently processed to separate into three clear classes: LOS, shadowed, and blocked for the full upper hemisphere (Akturan and Vogel, 1997). It is possible to associate to each class a given signal distribution in such a way that route or environment type signal statistics can be generated.

The main application of correlation information is for being used in combination with statistical models. One such model was proposed in Lutz (1996) where the original two-state model was converted to four-states: good–good, good–bad, bad–good, and bad–bad, and where the cross-correlation coefficient was used to drive the generation process (Figure 4.15).

A *physical–statistical* model for two satellites was developed in Tzaras et al. (1998) where a canonical street canyon was studied assuming a simple geometry with constant building heights and street width (Figure 4.16), yielding a model for ρ as a function of an azimuth spacing Δθ and two elevation angles (φ1, φ2). A similar approach based on an on–off model was proposed in

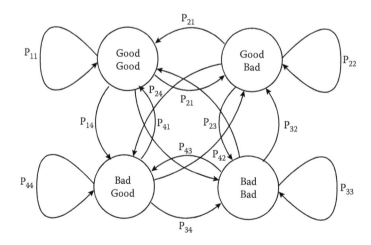

FIGURE 4.15　First-order Markov model for two satellites. (After Lutz, E., *International Journal of Satellite Communications*, 14, 333–339, 1996.)

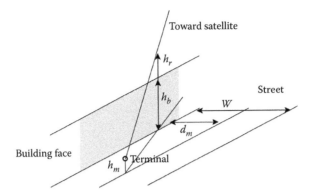

FIGURE 4.16 Physical–statistical model. (After Tzaras, C.S.R. Saunders, and B.G. Evans, *IEEE-APS Conference on Antennas and Propagation for Wireless Communications*, November 1–4, 1998.)

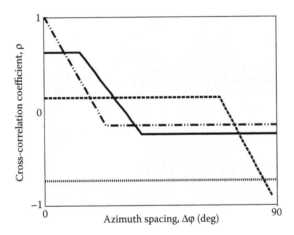

FIGURE 4.17 Shadowing cross-correlation model. (After Vazquez-Castro, M. et al. *Microwave and Optical Technology Letters*, 28(3), 160–164, 2001.)

Vazquez-Castro et al. (2001), where a three-segment model was proposed as illustrated in Figure 4.17. Further, in Vazquez-Castro et al. (2002), the model presented in Vazquez-Castro et al. (2001) was extended to include more realistic conditions using random building heights and widths following distributions from several European cities (London, Guildford, and Madrid).

As an example, a synthetic urban scenario analyzed using 2D Fresnel–Kirchhoff diffraction is presented in Figure 4.18a–c (Pérez-Fontán and Mariño Espiñeira, 2008) where the deterministically computed shadowing time series were cross-correlated. It is clear how the angular spacing has an impact on the correlation coefficient.

4.6 MODELING THE WIDEBAND LMS CHANNEL

As indicated earlier, the LMS channel has, in general, very small delay spread values. This signifies that current communication systems at the lower frequencies, L- and S-bands, do not require the use of wideband models for physical layer performance evaluation. The same may occur at Ku- or Ka-band where new communication systems are envisaged. The use of more directive antennas makes it more difficult to pick up long delayed echoes. In any case, there has been a limited interest in describing the wideband behavior of this channel for communication applications, in case larger

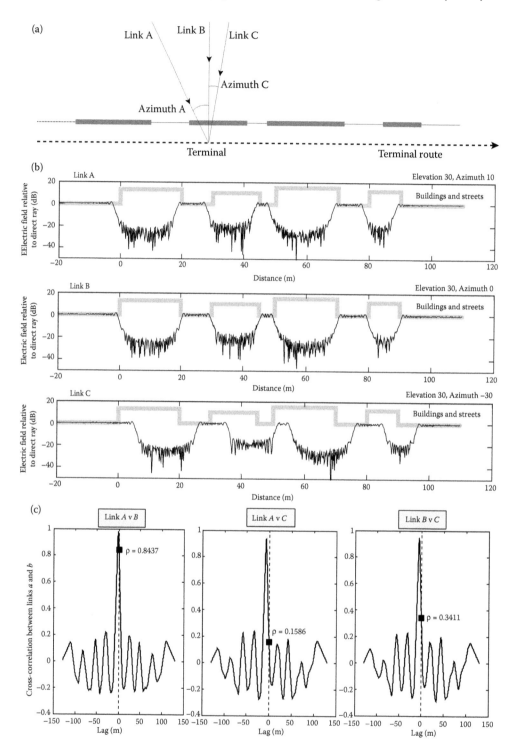

FIGURE 4.18 (a) Urban geometry, terminal route, and three satellites illuminating the area. (b) Time series for each of the three links. (c) Cross-correlation coefficients for different pairs of satellites. (After Pérez-Fontán, F. and P. Mariño Espiñeira, *Modeling of the Wireless Propagation Channel: A Simulation Approach with MATLAB*, John Wiley & Sons, Chichester, West Sussex, 2008.)

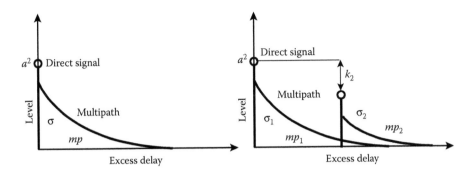

FIGURE 4.19 Observed delay spread modeling at Ka-band with nondirective antennas. (After Pérez-Fontán, F. et al. Statistical and physical–statistical modeling of the LMS, channel at Ku- and Ka-band, in *European Conference on Antennas and Propagation, EuCAP 2011*, Rome, Italy, April 11–15, 2011a.)

bandwidths are transmitted or less directive antennas are possible at the higher-frequency bands, as discussed in Pérez-Fontán et al. (2011a) where an omnidirectional antenna was used with a receiver operating at Ka-band. A Hercules aircraft flying in circles around the receiver at a relative short distance was used to simulate the satellite. This permitted the development of a wideband model with one or two exponentially decaying multipath clusters (Figure 4.19). Still very small delay spread values were recorded.

Coming back to the lower bands, the narrowband assumption is mostly correct for two-way communications or broadcast applications. In the case of satellite navigation systems, based on direct sequence spreading techniques, the echoes that count and may cause distance estimation errors are those within the duration of one transmitted chip of the ranging code. In this case, a wideband knowledge of the channel is relevant.

The time-spreading and time-varying nature of the channel are normally represented in the delay-time domain using a complex impulse response representation in its low-pass equivalent version, that is,

$$\tilde{h}(\tau,t) = \sum_{n=1}^{N} \tilde{a}_n(t)\delta(t-\tau_n(t)) \quad \text{with} \quad \tilde{a}_n(t) = a_n(t)\exp[-j2\pi f_c \tau_n(t)] \tag{4.11}$$

alternatively, in the frequency–time domain by

$$\tilde{H}(f,t) = \sum_{n=1}^{N} \tilde{a}_n \exp(-j2\pi f \tau_n(t)) \tag{4.12}$$

with complex amplitude $\tilde{a}_n(t)$ and delay τ_n for the nth signal component.

For simulation purposes, the time-varying impulse response is usually converted to tap delay-line (TDL) form taking into consideration the sampling requirements of the transmitted signal (Figure 4.20) (Tranter et al., 2004). The time-varying taps are assumed to be uncorrelated random fading processes whose average powers decay according to specific laws. In Figure 4.20, the path from the satellite to the terminal is represented by a delay, a loss, and a phase/frequency shift term. As the signal arrives in the local environment, it undergoes time spreading, Doppler spreading, and slow and fast fading. The decay with delay is represented by weights W_i, the fast fading and Doppler spreading by R_i, and the slow variations by S_i. The spacing $\Delta\tau$ between taps has to accommodate the transmitted signal bandwidth and the channel's Doppler spread.

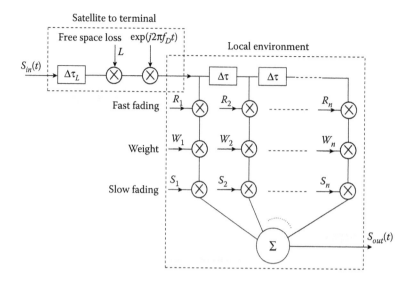

FIGURE 4.20 Wideband channel model with *n* fixed delays.

Jahn et al. (1996) proposed a measurement-based model for L-band where the various received components are spread out in delay, namely, direct path, near echoes, and far echoes (Figure 4.21). The *direct path* is subjected to shadowing and follows a Rice distribution in the LOS state or by a Suzuki distribution in the non-LOS state. Then, *near echoes* are produced in the neighborhood of the terminal with delays up to 600 ns and its number is Poisson distributed appearing with exponential arrivals in the delay axis while their decay is also exponential (linear in dB). Finally, *far echoes* were found to follow uniformly distributed delays up to a maximum value. Again their number is Poisson distributed. As in the near-echo case, their amplitudes follow a Rayleigh distribution.

A simpler model (Pérez-Fontán et al., 2001) extends a three-state model in the delay domain for L- and S-bands. That is, the multipath power is expanded according to exponential arrivals relative to the direct signal. The number of assumed rays was 50 but any number would suffice (depending on the application) while maintaining the total multipath power. The assumption made also gives rise to lineal power delay profiles in logarithmic units. Unlike in Jahn et al. (1996), no long echoes are considered.

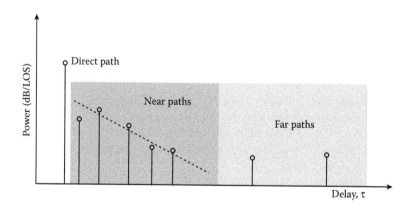

FIGURE 4.21 Definition of the echo characteristics for the proposed wideband LMS channel. (After Jahn, A., H. Bischl, and G. Heiss, Channel characterisation for spread spectrum satellite communications, in *IEEE 4th International Symposium on Spread Spectrum Techniques and Applications Proceedings*, 3, 1221–1226, September 1996.)

The usual approach followed in TDL models is that the different taps are uncorrelated. Moreover, the tap spacings are constant throughout the simulation process. This means that a specific echo whose delay changes in time across various taps is bunched together with other echoes making up a tap. This is not a good approach in the case of navigation given that the history of each echo has an impact in the estimation of the distance to a satellite. This means that TDL modes with constant delays may lead to a wrong representation of the channel. This means that a more appropriate approach is going back to representations of the channel based on the impulse response where the delays of the main contributions are allowed to vary and take up continuous valued delays. This is inherently linked with the physical–statistical models discussed earlier where individual delays can be tracked down through the entire simulation session. Similarly, models based on point scatterers permit this feature.

When processing wideband measurements, it is important to use advanced techniques that allow the tracking of individual echoes across consecutive measured snapshots (Jost et al., 2012). However, work toward the simplification of dense scattering environments is still necessary; otherwise, the simulation process can become unwieldy and time consuming.

4.7 MODELING THE SATELLITE-TO-INDOOR PROPAGATION CHANNEL

We now review some ideas proposed in regard to the satellite-to-indoor channel. Serving users inside buildings from a satellite is a very challenging issue given the limited margins already provided for outdoor users. However, it is still possible to reach into rooms near external building walls. Satellite-to-indoor measurement and modeling results can be found in Axiotis and Theologou (2003), Bodnar et al. (1999), Veltsistas et al. (2007), Kvicera and Pechac (2011), Jost et al. (2011), and Pérez-Fontán et al. (2008b, 2011b).

Several issues need modeling: the entry loss and the distribution of the received energy in the delay domain. Entry loss is the most critical of these effects given its very large values, while the delays are in the order of some tens of nanoseconds. The main contribution will be the unblocked direct signal illuminating the external walls of a building; reflections and diffractions off other building faces will show lower levels. If the direct signal is blocked, then indirect illuminating sources will play a role.

Entry loss is the result of very complex processes, including transmission through building materials and paths through different external and internal walls, the roof, the floor, through windows, glass facades, etc. where window glass can also be metal tinted, thus giving rise to strong attenuations. Moreover, satellite paths are very much dependent on the incidence angle: elevation and relative azimuth.

ITU-R Rec. P.1411-7 (2013) defined *building entry loss*, with short-range terrestrial paths in mind, as the excess loss due to the presence of a building wall (including windows and other features). It may be determined by comparing signal levels outside and inside the building at the same height. The building entry loss value has to be propagated further on within the building to account for the actual distances from the external wall to locations within the same room or within other rooms. Additional losses within the building itself are due to partitions and furniture in the rooms. The geometry of the path can in some cases be described by a single angle, the so-called grazing or entry angle instead of the relative azimuth and elevation. Further definitions on building entry and penetration loss are provided in Jost et al. (2013).

Channel models, as in the case of the LMS channel, can be classified as empirical, statistical, or analytical. An empirical approach was proposed in Pérez-Fontán et al. (2008b) for the entry loss, while a statistical model was proposed also in Pérez-Fontán et al. (2011b) for the spreading effects. The model is based on a combination of two existing models: one developed in the frame of the European COST 231 Action (COST 231 1999) and applied by Glazunov and Berg (2000) to the assessment of indoor wireless-to-satellite interference levels, and the other by Saleh and Valenzuela (1987), including its directional version (Spencer et al., 2000) (modified Saleh–Valenzuela model).

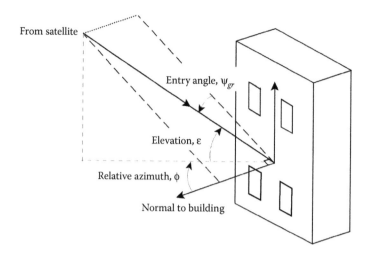

FIGURE 4.22 Outdoor-to-indoor propagation geometry.

The basic result of the developed model is a received echo structure (impulse response/scattering matrix) consisting of a maximum of five echo ensembles (four external walls and the roof). In reality, the observations reported indicated the presence of only one ensemble of echoes propagated into the building directly from the transmitter. Note that this ensemble further consists of clusters of echoes and individual echoes or rays. The assumed scenario is depicted in Figure 4.22.

The COST 231 model calculates the excess path loss through a given external wall to a point inside the building, a distance d meters away from it, using the expression (COST 231, 1999),

$$L_{Excess}(\text{dB}) = W_e + W_{Ge}(1 - \sin(\psi_{gr}))^2 + \max(L_1, L_2) \tag{4.13}$$

where $L_1 = W_i p$ and $L_2 = \alpha(d - 2)(1 - \sin(\psi_{gr}))^2$, and p is the number of internal walls traversed; d is the distance within the building up to the terminal; ψ_{gr} is the entry angle with respect to the external wall; W_e is the external wall perpendicular incidence attenuation; W_{Ge} is the additional loss due to incidence angles away from the normal to the wall; W_i is the loss through internal walls; and α is a loss per meter due to furniture and other indoor features.

Taking into account the definition given earlier, the building entry loss part of the COST 231 model would be

$$L_{Entry}(\text{dB}) = W_e + W_{Ge}(1 - \sin(\psi_{gr}))^2 \tag{4.14}$$

for terminals in rooms with external walls.

The entry angle seems to be the most important geometrical parameter and its influence has been quantified (Pérez-Fontán et al., 2008b) together with the influence of the relative incidence azimuth and elevation. Measurements of the entry loss, rather than following a $(1 - \sin(\psi_{gr}))^2$ trend with the grazing angle as in the previous baseline equation, have shown an evolution of the form

$$L_{Entry}(\text{dB}) = A - B\psi_{gr} \tag{4.15}$$

where ψ_{gr} is in degrees. Parameters obtained by measurement fitting were provided in Pérez-Fontán et al. (2011b).

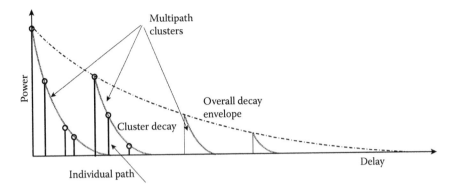

FIGURE 4.23 Proposed cluster structure (in linear units) in the Saleh–Valenzuela model.

For the wideband case, a modified version of the S&V model (Saleh and Valenzuela, 1987) (Figure 4.23) was used. The S&V model makes the assumption that echoes arrive in clusters, the overall impulse response being

$$h(\tau) = \sum_l \sum_k \beta_{kl} \exp(j\phi_{kl}) \, \delta(\tau - T_l - \tau_{kl}) \qquad (4.16)$$

The S&V wideband model was extended in Spencer et al. (2000) to include AoA. From measurements carried out in indoor environments, it was found how ray clusters are not only spread out in time/delay but also in AoA. Thus, the original Saleh–Valenzuela model was extended to include AoA, θ, as indicated by the equation

$$h(\tau, \theta) = \sum_l \sum_k \beta_{kl} \exp(j\phi_{kl}) \, \delta(\tau - T_l - \tau_{kl}) \, \delta(\theta - \Theta_l - \theta_{kl}) \qquad (4.17)$$

index l indicates cluster number and index kl indicates ray number k within a cluster l. Angle θ is measured here on the horizontal plane. A more complex model could also include the elevation angle. The model for the AoA, θ, is a zero-mean Laplace distribution with standard deviation σ, that is,

$$p(\theta_{kl}) = \frac{1}{\sqrt{2}\sigma} \exp(-|\sqrt{2}\theta_{kl}/\sigma|) \qquad (4.18)$$

and where θ_{kl} is uniformly distributed between 0 and 2π. A full parameterization of this model is provided in Pérez-Fontán et al. (2008b).

A more intensive, but narrowband, campaign was carried out in Kvicera and Pechac (2011), where a remote-controlled airship was used as a pseudosatellite. The empirical models developed for an NLOS, that is, with direct illumination of the facade where the receiver was located. The receiver was positioned indoors in front of a window.

An almost linear rise of penetration loss (dB) with elevation for the LOS case was observed and the following model was proposed:

$$L_{PL,LOS}(\text{dB}) = A + B\alpha \qquad (4.19)$$

with the loss in dB and the elevation angle, α, in degree. The terms A and B are model parameters. For the NLOS case, penetration loss first rises with elevation and then tends to be nearly constant at high elevation angles. The following model was proposed:

$$L_{PL,NLOS}(\text{dB}) = C - D(90 - \alpha)^E \tag{4.20}$$

again with the elevation angle, α in degree, and where C, D, and E are model parameters.

A statistical wideband propagation model for the satellite-to-indoor channel suitable for navigation applications was presented in Jost et al. (2013), where a deterministic part based on diffractions and reflections off windows and walls in the room where the user is located and the definition of various scatterer types extracted from measurements was proposed. The continuity of all ray trajectories, a key requirement for navigation applications is fulfilled for any closed trajectory of the user within the room.

4.8 SUMMARY

In this chapter, an extensive review of the approaches followed for modeling the LMS channel was presented. The text concentrated on providing the main ideas behind these models but not so much the details, which can be found in the references provided. The LMS channel has been mainly described using statistical models in which parameters are derived from experimental data. As the length of the terminal route increases, there is the need to use more complex approaches; thus, for short distances, a single distribution suffices, and then combinations of distributions are used and, finally, for long routes, maybe representing the duration of one call, the viewing of a news broadcast or the duration of a navigation session, states are used.

One approach that provides mode information is the so-called physical–statistical or virtual city approach where the environment is described in terms of a limited number of canonical elements such as screens, boxes, etc. They are randomly laid out based on statistical distributions such as building heights. These synthetic scenarios are analyzed using more or less complex deterministic models such as ray-tracing. They provide a deeper insight into the various channel phenomena, which are inherently directional, wideband, and polarimetric. Moreover, they directly contemplate the possibility of analyzing simultaneous multiple satellite links and also MIMO configurations.

Attention has also been paid to the modeling of satellite diversity issues through a statistical and a physical–statistical approach. Finally, the wideband case, though not very frequent in communications, is important in navigation applications. A short discussion on the satellite-to-indoor channel has also been provided.

REFERENCES

Abdi, A., W.C. Lau, M.-S. Alouini, and M, Kaveh, A new simple model for land mobile satellite channels: First- and second-order statistics, *Transactions on Vehicular Technology*, 2(3), 519–528, 2003.

Abele, A., F. Pérez-Fontán, M. Bousquet, P. Valtr, J. Lemorton, F. Lacoste, and E. Corbel, A new physical–statistical model of the land mobile satellite propagation channel, in *4th European Conference on Antennas and Propagation, EuCAP'2010*, Barcelona, Spain, April, 2010.

Aguado Agelet, F., A. Formella, J.M. Hernando Rabanos, F. Isasi de Vicente, and F. Pérez-Fontán, Efficient ray-tracing acceleration techniques for radio propagation modeling, *IEEE Transactions on Vehicular Technology*, 49(6), 2089–2104, 2000.

Ait-Ighil, M., J. Lemorton, F. Pérez-Fontán, F. Lacoste, P. Thevenon, C. Bourga, and M. Bousquet, SCHUN—A hybrid land mobile satellite channel simulator enhanced for multipath modelling applied to satellite navigation systems, in *EuCAP 2013, 7th European Conference on Antennas and Propagation*, Gothenburg, Sweden, April 8–12, 2013.

Akturan, R., An overview of the Sirius satellite radio system, *International Journal of Satellite Communications and Networking*, 26(5), 349–358, 2008.

Akturan, R. and W.J. Vogel, Optically derived elevation angle dependence of fading for satellite PCS, in *19th NASA Propagation Experimenters Meeting (NAPEX 19)*, Fort Collins, CO, August, 127–132, 1995.

Akturan, R. and W.J. Vogel, Path diversity for LEO satellite-PCS in the urban environment, *IEEE Transactions on Antennas and Propagation*, 45(7), 1107–1116, 1997.

Arndt, D., On channel modelling for land mobile satellite reception, PhD thesis, Technische Universität, Ilmenau, 2015.

Axiotis, D.I. and M.E. Theologou, An empirical model for predicting building penetration loss at 2 GHz for high elevation angles, *IEEE Antennas and Wireless Propagation Letters*, 2(1), 234–237, 2003.

Barrick, D.E., Rough surface scattering based on the specular point theory, *IEEE Transactions on Antennas and Propagation*, 16, 449–454, 1968.

Barts, R.M. and W.L. Stutzman, Modeling and simulation of mobile satellite propagation, *IEEE Transactions on Antennas and Propagation*, 40(4), 375–382, 1992.

Bodnar, Z., Z. Herczku, J. Berces, I. Papp, F. Som, B.G. Molnar, and I. Frigyes, A detailed experimental study of the LEO satellite to indoor channel characteristics, *International Journal of Wireless Information Networks*, 6(2), 79–91, 1999.

Braten, L., Satellite visibility in Northern Europe based on digital maps, *International Journal of Satellite Communications*, 18(1), 47–62, 2000.

Braten, L.E. and T. Tjelta, Semi-Markov multistate modeling of the land mobile propagation channel for geostationary satellites, *IEEE Transactions on Antennas and Propagation*, 50(12), 1795–1802, 2002.

Caldeirinha, R.F.S., Propagation modelling of bistatic scattering of isolated trees for micro- and millimeter wave urban microcells, *IEEE Personal, Indoor and Mobile Radio Communications (PIMRC)*, 1, 135–139, 2002.

Cheffena, M. and T. Ekman, Dynamic model of signal fading due to swaying vegetation, *EURASIP Journal on Wireless Communications and Networking, Advances in Propagation Modeling for Wireless Systems*, Special issue, 2009, Article ID 306876, 11, 2009.

Cheffena, M. and F. Pérez-Fontán, Land mobile satellite channel simulator along roadside trees, *IEEE Antennas and Wireless Propagation Letters*, 9, 748–751, 2010.

Cheffena, M. and F. Pérez-Fontán, Channel simulator for land mobile satellite channel along roadside trees, *IEEE Transactions on Antennas and Propagation*, 59(5), 1699–1706, 2011.

Cheffena, M., F. Pérez-Fontán, F. Lacoste, E. Corbel, H.-J. Mametsa, and G. Carrie, Land mobile satellite dual polarized MIMO channel along roadside trees: Modeling and performance evaluation, *IEEE Transactions on Antennas and Propagation*, 60(2), 597–605, 2012.

Corazza, G.E. and F. Vatalaro, A statistical model for land mobile satellite channels and its application to nongeostationary orbit systems, *IEEE Transactions on Vehicular Technology*, 43(3), 738–742, 1994.

COST 231. *Evolution of Land Mobile Radio (Including Personal) Communications*. Final Report. Available at http://www.lx.it.pt/cost231/final_report.htm, 1999.

Davarian, F., Channel simulation to facilitate mobile-satellite communications research, *IEEE Transactions on Communications*, 35(1), 47–56, 1987.

de Jong, Y.L.C. and M.H.A.J. Herben, A tree-scattering model for improved propagation prediction in urban microcells, *IEEE Transactions on Vehicular Technology*, 53, 503–513, 2004.

Degli-Esposti, V., F. Fuschini, E.M. Vitucci, and G. Falciasecca, Measurement and modelling of scattering from buildings, *IEEE Transactions on Antennas and Propagation*, 55(1), 143–153, 2007.

Diaz, L. and T. Milligan, *Antenna Engineering Using Physical Optics: Practical CAD Techniques and Software*, Artech House, Norwood, MA, 1996.

Dongya, S., R. Jian, Y. Yihuai, Q. Yong, C. Hongliang, and F. Shigang, The six-state Markov model for land mobile satellite channels, in *IEEE International Symposium on Microwave, Antenna, Propagation and EMC Technologies for Wireless Communications, 2005, MAPE 2005*, 2, 1619–1622, 2005.

Dottling, M., A. Jahn, D. Didascalou, and W. Wiesbeck, Two and three-dimensional ray tracing applied to the land mobile satellite (LMS) propagation channel, *Antennas and Propagation Magazine, IEEE*, 43(6), 27–37, 2001.

DVB BlueBook A120, *DVB-SH Implementation Guidelines*, www.dvb-h.org/, 2008-05.

ESTEC. Final report S-UMTS. *Preparation of Next Generation Universal Mobile Satellite Telecommunications Systems*. Technical report, ESA, ESTEC, 2000.

Evans, J.V., Satellite systems for personal communications, *Antennas and Propagation Magazine, IEEE*, 39(3), 7–20, 1997.

Fleury, B.H., M. Tschudin, R. Heddergott, D. Dahlhaus, and K.L. Pedersen. Channel parameter estimation in mobile radio environments using the SAGE algorithm, *IEEE Journal on Selected Areas in Communications*, 17, 434–450, 1999.

Franchi, A., A. Howell, and J.R. Sengupta, Broadband mobile via satellite: Inmarsat BGAN, in *IEE Seminar on Broadband Satellite: The Critical Success Factors—Technology, Services and Markets (Ref. No. 2000/067)*, London, UK, pp. 23/1–23/7, 2000.

Glazunov, A.A. and J.-E. Berg, *Building-Shielding Loss Model*, IEEE VTC Spring, Tokyo, Japan, 2000.

Hristov, H.D., *Fresnel Zones in Wireless Links, Zone Plate Lenses and Antennas*, Artech House, Norwood, MA, 2000.

Hwang, S.-H., K.-J. Kim, J.-Y. Ahn, and K.-C. Whang, A channel model for nongeostationary orbiting satellite system, in *47th IEEE Conference on Vehicular Technology*, Phoenix, Arizona, 1, 41–45, May 1997.

Ishimaru, A., *Wave Propagation and Scattering in Random Media*, Wiley-IEEE Press, Piscataway, NJ, 1999.

Israel, J., F. Lacoste, J. Mametsa, and F. Pérez-Fontán, A propagation model for trees based on multiple scattering theory, in *European Conference on Antennas and Propagation, EUCAP 2014*, April 6–11, the Hague, the Netherlands, 2014.

ITU-R Rec. P.531-12. *Ionospheric Propagation Data and Prediction Methods Required for the Design of Satellite Services and Systems*. Geneva, 2013.

ITU-R Rec. P.618-11. *Propagation Data and Prediction Methods Required for the Design of Earth–Space Telecommunication Systems*. Geneva, 2013.

ITU-R Rec. P.676-10. *Attenuation by Atmospheric Gases*. Geneva, 2013.

ITU-R Rec. P.681-8. *Propagation Data Required for the Design of Earth Space Land Mobile Telecommunication Systems*. Geneva, 2015.

ITU-R Rec. P.1411-7. *Propagation Data and Prediction Methods for the Planning of Short-Range Outdoor Radio Communication Systems and Radio Local Area Networks in the Frequency Range 300 MHz to 100 GHz*. Geneva, 2013.

Jahn, A., H. Bischl, and G. Heiss, Channel characterisation for spread spectrum satellite communications, in *IEEE 4th International Symposium on Spread Spectrum Techniques and Applications Proceedings*, 3, Mainz, Germany, pp. 1221–1226, September 1996.

Jahn, A. and E. Lutz, DLR channel measurement programme for low Earth orbit satellite systems, in *Third Annual International Conference on Universal Personal Communications, Record*, San Diego, California, pp. 423–429, September–October 1994.

Jahn, A., M. Sforza, S. Buonomo, and E. Lutz, Narrow- and wideband channel characterization for land mobile satellite systems: Experimental results at L-band, in *Proceedings of the 4th International Mobile Satellite Conference (IMSC '95)*, Pasadena, California, pp. 115–121, 1995.

Jost, T., G. Carrie, F. Pérez-Fontán, W. Wang, and U.-C. Fiebig, A deterministic satellite-to-indoor entry loss model, *IEEE Transactions on Antennas and Propagation*, 61(4), 2223–2230, 2013.

Jost, T., W. Wang, U.-C. Fiebig, and F. Pérez-Fontán, Comparison of L- and C-band satellite-to-indoor broadband wave propagation for navigation applications, *IEEE Transactions on Antennas and Propagation*, 59(10), 3899–3909, 2011.

Jost, T., W. Wang, U.-C. Fiebig, and F. Pérez-Fontán, Detection and tracking of mobile propagation channel paths, *IEEE Transactions on Antennas and Propagation*, 60(10), 4875–4883, 2012.

Kanatas, A.G. and P. Constantinou, Narrowband characterisation of the land mobile satellite channel: A comparison of empirical models, *European Transactions on Telecommunications*, 7(4), 315–321, 1996.

Karaliopoulos, M.S. and F.-N. Pavlidou, Modelling the land mobile satellite channel: A review, *Electronics Communication Engineering Journal*, 11(5), 235–248, 1999.

Karasawa, Y., K. Minamisono, and T. Matsudo, A propagation channel model for personal mobile-satellite services, in *Proceedings of Progress of Electromagnetic Research Symposium of the European Space Agency (ESA)*, Noordwijk, the Netherlands, pp. 11–15, July 1995.

Kvicera, M. and P. Pechac, Building penetration loss for satellite services at L-, S- and C-band: Measurement and modeling, *IEEE Transactions on Antennas and Propagation*, 59(8), 3013–3021, 2011.

Loo, C., A statistical model for a land mobile satellite link, in *ICC'84—Links for the Future: Science, Systems and Services for Communications, Proceedings of the International Conference on Communications*, Amsterdam, The Netherlands, 2, pp. 588–594, May 14–17, 1984.

Lutz, E., D. Cygan, M. Dippold, F. Dolainsky, and W. Papke, The land mobile satellite communication channel—Recording, statistics, and channel model, *IEEE Transactions on Vehicular Technology*, 40, 375–386, 1991.

Lutz, E., A Markov model for correlated land mobile satellite channels, *International Journal of Satellite Communications*, 14, 333–339, 1996.

Mehrnia, A. and H. Hashemi, Mobile satellite propagation channel, Part II-a new model and its performance, in *50th IEEE Conference on Vehicular Technology, VTC 1999—Fall*, Amsterdam, the Netherlands, 5, pp. 2780–2784, 1999.

Michalski, R.A., An overview of the XM satellite radio system, in *20th AIAA International Communication Satellite Systems Conference and Exhibit*, Montreal, Quebec, Canada, May 2002.

MiLADY. Project web page. http://telecom.esa.int/telecom/ www/object/index.cfm?fobjectid = 29020, 2010.

Milojevic, M., M. Haardt, E. Eberlein, and A. Heuberger, Channel modeling for multiple satellite broadcasting systems, *IEEE Transactions on Broadcasting*, 55(4), 705–718, 2009.

Ming, H., S. Dongya, C. Yanni, X. Jie, Y. Dong, C. Jie, and L. Anxian, A new five-state Markov model for land mobile satellite channels, in *8th International Symposium on Antennas, Propagation and EM Theory, ISAPE 2008*, Kunming, China, pp. 1512–1515, November, 2008.

Oestges, C. and D. Vanhoenacker-Janvier, A physical–statistical shadowing correlation model and its application to low-Earth orbit systems, *IEEE Transactions on Vehicular Technology*, 50(2), 416–421, 2001.

Parks, M.A.N., S.R. Saunders, and B.G. Evans, Wideband characterisation and modelling of the mobile satellite propagation channel at L- and S-bands, in *Tenth International Conference on Antennas and Propagation (Conf. Publ. No. 436)*, Edinburgh, UK, Vol. 2, pp. 39–43, 1997.

Patzold, M., U. Killat, and F. Laue, An extended Suzuki model for land mobile satellite channels and its statistical properties, *IEEE Transactions on Vehicular Technology*, 47(2), 617–630, 1998.

Pérez-Fontán, F., A. Abele, B. Montenegro, F. Lacoste, V. Fabbro, L. Castanet, B. Sanmartin, and P. Valtr, Modelling of the land mobile satellite channel using a virtual city approach, *EuCAP 2007*, in *The Second European Conference on Antennas and Propagation*, Edinburgh, UK, November 11–16, 2007b.

Pérez-Fontán, F., A. Mayo Bazarra, D. Marote Alvarez, R. Prieto-Cerdeira, P. Mariño, F. Machado, and N. Riera Diaz, Review of generative models for the narrowband land mobile satellite propagation channel, *International Journal of Satellite Communications and Networking*, 26(4), 291–316, 2008a.

Pérez-Fontán, F., G. Carrie, J. Lemorton, F. Lacostec, B. Montenegro-Villacieros, and D. Vanhoenacker-Janvier, An enhanced narrowband statistical land mobile satellite channel model for the S- and C-bands, *(Special issue: ESA Workshop on Radiowave Propagation—2011), Space Communications*, 22(2–4), 125–132, 2013.

Pérez-Fontán, F. and P. Mariño Espiñeira, *Modeling of the Wireless Propagation Channel: A Simulation Approach with MATLAB*. John Wiley & Sons, Chichester, West Sussex, 2008.

Pérez-Fontán, F., J. Pereda Gonzalez, M.J. Sedes Ferreiro, M. Vazquez-Castro, S. Buonomo, and J.P. Poiares-Baptista, Complex envelope three-state Markov model based simulator for the narrow-band LMS channel, *International Journal of Satellite Communications*, 15(1), 1–15, 1997.

Pérez-Fontán, F., V. Hovinen, M. Schonhuber, R. Prieto-Cerdeira, J.A. Delgado-Penin, F. Teschl, J. Kyrolainen, and P. Valtr, Building entry loss and delay spread measurements on a simulated hap-to-indoor link at S-band, *EURASIP Journal of Wireless Communications and Networking, Advanced Communication Techniques and Applications for High-Altitude Platforms*, Special issue, 2008b.

Pérez-Fontán, F., V. Hovinen, M. Schonhuber, R. Prieto-Cerdeira, F. Teschl, J. Kyrolainen, and P. Valtr, A wideband, directional model for the satellite-to-indoor propagation channel at S-band, *International Journal of Satellite Communications and Networking*, 29, 23–45, 2011b.

Pérez-Fontán, F., N. Jeannin, L. Castanet, H.J. Mametsa, F. Lacoste, V. Hovinen, M. Schonhuber, F. Teschl, and R. Prieto-Cerdeira, Statistical and physical–statistical modeling of the land mobile satellite, LMS, channel at Ku- and Ka-band, in *European Conference on Antennas and Propagation, EuCAP 2011*, Rome, Italy, April 11–15, 2011a.

Pérez-Fontán, F., I. Sanchez-Lago, R. Prieto-Cerdeira, and A. Bolea-Alamañac, Consolidation of a multi-state narrowband land mobile satellite channel model, in *The Second European Conference on Antennas and Propagation*, Edinburgh, UK, November, 1–6, 2007a.

Pérez-Fontán, F., M. Vazquez-Castro, S. Buonomo, J.P. Poiares-Baptista, and B.R. Arbesser-Rastburg, S-band LMS propagation channel behaviour for different environments, degrees of shadowing and elevation angles, *IEEE Transactions on Broadcasting*, 44(1), 40–76, 1998.

Pérez-Fontán, F., M. Vázquez-Castro, C. Enjamio Cabado, J. Pita Garcia, and E. Cubista, Statistical modeling of the LMS channel, *IEEE Transactions on Vehicular Technology*, 50(6), 1549–1567, 2001.

Prieto-Cerdeira, R., F. Pérez-Fontán, P. Burzigotti, A. Bolea-Alamañac, and I. Sanchez-Lago, Versatile two-state land mobile satellite channel model with first application to DVB-SH analysis, *International Journal of Satellite Communications and Networking*, 28, 291–315, 2010.

Rice, M. and B. Humpherys, Statistical models for the acts k-band land mobile satellite channel, in *47th IEEE Conference on Vehicular Technology*, Phoenix, Arizona, 1, May, 46–50, 1997.

Rieche, M., D. Arndt, A. Ihlow, F. Pérez-Fontán, and G. Del Galdo, Impact of driving direction on land mobile satellite channel modeling, in *European Conference on Antennas and Propagation, EUCAP 2014*, April 6–11, the Hague, the Netherlands, 2014.

Robert, P.P., B.G. Evans, and A. Ekman, Land mobile satellite communication channel model for simultaneous transmission from a land mobile terminal via two separate satellites, *International Journal of Satellite Communications*, 10(3), 139–154, 1992.

Rogers, N.C., A. Seville, J. Richter, D. Ndzi, N. Savage, R.F.S. Caldeirinha, A.K. Shukla, M.O. Al-Nuaimi, K. Craig, E. Vilar, and J. Austin. *A Generic Model of 1–60 GHz Propagation through Vegetation*. QINETIQ Report QINETIQ/KI/COM/CR020196/1.0, 2002.

Saleh, A.A.M. and R.A. Valenzuela, A statistical model for indoor multipath propagation, *IEEE Journal on Selected Areas in Communications*, SAC-5(2), 128–137, 1987.

Simunek, M., F. Pérez-Fontán, and P. Pechac, The UAV low elevation propagation channel in urban areas: Statistical analysis and time-series generator, *IEEE Transactions on Antennas and Propagation*, 61(7), 3850–3858, 2013.

Spencer, Q.H., B.D. Jeffs, M.A. Jensen, and A.L. Swindlehurst, Modeling the statistical time and angle of arrival characteristics of an indoor multipath channel, *IEEE Journal on Selected Areas in Communications*, 18(3), 347–360, 2000.

Suzuki, H., A statistical model for urban radio propagation, *IEEE Transactions on Communications*, 25(7), 673–680, 1977.

Tranter, W.H., K.S. Shanmugan, T.S. Rappaport, and K.L. Kosbar, *Principles of Communication Systems Simulation with Wireless Applications*, Prentice-Hall, Upper Saddle River, New Jersey, Professional Technical Reference, 2004.

Tzaras, C., S.R. Saunders, and B.G. Evans, A physical–statistical time-series model for the mobile-satellite channel, in *IEEE-APS Conference on Antennas and Propagation for Wireless Communications*, Waltham, Massachusetts, November, pp. 1–4, 1998.

Valtr, P. and F. Pérez-Fontán, Wave scattering from rough building surfaces in a virtual city model for the land mobile satellite propagation channel, *Microwave and Optical Technology Letters*, 53(10), 2310–2314, 2011.

Vatalaro, F., Generalised rice-lognormal channel model for wireless communications, *Electronics Letters*, 31(22), 1899–1900, 1995.

Vauzelle, R., Un modele de diffraction 3D dans le premier ellipside de Fresnel [A 3D diffraction model within the first Fresnel ellipsoid], These pour l'obtention du grade de Docteur, de l'Universite de Poitiers, Fevrier 18, 1994.

Vazquez-Castro, M., F. Pérez-Fontán, H. Iglesias-Salgueiro, and M.A. Barcia-Fernandez, A simple three-segment model for shadowing cross correlation in multisatellite systems in street canyons, *Microwave and Optical Technology Letters*, 28(3), 160–164, 2001.

Vazquez-Castro, M., F. Pérez-Fontán, and S.R. Saunders, Shadowing correlation assessment and modeling for satellite diversity in urban environments, *International Journal of Satellite Communications*, 20(2), 151–166, 2002.

Veltsistas, P., G. Kalaboukas, G. Konitopoulos, D. Dres, E. Katimertzoglou, and P. Constantinou, Satellite-to-indoor building penetration loss for office environment at 11 GHz, *IEEE Antennas and Wireless Propagation Letters*, 6, 96–99, 2007.

Vogel, W.J., J. Goldhirsh, and Y. Hase, Land-mobile satellite fade measurements in Australia, *Journal of Spacecraft and Rockets*, 29, 123–128, 1992.

Wakana, H. A propagation model for land mobile satellite communications, in *Antennas and Propagation Society International Symposium, AP-S, Digest*, 3, 1526–1529, 1991.

Xie, Y. and Y. Fang, A general statistical channel model for mobile satellite systems, *IEEE Transactions on Vehicular Technology*, 49(3), 744–752, 2000.

5 Propagation Effects on Satellite Navigation Systems

Roberto Prieto-Cerdeira

CONTENTS

5.1 INTRODUCTION

Navigation is related to resolving how to travel from one place to another. Throughout history, finding accurate methods of positioning and navigation has been a major challenge for scientists, astronomers, and mariners, and a mechanism of great discovery. Celestial navigation allowed latitude determination through the observation of angular measurements to the sun or known celestial bodies. The longitude determination problem was far more challenging. It was not until the eighteenth century when ships would be able to determine longitude more accurately with the use of chronometers in order to keep track of the time at departure port, which together with the actual time, would allow the estimation of the time difference into geographical separation (Sobel, 1998). The advent of radio navigation (using radio signals for determining position) during the twentieth century, proved valuable for a number of new applications including flight navigation. Several terrestrial radio navigation techniques were developed and many exist today, employing angular, range, or Doppler measurements. In the 1960s, the first satellite navigation system (Transit) was the pioneer of the current Global Navigation Satellite Systems (GNSS) (Parkinson et al., 1995).

Modern GNSS such as the initial GPS from the United States and GLONASS from Russia, and more recent systems such as the European Galileo and Chinese Beidou, are based on satellite constellations in medium Earth orbit (Beidou, also including satellites in geostationary [GEO] and inclined geosynchronous orbits [IGSO]). These satellites broadcast ranging signals in the L-band. Other regional GNSS from India and Japan include navigation and/or augmentation satellites in GEO and/or IGSO (with data and ranging signals also in the L-band, and S-band for the Indian IRNSS system). The core of these systems is the availability of time typically provided by means of atomic clocks on the satellites which allows for accurate ranging measurements at the receiver. With the availability of more than three satellites in view, it is possible to resolve for position, usually one more is required for correcting clock errors at the user receiver, and with more satellites it is possible to compensate for geometry and other limiting factors.

Satellite-based augmentation systems (SBAS) augments GNSS, providing real-time integrity and accuracy for safety-of-life applications (civil aviation as the main application but also maritime, land mobile, etc.). They include the American WAAS, the European EGNOS, the Indian GAGAN, and the Japanese MSAS belonging to a network of interoperable SBAS systems complying with standards defined by the International Civil Aviation Organization (ICAO).

Signals from the modern GNSS are based on direct-sequence spread spectrum (DSSS), based on high symbol rate pseudorandom sequences (PRNs) modulated with a given spreading shape. A spreading symbol is known as a chip, its length as chip period, and the spreading sequence rate as chip rate. The PRNs are reproducible sequences that intend to present autocorrelation properties of true infinite random sequences (total correlation for the sequence perfectly aligned with a replica of itself and decaying rapidly for the replica shifted by one chip; total uncorrelation between two different sequences). The PRNs are usually modulated with data (navigation message) at low rates, but also pilot sequences (with no data modulated) are included in recent signals.

Spread-spectrum signals offer three main advantages for satellite navigation: "precise ranging; access to different satellites, each one broadcasting a known PRN (CDMA); and, rejection of narrowband interference by the bandwidth spreading operation" (Kaplan and Hegarty, 2006). Ranging is performed by calculating the transit time of the signal from the satellite to the receiver using the autocorrelation function of a broadcast PRN by a satellite and a locally generated replica in the receiver. True ranges are contaminated by a number of errors such as ionospheric and tropospheric propagation delays or clock errors, thus calling pseudorange to the range estimates. Apart from code pseudoranges, GNSS receivers may use carrier phase for positioning, which is more precise than the code due to more reduced noise but it requires solving the ambiguity of the number of cycles between satellite and receiver. A simplified diagram of the code and carrier phase ranging system including satellite and receiver is included in Figure 5.1.

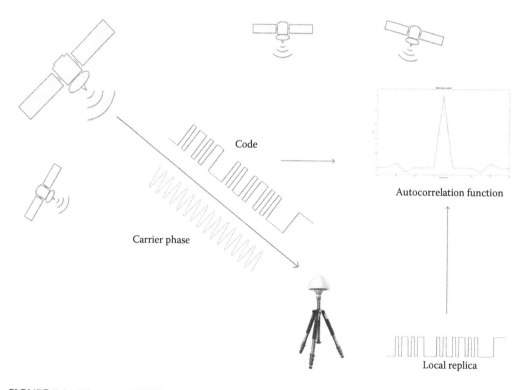

FIGURE 5.1   Diagram of GNSS code and carrier phase ranging.

The GPS coarse acquisition (C/A) civil signal is based on Gold codes modulated onto a BPSK rectangular spreading shape with a chip rate of 1.023 Mcps (chip length of 977.5 ns and main lobe bandwidth of 2.046 MHz). The code period for the C/A codes is 1 ms. Data is BPSK modulated on top of the spreading signal at 50 bps. The C/A signal is modulated into the L1 (1575.42 MHz) carrier frequency. The precise (P) military signal is similar to C/A but with a chip rate 10 times larger, much longer code period, and encrypted codes accessible only by authorized users. The P signal is modulated in L1 and in L2 (1227.6 MHz) frequencies. For Galileo and GPS modernization, new signals have been selected with a variety of bandwidths, sequence properties, code periods, with data modulation at various data rates and without data (pilot signals), etc. As an example, binary offset carrier (BOC) modulations have been proposed. BOC signals include spreading shapes formed by a rectangular signal multiplied by a square wave subcarrier allowing more flexibility on the power distribution within the available frequency spectrum regulated for satellite navigation. Also, civil signals at different carrier frequencies have been introduced to minimize the impact of ionospheric propagation group delay error on the ranging signals and to increase robustness of the system.

For GNSS, different propagation mechanisms (diffraction, reflection, and scattering) are encountered by the signal broadcast from the satellite affecting receiver performance. The propagation effects that impact performance (accuracy, availability, integrity and/or continuity of service) are due to Earth's atmosphere and characteristics of the local environment of the receiver. In this sense, Earth's atmosphere can be classified into the troposphere, whose main effect is a group delay in the navigation signal due to water vapor and the gas components of the dry air, which is non-dispersive (independent of frequency); and the ionosphere, the ionized part of the atmosphere, that induces a dispersive group delay that is significantly larger than the one from the troposphere. Other ionospheric effects such as scintillations and refraction may be present. The local environment may affect the navigation signal in various ways such as fading due to shadowing, the signal is scattered and/or diffracted by obstacles (buildings, vegetation, etc.) between satellite and receiver; multipath, where replicas of the signal arrive to the receiver with a certain delay and phase due to reflection and diffraction with surrounding objects. Ambient noise and interference, although not considered radiowave propagation effects, have an impact on navigation performance.

Applications based on GNSS positioning for difficult environments, such as urban canyons, indoor environments, or forests, requires a rigorous modeling of the electromagnetic radio channel in order to be able to assess receiver performances in controlled laboratory conditions with respect to their target conditions.

In addition to the effects on pseudorange and carrier phase of navigation signals, radiowave propagation affects many aspects of a GNSS mission, including TT&C and mission uplink communications, two-way laser and S-band ranging, data dissemination, or search-and-rescue. This chapter focuses only on propagation effects on navigation signals.

In order to evaluate how propagation affects positioning and navigation, it is important to understand how propagation affects the primary GNSS observables code and carrier phase. The basic GNSS observables for a given signal are the code observable

$$R_{Pn}^j = \rho^j + c \cdot (dt - dt^j) + T^j + rel^j + \alpha_n \cdot I^j + K_{Pn,Rx} + K_{Pn}^j + M_{Pn}^j + \varepsilon_{Pn}^j \tag{5.1}$$

and the phase observable

$$\varphi_{Ln}^j = \rho^j + c \cdot (dt - dt^j) + T^j + rel^j - \alpha_n \cdot I^j + k_{Ln,Rx} + k_{Ln}^j + \lambda_{Ln} N_{Ln}^j + w_{Ln}^j + m_{Ln}^j + \varepsilon_{Ln}^j \tag{5.2}$$

expressed here in meters, where ρ^j is the true range between the receiver and satellite j, c is the speed of light, $(dt - dt^j)$ is the clock error between satellite and receiver clocks, T^j is the tropospheric delay, rel^j is the relativistic effect, $\alpha_n \cdot I^j$ is the ionospheric delay (an advance in carrier phase, notice the different sign), $K_{Pn,Rx} + K_{Pn}^{j}$ are code hardware biases at receiver and satellite, $k_{Ln,Rx} + k_{Ln}^{j}$ are carrier phase hardware biases at receiver and satellite, respectively, M_{Pn}^{j} is the code multipath error and m_{Ln}^{j} carrier phase multipath error, ε_{Pn}^{j} is code noise and ε_{Ln}^{j} carrier phase error, $\lambda_{Ln}N_{Ln}^{j}$ is carrier phase ambiguity, and w_{Ln}^{j} is carrier phase wind-up. In these formulas, the components directly affected by propagation aspects are troposphere and ionosphere delay, multipath, and in the case of fading, noise may increase. These components will be described and models for each of them presented in the next sections.

For a more detailed description of GNSS system, signals, and data processing, there are various texts available, including Kaplan and Hegarty (2006) and Sanz Subirana et al. (2013).

5.2 IONOSPHERIC EFFECTS

The ionosphere is a region of weakly ionized gas in Earth's upper atmosphere, between the thermosphere and the exosphere, lying between about 50 kilometers up to several thousand kilometers from Earth's surface (Davies, 1990). The ionosphere affects radio wave propagation in different ways such as refraction, absorption, Faraday rotation, group delay, time dispersion, and scintillations, and these effects are dispersive. Faraday rotation, which is the rotation of the polarization due to the interaction of the electromagnetic wave with the ionized medium in Earth's magnetic field along the path (ITU-R, 2013), is often neglected for satellite navigation and satellite communications in the L-band through the use of circular polarization.

Most ionospheric propagation effects are directly related to the level of ionization in the ionosphere, and in particular to electron density. The total electron content (TEC) is the electron density integrated along a slant (or vertical) path between a satellite and a receiver. It is expressed in TEC units (TECu) where 1 TECu = 10^{16} electrons/m^2.

Ionospheric electron density and, in general, all ionospheric effects depend on different factors such as the time of day, location, season, solar activity (which is related to the epoch within the solar cycle), or level of disturbance of the ionosphere, such as those happening during geomagnetic storms. In the same manner, on a time scale of the order of several months, solar activity follows a periodic 11-year cycle. The level of solar activity (and hence the solar cycle) is usually represented by solar indices such as the sun spot number (SSN) or the solar radio flux at 10.7 cm (F10.7), and the level of geomagnetic activity is often represented by geomagnetic indices such as Ap, Kp, or Dst. The evolution of solar and geomagnetic indices including the solar cycle 23 and 24 (January 1995–June 2015) is depicted in Figure 5.2.

5.2.1 Ionospheric Delay Error

The ionosphere is a dispersive medium and starting from the propagation model of an electronic wave in a medium, the ionosphere delay can be represented in its analytical form as the sum of contributions related to the signal carrier frequency. The refractive index and thus the delay are frequency dependent, in particular, the higher the frequency, the smaller the ionosphere effect. In general, an important characteristic of the ionosphere delay is that it has opposite signs in carrier phase and pseudorange observables. In particular, the ionosphere refractive index can be expressed as a series expansion:

$$n_{gr} = 1 + \frac{a_1}{f^2} + \frac{a_2}{f^3} + \frac{a_3}{f^4} + \cdots \tag{5.3}$$

where the term a_1 (first-order term) is the most important parameter and it is directly correlated to the number of free electrons in the signal path (TEC). Higher-order terms that usually account for differences at the millimeter level and up to few centimeters in the worst case (Hernández-Pajares

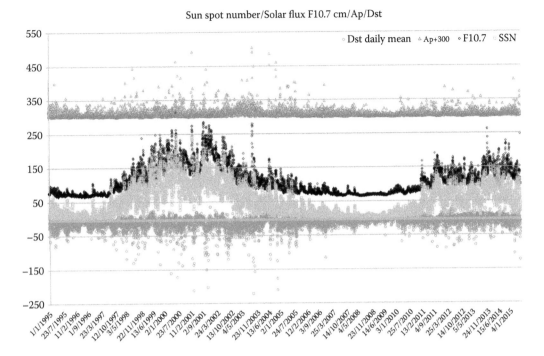

FIGURE 5.2 Evolution of solar (sun spot number and solar flux F10.7) and geomagnetic indices (Ap and Dst) between January 1995 and June 2015—parts of solar cycle 23 and 24. (Data from NOAA/NGDC, NOAA/SPWC, and WDC for Geomagnetism, Kyoto.)

et al., 2014) should be considered only for high-precision applications where all other error sources of similar or higher orders of magnitude may be estimated as well. For more details on effects and modeling of higher-order effects, see Bassiri and Hajj (1993), Petrie et al. (2010), and Hernández-Pajares et al. (2014). Neglecting higher-order errors, the group delay experienced by an electromagnetic wave propagating through the ionosphere is directly proportional to the TEC (neglecting) as follows:

$$d_{Igr} = \int_{path} (n_{gr} - 1) \cdot dl = \frac{40.3}{f^2} \cdot \int_{path} N \cdot dl = \frac{40.3}{f^2} \cdot sTEC \qquad (5.4)$$

where d_{Igr} is the group delay error (m), n_{gr} is the group refractive index in the ionosphere, f is frequency (Hz), N is electron density (electrons/m³), $sTEC$ is slant TEC, and path is the propagation path between the receiver and the satellite. For phase delay, the expression for group refractive index should be replaced by the phase refractive index considering its relation $n_{gr} = n + f(\partial n + \partial f)$. The first-order ionospheric phase delay experienced by the signal is given by

$$d_I = \int_{path} (1 - n) \cdot dl = -\frac{40.3}{f^2} \cdot sTEC \qquad (5.5)$$

It is observed that the phase is advanced by the same amount as the group delay.

At the GPS L1 frequency (1575.42 MHz), the group delay is approximately 16 cm for 1 TECu. For oblique paths between satellite and receivers, the TEC is designated as sTEC in comparison to vertical TEC (vTEC) for zenith paths. A simple model to convert the vTEC value into sTEC for a given elevation angle follows the assumption that all the ionospheric electron density content is concentrated in a single thin layer at a given height close to the height of maximum electron density. The conversion uses a mapping function (also known as obliquity function or slant conversion factor) F_{IPP} as follows:

$$sTEC(\varepsilon) = F_{IPP}(\varepsilon) \cdot vTEC \qquad (5.6)$$

where ε is the elevation angle of the station–satellite ray link. Notice that the ionospheric pierce point (IPP) is the coordinate point where the station–satellite ray link intersects with the ionospheric layer.

The most common ionospheric mapping function is the geometric approximation (see geometry in Figure 5.3). After doing the necessary mathematical arrangements, one can derive the following expression for the mapping function:

$$F_{IPP}(\varepsilon) = \frac{1}{\cos(\alpha)} = \frac{1}{\sqrt{(1 - ((R_e/(R_e + h_{ion})) \cdot \cos(\varepsilon))^2)}} \qquad (5.7)$$

where R_e is the radius of Earth and h_{ion} is the height of the ionospheric layer (e.g., 350 km).

In order to estimate the ionospheric delay contribution using GNSS observations, dual-frequency observations are required since the ionosphere is a dispersive media. This implies that the observations on different frequencies have to be combined in order to retrieve the ionospheric delay. In this context, the available observables are those in code and phase. These observables have their advantages and drawbacks that will be the issue that will drive their performance.

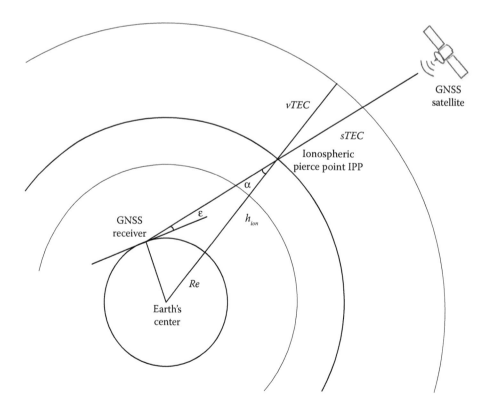

FIGURE 5.3   Scheme to derive the geometric mapping function.

As already mentioned, the code observable is more noisy and less precise than the phase one but unambiguous. Thus, combining Equations 5.1 and 5.2 for two different frequencies, the expressions for code and phase for the ionospheric combination or *geometry-free combination* for L1 and L2 frequencies are

$$R_{PI}^j = R_{P2}^j - R_{P1}^k = \alpha_I I^j + K_{PI,Rx} + K_{PI}^j + M_{PI}^j + \varepsilon_{PI}^j \tag{5.8}$$

$$\phi_{LI}^j = \phi_{L1}^j - \phi_{L2}^j = \alpha_I I^j + k_{LI,Rx} + k_{LI}^j + (\lambda_{L1} N_{L1}^j - \lambda_{L2} N_{L2}^j) + w_{LI}^j + m_{LI}^j + \epsilon_{LI}^j \tag{5.9}$$

where $\alpha_I I^j = (\alpha_2 - \alpha_1) I^j$ is the ionospheric delay in meters of LI, $K_{PI,Rx}$ and K_{PI}^j are the code interfrequency biases for the receiver and the satellite, respectively (also known as differential code biases), and similarly $k_{LI,Rx}$ and k_{LI}^j for the interfrequency biases for phase. The use of phase observables is more complex since it implies the calculation of an ambiguity term for each satellite–station arc, which increases the numbers of unknowns.

It is also possible to obtain ionospheric estimation through vTEC global ionospheric maps (GIMs) available from one of the International GNSS Service (IGS), Ionosphere Associate Analysis Centers (IAACs), centers delivering ionospheric products, CODE (Center for Orbit Determination in Europe, University of Berne, Switzerland), ESA/ESOC (European Space Agency/European Space Operation Center of ESA, Darmstadt, Germany), JPL (Jet Propulsion Laboratory, Pasadena, USA), and UPC (Technical University of Catalonia/gAGE, Barcelona, Spain) (Hernández-Pajares et al., 2009). They are provided in IONEX format (Schaer et al., 1998). Nowadays, IGS also provides final and rapid ionospheric products in IONEX formats obtained through the combination of GIMs from the IAACs. The GIMs files provide global maps with a temporal resolution of 2 h and a spatial resolution of 5° in longitude and 2.5° in latitude. In a daily file, there are 13 vTEC maps with 13 vTEC root mean square (RMS) maps in order to provide information about the accuracy of the measurement. Final products are provided with a latency of several days and rapid products with slightly lower accuracy are provided with a latency of roughly 24 h. Apart from IGS, other institutions provide global and regional GIMs IONEX files with various accuracies, update rates, and latencies, for example, generating ultrarapid vTEC map products with 15 min temporal resolution or below and latencies within a few hours (Hernández-Pajares et al., 2015; European Space Agency (ESA), 2015; Deutsches Zentrum für Luft- und Raumfahrt e.V. (DLR), 2015), and even two-layer vTEC models maps are becoming available (Rovira-Garcia et al., 2015). Figure 5.4 shows an example of a global map of vTEC for a solar maximum condition from IGS final ionospheric products. The figure shows that in the region around the geomagnetic equator, the ionosphere has the largest vTEC values and the strongest gradients, whereas in the mid-latitude regions, the daytime vTEC values are less than half the value found in the equatorial anomaly region. The vTEC shape in the picture moves approximately westwards along lines of geomagnetic latitude at Earth's rotation rate of 15° per hour.

It is important to mention that the accuracy of GIMs is limited by the number of stations and the time and space resolution, and although they represent an invaluable data source, they cannot be considered an error-free reference of the ionosphere. In order to estimate accurate vertical or slant TEC, it must be considered that errors due to mapping function, time interpolation, and space interpolation may occur, and it will be particularly pronounced in areas where no IGS stations are available and interpolation is used (e.g., oceans). When data to a larger network of stations is available, three-dimensional tomographic approaches (Mitchell and Spencer, 2003; Hernández-Pajares et al., 1999; Rovira-Garcia et al., 2015) may provide a more accurate reference ionosphere. In the cases where TEC estimation is required and single stations are available or the network is too sparse, one approach for estimating the ionosphere exploits dual-frequency carrier phase measurements together with GIMs (Montenbruck et al., 2014; Prieto-Cerdeira et al., 2014). The GIMs are used

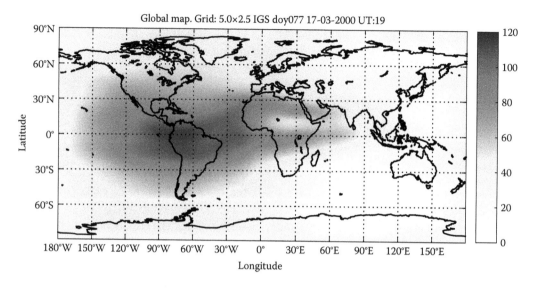

FIGURE 5.4   Global map of observed vertical TEC for 17-03-2000 at 19UT.

to level the geometry-free combination of carrier phases. The alignment of the ionospheric carrier phase combination with a previously computed sTEC (from a GIM) has the advantage of avoiding local errors affecting the pseudorange measurements, such as multipath, which produces significant errors in the pseudorange alignment procedure.

The ionosphere can be categorized in three main regions according to characteristics and intensity of effects. The boundaries are defined by geomagnetic latitude. They are equatorial regions, mid-latitude regions, and auroral/polar regions. The boundaries are roughly represented in Figure 5.5 for nominal geomagnetic activity.

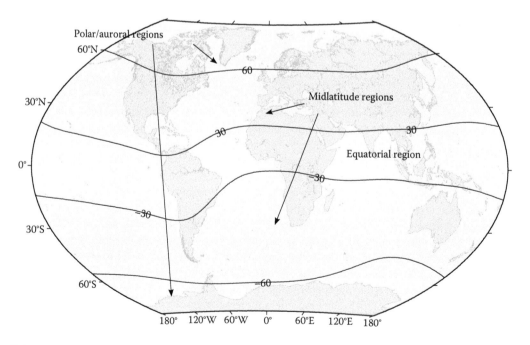

FIGURE 5.5   Main areas of ionospheric activity.

For equatorial or low-latitude regions, for instance, the ionospheric effects are considerably stronger. The regions around the geomagnetic equator (equatorial anomaly, located at around ±15–20°) present the largest values of TEC and also the largest ionospheric disturbances and irregularities. These high TEC values are produced by the electromagnetic ($E \times B$) drift force that cause a fountain effect transporting the plasma away from the magnetic equator toward higher latitudes peaking at the anomaly "crests" (SBAS Ionospheric Working Group, 2003). During post-sunset hours, large depletions or "bubbles" of electron density are observed. They can be large in absolute value but small in longitudinal extent; thus, they may go undetected in the reference station measurements (Klobuchar et al., 2002; Doherty et al., 2002). These depletions are associated with small-scale irregularities that produce intense amplitude and phase scintillation effects. Also, large spatial and temporal electron density gradients are observed in low-latitude regions.

High-latitude regions (auroral/polar regions) are considered for geomagnetic latitudes larger than around 65°. The ionosphere of the auroral regions normally causes lower ionospheric delays than that of the mid-latitude regions; however, the variability of the auroral ionosphere tends to be greater than that of the mid-latitude regions. The polar cap (above ~75°) is a small size area, which presents large TEC variability (SBAS Ionospheric Working Group, 2003); within this area, the polar cusp is the area where Earth's magnetic field is reduced, creating a cone where barely no magnetic activity is observed; the behavior is difficult to predict (however, very few satellite paths would cross the ionosphere through the cusp). In the polar cap, TEC daily dependencies are less predictable and are often driven by polar cap patches (structures with plasma density enhancements) (Weber et al., 1986; Prikryl et al., 2010).

During geomagnetic disturbed periods, severe effects may be observed at mid-latitudes since the boundaries of the mid-latitude regions are shortened from auroral and equatorial areas.

Although the differences between various regions are related to large- and small-scale characteristics, in order to illustrate such differences, a long-term large-scale characterization of the different regions is presented based on different percentiles from vTEC values obtained from GIMs over Solar Cycle 23 obtaining the approximate percentiles shown in Table 5.1.

Carrier smoothing is used in GPS receivers to reduce the effects of multipath and thermal noise (effects described at the end of this chapter). The noisy code measurements are used to estimate the bias on more precise but ambiguous carrier measurements by means of averaging with a Hatch filter (Hatch, 1983). However, the code–carrier divergence, that is, the effect on which the code is delayed proportionally to the TEC and the carrier is advanced by nearly the same amount, limits the use of the smoothing method by introducing a bias when large temporal ionospheric gradients are encountered due to the fact that the ionospheric delay over the code is assumed to be constant during the averaging period. The filter bias is equal to negative twice the ionospheric rate of change, times the time constant (Walter et al., 2004). The ionospheric rate of change during nominal periods in mid-latitudes is below 1 mm/s. In solar maximum conditions and based on statistics over long observation periods, rates of change of 8 mm/s can be considered a reasonable bound for quiet solar maximum days, and the largest observed gradient in Continental US (CONUS) is 150 mm/s

---

TABLE 5.1

vTEC Reference Values (in TECu) for Different Percentiles and Latitudes Based on Analysis of IGS GIMs Data during Solar Cycle 23

	68th	90th	95th	99th
High latitudes	18	28	35	50
Mid-latitudes	25	50	60	90
Low latitudes	45	80	95	120
Global	25	60	65	105

(Walter et al., 2004). For a 100-s smoothing, required for SBAS receivers, a gradient of 8 mm/s is equivalent to a filter bias of −1.6 m, and 150 mm/s to −30 m, respectively.

For GNSS dual-frequency receivers, an *ionospheric-free linear combination* may be used in order to remove the first-order ionospheric delay for code and phase, respectively:

$$R^j_{IonoFree} = \frac{f^2_{P1} R^j_{P1} - f^2_{P2} R^j_{P1}}{f^2_{P1} - f^2_{P2}} \tag{5.10}$$

$$\phi^j_{IonoFree} = \frac{f^2_{L1} \phi^j_{L1} - f^2_{L2} \phi^j_{L2}}{f^2_{L1} - f^2_{L2}} \tag{5.11}$$

This combination eliminates the ionospheric delay contribution at the expense of increased noise and multipath error. The increase depends on the frequency separation of the two signal frequencies used. Given the assumption that noise at the two frequencies is independent, the noise variance of the ionospheric-free combination is calculated as follows:

$$\sigma^j_{IonoFree^2} = \frac{f^4_{P1} \sigma^j_{P1^2} - f^4_{P2} \sigma^j_{P2^2}}{\left(f^2_{P1} - f^2_{P2}\right)^2} \tag{5.12}$$

and assuming that the noise in both frequencies are identically distributed

$$\sigma^j_{IonoFree^2} = \frac{f^4_{P1} - f^4_{P2}}{\left(f^2_{P1} - f^2_{P2}\right)^2} \sigma^{j^2}_P \tag{5.13}$$

this relation is often given related to factor $\gamma = f^2_{P1}/f^2_{P2}$ as follows:

$$\sigma^j_{IonoFree} = \sigma^{j^2}_P \sqrt{\frac{\gamma^2 + 1}{(\gamma - 1)^2}} = K_{noise} \cdot \sigma^{j^2}_P \tag{5.14}$$

As examples, the value of K_{noise} is equal to 2.98 for GPS L1–L2 combination and 2.59 for Galileo E1–E5a or GPS L1–L5 combinations.

For GNSS single-frequency receivers, correction algorithms based on empirical or climatological models are driven by broadcast parameters in the navigation message. For the GPS system, the ionospheric correction algorithm (ICA) (Klobuchar, 1987) uses eight coefficients to describe the ionosphere, which is represented as a thin-shell model. This model performs rather well in mid-latitude regions but the representation of the equatorial anomaly is limited. The GPS ICA algorithm was designed to remove about 50% of the ionospheric error although larger correction capabilities are usually observed during nominal periods. For Galileo, a new algorithm has been developed using NeQuick G model, an adaptation of the NeQuick empirical climatological electron density model (Hochegger et al., 2000; ITU-R, 2013) suitable for trans-ionospheric applications, which generates electron density for given space, time, and solar activity conditions from a minimum set of anchor point characteristics. This model is driven by an *effective ionization level* broadcast in the navigation message. For further details of the Galileo single-frequency algorithm and its performance, see European Union (2015) and Prieto-Cerdeira et al. (2014). Other correction models available to GNSS receivers are based on correction/augmentation data from wide-area differential networks.

The correction model for SBAS such as WAAS or EGNOS is defined in MOPS DO-229 (RTCA, 2006). The MOPS ionospheric model represents the ionosphere as a *thin shell* at the height of highest electron density (this is assumed to be a sphere with a radius of 6728 km, equivalent to 350 km above Earth's surface). The information on discrete points (ionospheric grid points [IGPs]) of the shell on a regular grid, the ionospheric grid, is provided to SBAS users. The IGPs are spaced at 5° latitude/longitude for mid- and equatorial latitudes. The ionospheric corrections at the IGPs are broadcast to the user as part of the SBAS messages in the form of a grid ionospheric vertical delay (GIVD) for the ionospheric delay corrections, and integrity information through the grid ionospheric vertical error indicator (GIVEI), which is an index that characterizes the total error bound from all possible error sources: measurement, estimation, slant-to-vertical correction, spatial/temporal interpolation, etc.

As a general rule of thumb, during solar maximum periods, ionization levels are larger and therefore absolute parameters are also larger. On the other hand, during geomagnetic disturbed periods, resulting from the interaction between space weather solar events and Earth's magnetic field, the ionosphere may deviate from the typical (geographical, diurnal, and seasonal) behavior. Geomagnetic storms are more frequent during active periods of the solar cycle but their strength is not necessarily correlated with solar activity peaks (this can be observed in Figure 5.2 comparing large values of Ap or low values of Dst with respect to SSN and solar flux). The ionospheric effects on performance observed by GNSS receivers tend to increase with solar or geomagnetic activity, respectively, but looking only at solar or geomagnetic global indices (SSN, F10.7, Ap/Kp, Dst) is not an unambiguous direct indicator that GNSS receivers may be significantly affected by the ionosphere. For that reason, a different regional parameter has been developed specifically for GNSS: the *Along Arc TEC Rate* (AATR index) has proven to be an effective independent indicator of ionospheric activity that degrades GNSS performance (Sanz Subirana et al., 2014), with

$$AATR_i = \frac{\Delta sTEC}{(M(\varepsilon))^2 \Delta t} \tag{5.15}$$

where *sTEC* is slant total electron content, t is time, Δt ranges between 30 and 60 s, and $M(\varepsilon)$ is the mapping function with respect to elevation ε. The AATR index at a given location is defined as the hourly RMS of all the individual instantaneous $AATR_i$ for all satellites in view. The AATR index over several stations in Europe during the St Patrick's storm day in 2015 is presented in Figure 5.6 showing the impact on stations in Scandinavia and Iceland, which was observed as reduced availability of the EGNOS system during that period.

5.2.2 Ionospheric Scintillations

Ionospheric amplitude and phase scintillations are the other main ionospheric effects on GNSS signals. They are observed as rapid fluctuations of amplitude and/or phase of the radio wave signal, respectively, caused by refraction and diffraction due to small-scale irregularities that modify the ionospheric refractive index. Strong scintillations can induce cycle slips and loss of lock in GNSS receivers on one or more signals simultaneously. Scintillations depend on the location, time of day, season, solar, and geomagnetic activity. An example of ionospheric scintillations in a station in Tahiti, French Polynesia is presented in Figure 5.7 where for the period of observation, about half of the satellites in view for GPS, GLONASS, and Galileo constellations are affected by scintillations. During nominal conditions, strong levels of scintillation are rarely observed in mid-latitudes, but they may be encountered daily during postsunset hours in low-latitude regions. A graphical representation of the depth of scintillation, geographical distribution, and time-of-day occurrence for two levels of solar activity is presented in Figure 5.8. Low-latitude scintillation is seasonally dependent and is limited to local night time hours (SBAS Ionospheric Working Group, 2010). The high-latitude region can also encounter significant scintillation and it is strongly dependent on

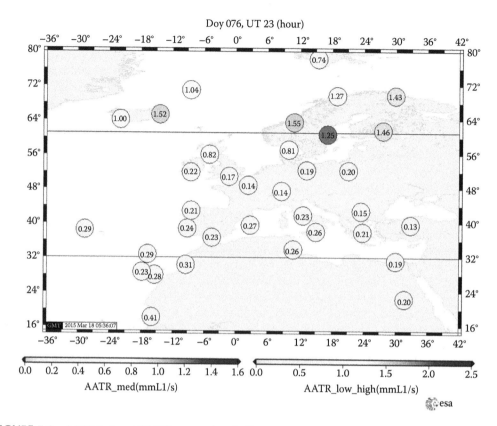

FIGURE 5.6   AATR index at 23 UT over stations in Europe during St Patrick's Day storm 2015. (Courtesy of R. Orus/ESA.)

geomagnetic activity levels, but can occur in all seasons and is not limited to local night time hours (SBAS Ionospheric Working Group, 2010).

The most commonly used parameter to characterize intensity fluctuations is the scintillation index S4 (ITU-R, 2013), defined by the equation

$$S4 = \left(\frac{\langle I^2 \rangle - \langle I \rangle^2}{\langle I \rangle^2} \right)^{1/2} \tag{5.16}$$

where I is intensity (proportional to the square of the signal amplitude) and $\langle \ \rangle$ denotes averaging, usually over a period of 60 s. Likewise, phase scintillations are often characterized by the standard deviation of phase variations, the phase scintillation index σ_φ

$$\sigma_\varphi = \sqrt{\langle \varphi^2 \rangle - \langle \varphi \rangle^2} \tag{5.17}$$

where φ is carrier phase in radians and $\langle \ \rangle$ denotes averaging, usually over a period of 60 s. The evolution of the S4 index for all GPS and Galileo satellites in view from a station in Cape Verde from 19–21 h local time on March 15, 2013 is presented in Figure 5.9, it is observed that only a few satellites are affected.

Scintillation strength may, for convenience, be classified into three regimes, see ITU-R (2013) for amplitude scintillation thresholds; for phase scintillation, various thresholds are proposed in Doherty et al. (2003) and Rodrigues et al. (2004).

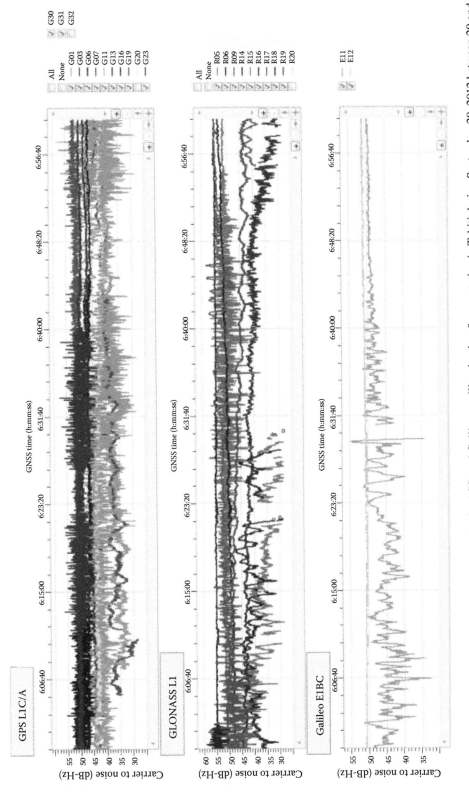

FIGURE 5.7 Carrier-to-noise-density ratio for GPS, GLONASS, and Galileo satellites in view from a station in Tahiti during September 28, 2012 between 20 and 21 local time. About half the satellites are affected by ionospheric scintillations.

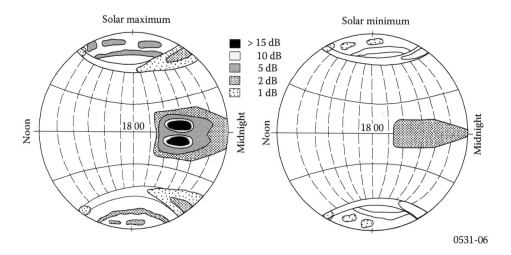

FIGURE 5.8 Depth of amplitude scintillation fading, geographical distribution, and time-of-day occurrence at the L-band for two levels of solar activity. (Courtesy of ITU-R, Ionospheric Propagation Data and Prediction Methods Required for the Design of Satellite Services and Systems, Recommendation ITU-R P. 531-12. P Series Radiowave Propagation. ITU, 2013.)

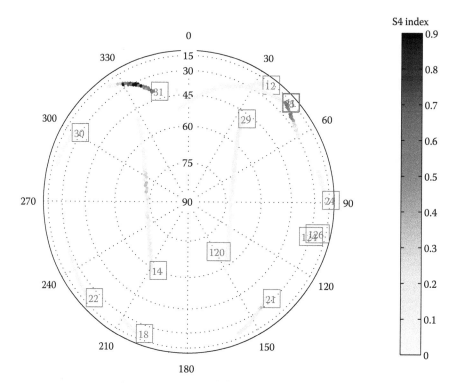

FIGURE 5.9 S4 index over Cape Verde station on March 15, 2013 at 19–21 h LT. Scintillation is observed over GPS satellite PRNs 14, 21, and 31 and Galileo satellite PRN 81.

Scintillation Regime	Amplitude Scintillation	Phase Scintillation (rad)
Weak	S4 < 0.3	$\sigma_\varphi < 0.25$–0.3
Moderate	$0.3 \le S4 \le 0.6$	0.25–$0.3 \le \sigma_\varphi \le 0.5$–0.7
Strong	S4 > 0.6	$\sigma_\varphi > 0.5$–0.7

The scintillation index S4 is related to the peak-to-peak fluctuations of the intensity. The exact relationship depends on the distribution of the intensity. The probability distribution function that best describes the intensity distribution is the Nakagami-m (Nakagami, 1960), particularly for weak and moderate scintillations. The probability density function of a Nakagami-m is given by

$$f_{\text{Nakagami}}(r) = \frac{2m^m}{\Gamma(m) \cdot \Omega^m} \cdot r^{2m-1} \cdot \exp\left(\frac{-m \cdot r^2}{\Omega}\right) \tag{5.18}$$

where r is the envelope (amplitude) of the received complex signal, $\Gamma(x)$ is the gamma function, Ω is the average power (usually normalized to unity), and m is the Nakagami-m coefficient, which is related to the scintillation index S4 by

$$m = \frac{1}{S4^2} \tag{5.19}$$

In addition to the amplitude and phase scintillation indices, the power spectral density (PSD) needs to be specified for a full characterization. Both amplitude and phase scintillation follow an inverse power law relationship following Tf^{-p}, where T is the *spectral strength* at 1 Hz and p is the *spectral index*. For amplitude scintillations, lower frequencies are attenuated with respect to phase scintillations. For more details on ionospheric scintillation spectra, see Rino (1979) and Knight and Finn (1998). Spectral indices ranging from 1 to 6 have been observed (ITU-R, 2013). Relevant spectral components usually appear up to 10–20 Hz, which constrain the minimum sampling frequency of instruments measuring scintillation to sampling rates larger than 40 Hz (commonly 50 Hz for GNSS-based instruments in the L-band).

For frequency scaling, typically, the following relation on S4 is used:

$$S4(f_2) = S4(f_1)\left(\frac{f_2}{f_1}\right)^{-n} \tag{5.20}$$

with $n = 1.5$ recommended for L-band frequencies. In Wheelon (2003), values of n derived from satellite measurement data between several pairs of frequencies from 30 MHz up to 6 GHz are presented ranging from 1–2. This relationship is valid particularly for weak scattering assumptions (higher elevations, low S4 values). For high S4 values, the relation saturates with n approaching 0.

As previously mentioned, scintillation effects differ at low and high latitudes, and are not observed at mid-latitudes except for strong geomagnetic storms. At high-latitudes, the effect is mainly driven by the interaction of solar radiation with Earth's magnetic field, and therefore, more dependent on geomagnetic activity, whereas at low latitudes it is driven by solar activity and therefore linked to solar cycle and seasons.

In general, scintillation activity is correlated with solar activity, increasing during solar maximum periods and with also seasonal dependencies. According to SBAS Ionospheric Working Group (2010), "from South America through Africa to the near East" indicates "the presence of scintillation in all seasons except the May–July period." On the contrary, "in the Pacific sector, scintillations are observed in all seasons except the November–January period." The dependence of scintillation with season for different longitudes was reported in (Aarons, 1993). For equatorial region during the solar maximum period and worst season, strong scintillations might be expected to appear up to around 20% of the time between sunset and midnight.

The spatial distribution of scintillation phenomena is characterized by randomly distributed patches that vary in size and drift speed (ranging from 50 to 150 m/s) (SBAS Ionospheric Working Group, 2010). The spatial distribution of patches is not homogeneous and therefore satellites at different orientations in the sky will observe time-varying scintillation levels with a somewhat arbitrary spatial correlation. This patchy pattern may manifest on very-large-scale structures, and can also cause significant variation in ionospheric delay and a loss of lock on the signal, whereas other smaller structures may not cause loss of the signal, but still can affect the integrity of the signal by producing ranging errors. Finally, owing to the patchy nature of irregularity structures, other satellites could remain unaffected. The correlation of deep signal fades is considered low between different satellites (below 15%) as reported in Seo et al. (2011). For high-latitude scintillations, "the irregularities move at speeds up to ten times larger in the polar regions as compared to the equatorial region. This means that larger sized structures in the polar ionosphere can create phase scintillation and that the magnitude of the phase scintillation can be much stronger" (SBAS Ionospheric Working Group, 2010).

Moderate and strong scintillations affect the performance of GNSS receivers, inducing tracking errors and potentially cycle slips for carrier-phase tracking or even loss of lock. Amplitude scintillation can create deep signal fades that interfere with a user's ability to receive GNSS signals whereas phase scintillation describes rapid fluctuations in the observed carrier phase obtained from the receiver's phase lock loop (PLL). Existing analyses are usually performed using ionospheric models and data for nominal averaged situations; however, the effects of strong scintillation cases on GNSS are not fully characterized.

Tracking performance under scintillation conditions depends on intensity of scintillation but also on tracking capabilities of the receiver. Closed-form analytical expressions of the tracking error variance at the output of classical PLL in GPS L1, L2, and L5 receivers have been derived in Knight and Finn (1998), Conker et al. (2003) and Kim et al. (2003) as:

$$\sigma^2_{\phi_\varepsilon} = \sigma^2_{\phi_S} + \sigma^2_{\phi_T} + \sigma^2_{\phi_{OSC}} \tag{5.21}$$

where $\sigma^2_{\phi_S}$, $\sigma^2_{\phi_T}$, and $\sigma^2_{\phi_{OSC}}$ are phase scintillation, thermal noise, and oscillator noise, respectively, and thermal noise expression depends on S4, C/No, predetection integration time, and PLL bandwidth. This formulation allows the estimation of tracking behavior under a certain number of scenarios but it has to be used with caution as it does not take into account explicitly the time correlation of the channel, or in other words, considers that throughout the receiver integration time the C/No is stationary, which is not necessarily the case under severe scintillation conditions. Also, the expression presents singularities at S4 = 0.707 and S4 = 1, therefore, it can be used for moderate and low scintillations only.

Various models exist for predicting and simulating nominal ionospheric scintillation characteristics. One of them is the WBMOD (Secan et al., 1995), which is "an empirical model which produces statistical predictions of trans-ionospheric activity as a function of radio frequency, day of year, time of day, ionospheric latitude and longitude, SSN, and magnetic activity." It is a climatological model, that is, it predicts average behavior, rather than instantaneous behavior (Cervera et al., 2001). There are other analytical models, such as the Cornell scintillation model (Humphreys et al., 2009), that are able to generate complex time series of scintillation from selected S4 index and decorrelation time, based on the generation of filtered white noise, with characteristics derived from a library of scintillation observations, but without establishing dependencies on location, time, etc. Another theoretical model, the global ionospheric scintillation model (GISM) (ITU-R, 2013; Béniguel and Hamel, 2011), is based on a multiple-phase screen technique, which involves resolution of the propagation equation for a medium divided into successive layers, each of them acting as a phase screen. This model provides reliable results for equatorial scintillations provided that the medium is

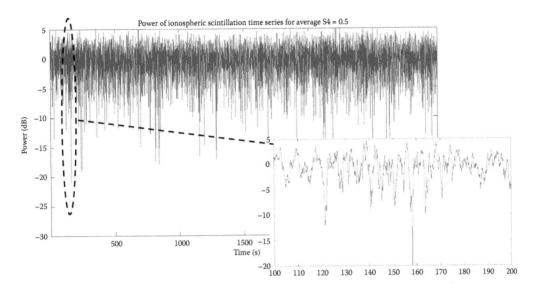

FIGURE 5.10   Intensity of ionospheric scintillation time series for average S4 = 0.5.

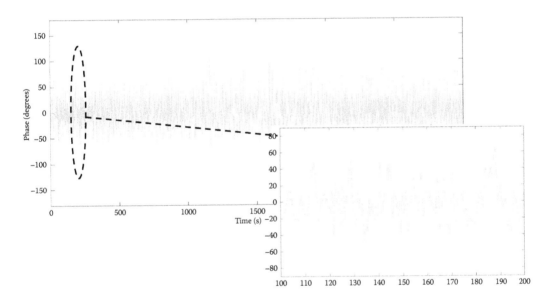

FIGURE 5.11   Phase of ionospheric scintillation time series for average S4 = 0.5.

accurately specified. It is driven by a background electron density climatological model (NeQuick). GISM is able to predict S4 and σ_φ indices for given locations, geometries, times, and SSN but also complex time series of scintillation for isolated events. Examples of intensity and phase scintillation times series from GISM are presented in Figures 5.10 and 5.11, respectively.

5.3   TROPOSPHERIC EFFECTS

The neutral (non-ionized) part of Earth's atmosphere lies from Earth's surface up to about 50 km, the troposphere being its bottom layer, ranging up to 6–10 km in the poles and 15–20 km in the equator, depending on the season. The troposphere is responsible for most of the effects due to the neutral atmosphere observed in GNSS signals. It is mainly composed of dry air molecular gases and water vapor. The main observed effect on GNSS signals is a propagation delay due to the refractivity of the medium, which results in propagation phase and group delay, which accounts for up to several meters, if not corrected, and path bending (below a few centimeters). Opposite to the frequency dependence of ionospheric delays, the troposphere is a non-dispersive media; thus, the delay is constant to all GNSS frequencies. Moreover, the group and phase tropospheric delays have the same magnitude and sign.

The tropospheric delay is the result of two components: delay due to dry air, which accounts for about 90% of the total delay and which is more predictable (due to the constant ratio of its constituents) and correctable; and delay due to water vapor, which is more variable and difficult to predict. Thus, the total tropospheric delay in the slant direction to a GNSS satellite—slant path delay (SPD)—can be divided into a hydrostatic and a wet component (Martellucci and Prieto-Cerdeira, 2009):

$$SPD(\varepsilon) = m_h(\varepsilon) \cdot ZHD + m_w(\varepsilon) \cdot ZWD \tag{5.22}$$

where SPD is slant path delay (m), as a function of elevation angle ε, ZHD and ZWD are zenith hydrostatic delay and zenith wet delay (m), and $m_h(\varepsilon)$, $m_w(\varepsilon)$ are the mapping functions to project from zenith to slant, for the hydrostatic and the wet path delay, as a function of elevation angle ε. Ray bending is usually accounted for by modern mapping functions. ZHD and ZWD are related to the air refractivity N (refractivity is related to air refractive index as $N = 10^{-6}(n - 1)$), which is determined by

$$N = N_d + N_v = K_1 \frac{(P-e)}{T} + K_2 \frac{e}{T} + K_3 \frac{e}{T^2} = k_1 \frac{(P-e)}{T} Z_d^{-1} + \left(k_2 \frac{e}{T} + k_3 \frac{e}{T^2} \right) Z_v^{-1}$$

$$N = N_h + N_w = k_1 \frac{P}{T} Z_d^{-1} + \left((k_2 - 0.622 k_1) \frac{e}{T} + k_3 \frac{e}{T^2} \right) Z_v^{-1} \tag{5.23}$$

where N_d is dry air refractivity, N_v is water vapor refractivity, P is air total pressure (hPa), T is temperature (K), e is partial pressure of water vapor (hPa), K_1, K_2, and K_3 are air refractivity coefficients (derived empirically), and Z_d and Z_v are the compressibility factors of dry air and water vapor (and k_1, k_2, and k_3 are the air refractivity coefficients when compressibility is considered). As observed in the second part of Equation 5.23, refractivity can also be expressed in its hydrostatic and non-hydrostatic/wet components N_h and N_w. In Martellucci and Prieto-Cerdeira (2009), experimental values of air refractivity coefficients from different authors are given (k_1 ranges from 77.6 and 77.7 [K/hPa], k_2 ranges from 64 and 72 [K/hPa], and k_3 ranges from 3.70 and 3.78 [10^5 K^2/hPa]). The uncertainty of air refractivity parameters, with particular regard to K_3, produces an RMS fluctuation of the ZTD of about 6 mm on a global scale.

The relation between zenith total delay and refractivity is given as follows:

$$ZTD = ZHD + ZWD = 10^{-6} \int_{h_0}^{\infty} [N_h(h) + N_w(h)] dh \tag{5.24}$$

Regarding mapping functions, $m_h(\varepsilon)$, $m_w(\varepsilon)$, various models exist from simply the secant of the zenith angle or simple approximations of that secant, to models based on continued fractions (such as the one in Niell, 1996) to mapping functions based on ray tracing along profiles from numerical weather prediction (NWP) models.

Blind tropospheric correction models do not use any input about the actual state of the atmosphere but only climatological maps (and therefore, do not need any broadcast or external data). One example is the tropospheric blind *RTCA model* for SBAS as defined in RTCA (2006) that is based on Equation 5.22, where

$$m(\varepsilon) = m_h(\varepsilon) = m_w(\varepsilon) = \frac{1.001}{\sqrt{0.002001 + \sin^2(\varepsilon)}} \tag{5.25}$$

$$ZHD = \left(1 - \frac{\beta H}{T}\right)^{(g/R_d\beta)} \cdot ZHD_0$$

$$ZWD = \left(1 - \frac{\beta H}{T}\right)^{((\lambda+1)g/R_d\beta)-1} \cdot ZWD_0 \tag{5.26}$$

$$ZHD_0 = \frac{10^{-6} k_1 R_d P}{g_m}$$

$$ZWD_0 = \frac{10^{-6} k_2 R_d}{g_m(\lambda+1) - \beta R_d} \cdot \frac{e}{T} \tag{5.27}$$

with $k_1 = 77.604$ K/mbar, $k_2 = 382{,}000$ K^2/mbar (notice the different definition and units for k_1 and k_2 with respect to Equation 5.23), $R_d = 287.054$ J/(kg · K), $g_m = 9.784$ m/s^2, and the five meteorological parameters: pressure (P [mbar]), temperature (T [K]), water vapor pressure (e [mbar]), temperature lapse rate (β [K/m]), and water vapor "lapse rate" (λ [dimensionless]) are calculated using

$$\xi(\phi, D) = \xi_0(\phi) - \Delta\xi(\phi) \cdot \cos\left(\frac{2\pi(D - D_{min})}{365.25}\right) \tag{5.28}$$

where ξ represents each of the five parameters, as a function of the receiver latitude, ϕ, and day of year, D, $D_{min} = 28$ for northern latitudes, $D_{min} = 211$ for southern latitudes, and ξ_0, $\Delta\xi$ are the average and seasonal variation values for the particular parameter at the receiver's latitude, tabulated for five different latitude thresholds in Table A-2 of RTCA (2006): for latitudes $|\phi| \leq 15°$ and $|\phi| \geq 75°$, values for ξ_0 and $\Delta\xi$ are taken directly from the table, and for latitudes in the range $15° < |\phi| < 75°$, values for ξ_0 and $\Delta\xi$ at the receiver's latitude are each precalculated by linear interpolation between values for the two closest latitudes $[\phi_i, \phi_{i+1}]$ in the table.

The *ESA Galileo tropospheric correction* blind model (Martellucci and Prieto-Cerdeira, 2009) introduces improved accuracy at the cost of additional complexity in terms of model parameters. For ZHD it uses the Saastamoinen model (Saastamoinen, 1972):

$$ZHD(\phi, h, P) = 10^{-6} \frac{k_1 R_d P}{g_m(\phi, h)} = \frac{0.0022767 \cdot P}{1 - 0.00266 \cdot \cos 2\phi - 0.00028 \cdot h} \tag{5.29}$$

with ϕ being ellipsoidal latitude, h surface height above the ellipsoid (km) and P total surface pressure (hPa), $k_1 = 77.604$ (K/hPa), $R_d = 287.0$ J/(kg K) is the ratio between molar gas constant and dry air molar mass, $g_m(\phi, h) = 9.784 \cdot (1 - 0.00266 \cdot \cos 2\phi - 0.00028 \cdot h)$ (m/s^2).

The ZWD is approximated by

$$ZWD(\phi, h, e_0, T_{m0}, \lambda) = 10^{-6} \frac{R_d}{g_m(\phi, h)} \cdot \frac{k_2}{(\lambda+1)} \cdot \frac{e_0}{T_{m0}} \tag{5.30}$$

with e_0 being water vapor pressure at Earth's surface (hPa), T_{m0} mean temperature of the water vapor above the surface (K), λ vapor pressure decrease factor and $k_2 = 370{,}100$ (K²/hPa), and the equation assumes the following conditions for receivers located at height from the mean sea level h (km), different than the surface height h_0, the values of the input meteorological parameters can be extrapolated from surface values, T_{m0} (K), e_0 (hPa), and P_0 (hPa), using the following equations:

$$T_m(h, T_{m0}, \alpha_m) = T_{m0} - \alpha_m \cdot (h - h_0)$$

$$P(h, P_0, T_0, \alpha) = P_0 \left[1 - \frac{\alpha(h - h_0)}{T_0} \right]^{(g/R_d'\alpha)} \tag{5.31}$$

$$e(h, e_0, \lambda) = e_0 \cdot \left[\frac{P(h)}{P_0} \right]^{\lambda+1}$$

where the required additional input parameters are α, lapse rate of air temperature (K/km), which can derived with the following equation:

$$\alpha \cong 0.5 \cdot \left[\frac{(\lambda+1) \cdot g}{R_d'} - \sqrt{ \frac{(\lambda+1) \cdot g}{R_d'} \left[\frac{(\lambda+1) \cdot g}{R_d'} - 4\alpha_m \right] } \right] \tag{5.32}$$

T_0, air temperature at surface (K), is derived as

$$T_0 = \frac{T_{m0}}{1 - (\alpha R_d' / (\lambda+1)g)} \tag{5.33}$$

where $R_d' = R_d/1000 = 0.287$ (J/(g K)) and α_m is lapse rate of the mean temperature of water vapor (K/km).

It is to be noted that the altitude above mean sea level, h, is different from the height above an assumed ellipsoid as calculated by navigation receivers. The derivation of the altitude above mean sea level, h, from height above the ellipsoid is provided in RTCA (2006) Annex H, page H-2.

For parameters affected only by seasonal fluctuations, air total pressure, P_0, mean temperature lapse rate, α_m, and decrease factor of water vapor pressure, λ, the value of each parameter X_i is given by

$$X_i(D_y) = a1_i - a2_i \cos\left[2\pi \frac{(D_y - a3_i)}{365.25} \right] \tag{5.34}$$

where $a1_i$ is the average value of the parameter, $a2_i$ is the seasonal fluctuation of the parameter, $a3_i$ is the day of the minimum value of the parameter, and D_y is the day of the year [1..365.25].

For parameters T_{m0} characterized both by seasonal and diurnal fluctuations (mean temperature of moist air at surface reference height, T_{m0}, and water vapor pressure at reference height, e_0), the value of the parameter X_i is given by

$$X_i(D_y, H_d) = a1_i - a2_i \cos\left[2\pi \frac{(D_y - a3_i)}{365.25} \right] - b2_i(D_y)\cos\left[2\pi \frac{(H_d - b3_i(D_y))}{24} \right] \tag{5.35}$$

where $a1_i$, $a2_i$, $a3_i$, and D_y are equivalent to Equation 5.34, H_d is hour of the day [0..24), $b2_i$ is the daily fluctuation, calculated by linear interpolation between the average values of daily fluctuations during nearest months to day D_y, $b3_i$ is the hour of the day at which the minimum value occur, by linear interpolation between the average values of daily fluctuations during nearest months to day D_y.

The values of the parameters of this harmonic model, $a1$, $a2$, $a3$, $b2$, and $b3$, for each meteorological parameter, are contained in digital maps with a resolution of 1.5 × 1.5 (deg), which were derived by fitting the harmonic model to monthly and hourly climatological values derived from the ECMWF ERA15 database (covering the period from 1979 to 1993). The model has been validated using an independent ECMWF global product covering the period from March to October 2002 with a spatial resolution of 1 × 1°.

The ESA Galileo tropospheric correction model uses the Niell mapping function (Niell, 1996) for elevation angles, ε, not lower than 3 (deg):

$$m_h(\varepsilon) = m(\varepsilon, c_{h1}, c_{h2}, c_{h3})$$

$$m_w(\varepsilon) = m(\varepsilon, c_{w1}, c_{w2}, c_{w3})$$

(5.36)

$$m(\varepsilon, a, b, c) = \frac{(1 + (a/(1 + (b/(1 + c)))))}{(\sin\varepsilon + (a/(\sin\varepsilon + (b/(\sin\varepsilon + c)))))}$$

(5.37)

The coefficients of the hydrostatic delay mapping function, c_{h1}, c_{h2}, and c_{h3} are tabulated depending on the latitude, *lat*, and characterized by a seasonal fluctuation such as that in Equation 5.34.

The ESA Galileo tropospheric correction overall RMS errors of ZTD, ZHD, and ZWD are 4.5, 1.6, and 4.4 cm, respectively. An example of the resulting zenith total delay output is presented in Figure 5.12.

In order to improve the accuracy significantly, blind models can also operate using real-time local meteorological data or model parameters from an external provider to replace data from climatological maps; these modes usually referred to as *local* or *site augmented modes*. The parameters from external providers may be derived from data produced by NWP systems, GNSS-based techniques (e.g., estimation from precise point positioning techniques), or remote sensing such as ground microwave radiometers.

Recent research has worked into more advanced mapping functions derived from NWP, for example, the Vienna mapping function VMF1 (Boehm et al., 2006), or the direct use of NWP in network-based precise positioning techniques.

Tropospheric delay from GNSS observations is often used for weather forecasting, both through ground-based integrated water vapor (IWV) predictions or space-based radio occultation profiles; furthermore, they can be assimilated into NWP systems for better spatial and temporal resolution.

5.4 MULTIPATH, SHADOWING, AND BLOCKAGE

5.4.1 GENERAL INTRODUCTION AND MULTIPATH IN REFERENCE STATIONS

Multipath propagation introduces code and phase errors on GNSS receivers through the interaction of the radiowave signal broadcast by the satellites with the local environment around the receiver. The propagation mechanisms that give rise to multipath are reflection/scattering and diffraction. At the receiver, the multipath and noise code errors depend not only on the actual multipath environment but also on receiver characteristics, antenna type, and possible multipath protection mechanisms. Shadowing is blockage produced by objects (buildings, trees, etc.) located between the line of sight (LOS) of the satellite and receiver.

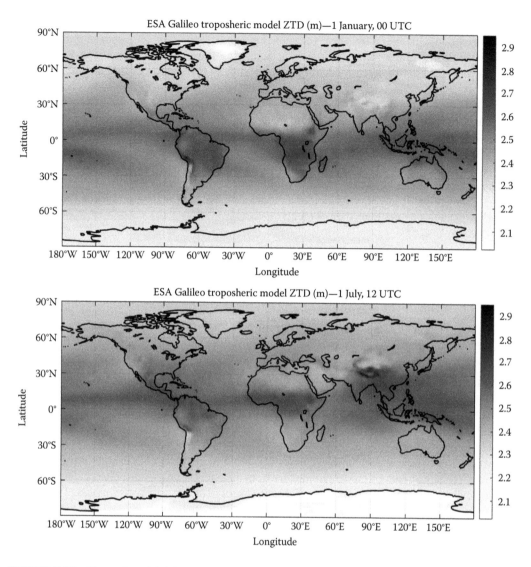

FIGURE 5.12 Maps of worldwide distribution of the model zenith total delay, at reference height over land and at mean sea level over sea, calculated for two different days of the year.

For code multipath, within the GNSS receiver signal processing module, acquisition of the signal is performed using locally generated replicas of the PRN codes and searching both in the code phase and Doppler shift domains for the right matching of the incoming sequence. This matching is given by the maximum of the cross-correlation of the local and incoming PRNs. This code phase is used to estimate the transit time of the signal from the satellite and thus the pseudorange. The cross-correlation of different PRN sequences is ideally zero and the autocorrelation of the same PRN is not zero only within one chip period of the autocorrelation function. In reality, PRNs are finite length and not perfectly random and non-negligible sidelobes plus finite bandwidth smoothes the sharp edges of the autocorrelation function. Once the signal is acquired, it must be tracked in order to maintain synchronization both in code phase using a delay lock loop (DLL) and in Doppler (carrier phase) using usually a frequency lock loop (FLL) or PLL. The DLL is driven by the output of a discriminator, which has the target to identify the maximum peak of the correlation function. The classical early-minus-late power non-coherent discriminator subtracts the power of an early correlation (the power of the complex correlation separated in in-phase and quadrature components) advanced by one chip from the power of a late correlation also delayed by one chip. A coherent discriminator performs the subtraction coherently using the carrier phase tracking of a prompt (unshifted) replica. Tracking is provided by maintaining the discriminator output to zero for negative slope of the discriminator function. In the presence of multipath, the autocorrelation function is distorted, shifting the zero-crossing of the discriminator function and thus creating an error on the pseudorange estimation. This is illustrated in Figure 5.13, where the autocorrelation function of a direct signal sequence together with a specular multipath signal is presented.

For a simple metric for the evaluation of multipath for a given signal and receiver discriminator, the multipath code phase error envelope has been extensively used in the literature. This error envelope indicates the error bounds of a single specular multipath ray with a fixed power relative to the direct signal for any relative delay, usually up to a maximum delay of 1.5, the chip period considering that longer delays do not affect the pseudorange estimation (true for negligible sidelobes in the autocorrelation function). A method to derive the multipath error envelope for other signals is given by applying an inverse Fourier transform to the analytical expression of the PSD of the signal, in order to obtain the autocorrelation function, and then deriving the discriminator function for the selected type of DLL. Figure 5.14 shows the code phase multipath error envelope for BPSK(1) and BOC(1,1) signal modulations with a signal-to-multipath ratio (SMR) of 6 dB, bandwidth of 15 MHz, and two discriminator spacings. It is evident that BOC(1,1) is less affected by multipath than BPSK(1). Higher-order BOC modulations present even sharper correlations with lower multipath noise, but with the presence of secondary peaks that may result in false locks inducing a permanent bias until recovery to the main peak is achieved.

The maximum multipath code tracking errors may be largely reduced using more advance discriminators. One of the most common techniques is based on the use of a narrow correlator using a shorter early–late spacing (Van Dierendonck et al., 1992). However, short delay multipath, around a few meters, is often difficult to mitigate. There are other more sophisticated discriminator schemes

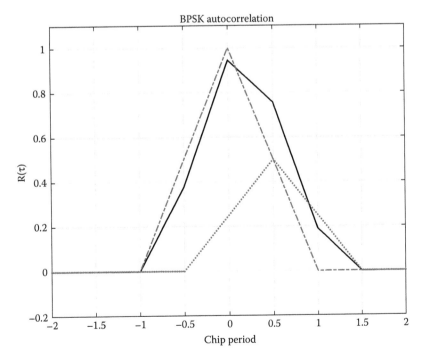

FIGURE 5.13 Autocorrelation function of BPSK PRN sequence in the presence of a specular multipath at half chip delay.

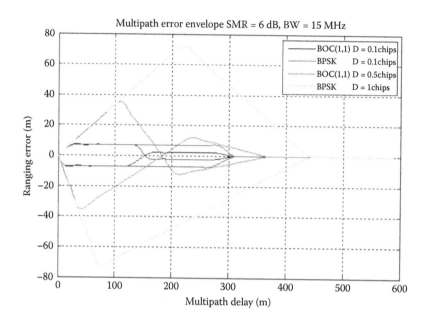

FIGURE 5.14   Code phase multipath error envelope for different signals and discriminator spacings.

(based on maximum likelihood estimators, particle filters, multiple correlators, etc.), for multipath mitigation available in the literature and implemented in commercial receivers.

For GNSS fixed reference stations, code multipath usually has a slowly varying component, introducing correlated noise in the observations. Its effects depend on the environment, antenna placement, and antenna design quality. Antenna plays a major role, through the design of adequate pattern for the desired signal above ground (depending on the application elevation masks of 5° or 10° are employed for avoiding very noisy signals), but also through a careful design and characterization of the co-polar and cross-polar gain below ground. Ground reflections will enter the antenna from negative elevation angles, and depending on the type of terrain, roughness, geometry, and elevation angle with respect to Brewster angle, a coherent reflected will be more co-polar right-hand circularly polarized (RHCP) (usually for low elevation angles) or more cross-polarized left-hand circularly polarized (LHCP) (higher elevation angles).

Figure 5.15 shows the situation in the ideal case of open air sky with a flat surface. Antenna choke rings or antenna multipath protection are often used to minimize the effects of reflections on the surrounding grounds. Figure 5.16 shows the antenna pattern of a GNSS antenna as measured in an anechoic chamber. The effects of buildings or other objects located higher than the antenna are more difficult to mitigate.

Code multipath effects (together with noise) are often characterized following the ionospheric-free code minus carrier (CmC) technique, which for L1 carrier frequency is given by

$$M_{P1}^i(t) + \varepsilon_{P1}^i(t) + \lambda_{L1}N_{L1}^i = R_{P1}^i(t) - \varphi_{L1}^i(t) - 2 \cdot \left(\frac{(f_{L2})^2}{(f_{L1})^2 - (f_{L2})^2} \right) \cdot (\varphi_{L1}^i(t) - \varphi_{L2}^i(t)) \qquad (5.38)$$

where M_{P1}^i denotes the multipath on the code phase estimate for L1 frequency and satellite i (in meters), ε_{P1}^i represents the receiver code noise at L1 (in meters), $\lambda_{L1}N_{L1}^i$ is the (unknown) carrier phase ambiguity, R_{P1}^i and φ_{L1}^i denotes code and phase observables, respectively, and f_{L1} and f_{L2} represent the center frequency of L1 and L2, respectively. This method assumes that carrier phase noise,

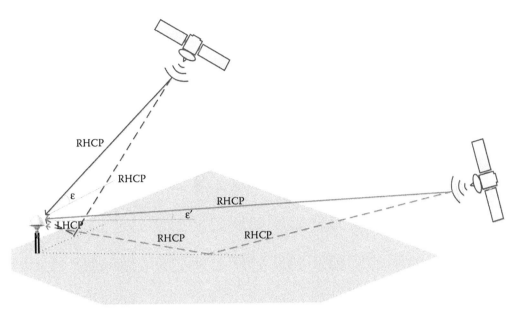

FIGURE 5.15 Geometry of direct and multipath signals from a flat surface indicating the effects on polarization depending on elevation angle.

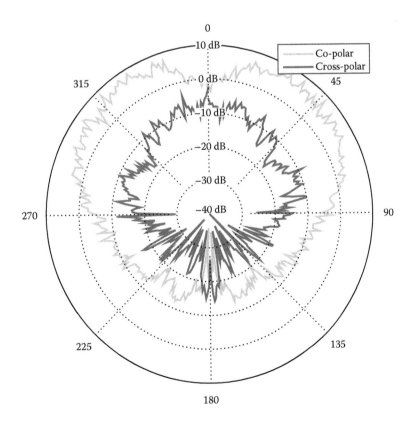

FIGURE 5.16 Co-polar (RHCP) and cross-polar (LHCP) patterns for a single azimuth planar cut and all elevations for a typical GNSS antenna pattern as measured from an anechoic chamber.

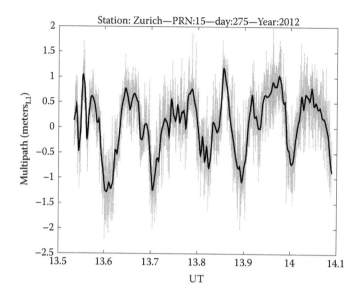

FIGURE 5.17 Multipath error (smoothed and non-smoothed) for a station in Zurich for a satellite strongly affected by correlated multipath.

multipath, and wind-up are negligible compared to code multipath and noise. During a continuous period of tracking satellite i, N is considered constant as long as no cycle slips have occurred. Therefore, the multipath + noise estimate is derived from the equation above by subtracting the mean value over a period determined by the epochs of start of tracking, end of tracking, and cycle slips, if any. This step also removes any multipath error (bias) that is constant over the whole continuous arc, but the method is able to characterize rapid random multipath and slow-fading multipath of medium time scales from seconds to several tens of minutes (provided that small arcs are rejected). Such effect is not considered to be persistent in the majority of arcs; therefore, is a sporadic and not a driving factor, and in the analysis it is assumed that the average of such long arc bias for a given station is zero mean. An example of such characterization for a station in Zurich is presented in Figure 5.17. Such multipath effects tend to repeat when the geometry for the satellite station pass repeats; this is, for instance, observed after one sidereal day for GPS and 10 days for Galileo.

The carrier phase error due to single ground reflection multipath in a flat terrain can be expressed as (Georgiadou and Kleusberg, 1988)

$$m_{Ln}^{j} = \frac{\lambda_{Ln}}{2\pi} \tan^{-1}\left(\frac{\alpha \sin((4\pi/\lambda_{Ln})h\sin(\varepsilon^{j}))}{1 + \alpha \sin((4\pi/\lambda_{Ln})h\sin(\varepsilon^{j}))} \right) \tag{5.39}$$

where m_{Ln}^{j} is expressed in meters, λ_{Ln} is the wavelength, α is an attenuation factor related to the ground reflection coefficient and antenna gain, h is the antenna height, and ε^{j} is the elevation angle to satellite j. The maximum of the multipath carrier phase error is $\lambda_{Ln}/4$ corresponding to ~4.8 cm in $L1$ frequency. Owing to the changing geometry among satellite, reflector, and antenna, over the satellite pass, the carrier phase multipath error gets a periodic character with frequency depending on antenna height, wavelength, elevation angle, and its rate of change over time.

5.4.2 LAND MOBILE EFFECTS

The land mobile covers vehicular and pedestrian dynamic users in various environments: for instance, open areas, tree-shadowed roads, suburban areas, or urban canyons. Different propagation mechanisms such as diffraction, reflection, and scattering in the local environment of satellite navigation land mobile receivers influence the navigation signal observed by the receiver in several ways: variations of amplitude (fading) and phase of the direct or LOS signal and the appearance of replicas of the transmitted signal arriving at the receiver. Multipath is related to the phenomena where electromagnetic signals arrive to a given location (user receiver) after interacting with various propagation paths (direct paths, and reflected, diffracted, or refracted paths). Shadowing/blockage is the signal attenuation due to partial/complete obstruction of the direct path.

In telecommunications, channels are considered narrowband if the channel is frequency non-selective, that is, when multipath affects all frequencies in the signal bandwidth in the same way. On the other hand, wideband channels are those where the channel is frequency selective for which the signal is affected by multipath differently for different frequencies in the bandwidth of interest. Narrowband channels are usually modeled as baseband-normalized complex time series representing the variations (fading) of the direct path. For wideband channels, the channel is often characterized as baseband-normalized complex time series of a channel impulse response (CIR) where each delta impulse represents a propagation path with given amplitude and phase and arriving at a given delay (some models also incorporate angle-of-arrival information). Considering bandwidth for modulated data, GNSS channels are categorized as narrowband. However, for ranging, considering how spread spectrum is used, the autocorrelation function distortion and its effects on the time synchronization (delay) are the key parameters; therefore, the autocorrelation is the basis pulse function for the channel, and tracking error is the figure of merit. The main difference in the required characteristics of multipath channel models for GNSS code-based ranging, compared to satellite communications is that in the former, it is critical to adequately represent the distortion of the signal autocorrelation function accurately synchronized in the delay access that will drive the tracking error depending on the type of receiver discriminator; whereas, in the latter, the important characteristics are the distortion of the transmitted signal pulses in order to be able to recover the transmitted data. In the latter, a proper energy distribution and time variation is sufficient even if the tap delays are fixed over time; however, for GNSS, the autocorrelation distortion strongly depends on the tap delay and therefore the delays need to be time-varying according to reality and modeled accurately. In this respect, channels that are able to describe the autocorrelation function in as an accumulated set of replicas with its phase, delay, and amplitude, which is the same type of model as the wideband models for telecommunications.

Tap-delay lines are often used to model wideband channels mathematically, considering the channel as a time-variant filter:

$$h(\tau,t) = a_0(t) \cdot \exp(j\phi_0(t)) + \sum_{k=1}^{N-1} a_k(t) \cdot \exp(j\phi_k(t)) \cdot \delta(\tau - \tau_k) \tag{5.40}$$

where a_i and ϕ_k are respectively the amplitudes and phases of the k-th tap, τ_k is its delay, and $k = 0$ is the LOS path. Depending on the particular model, a tap may incorporate the energy from a group of individual specular scatterers plus diffuse scattering, a cluster of reflections with related characteristics, or even an individual path if N is sufficiently large and both N and τ_k are time-varying. The instance of the filter for a given time t_i is usually called a CIR. An example of a typical CIR snapshot for GNSS is presented in Figure 5.18. Each tap component also varies according to a certain Doppler fading bandwidth related to the relative Doppler for the accumulated energy components contributing to a given tap, in addition to the Doppler shift imposed by satellite and receiver movement.

On the other hand, for the analysis of GNSS carrier-phase-based ranging, and also to analyze other effects at the GNSS receiver level, which are largely driven by the variations of the received carrier-to-noise power density (C/N0), such as time to first fix, loss of lock, reacquisition, etc., it

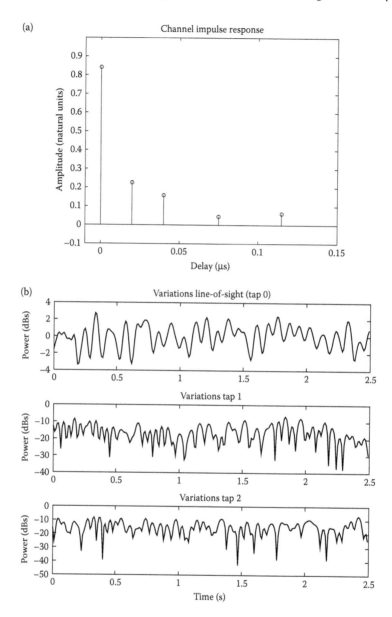

FIGURE 5.18 Example of channel impulse response (CIR) with line-of-sight and four taps: (a) instantaneous snapshot and (b) power variation over time for line of sight and taps 1 and 2.

is usually sufficient to use narrowband models, accurately representing variations of direct signal (including shadowing and accumulated energy variations due to multipath).

The adequacy of various types of land mobile models for the assessment of GNSS at different levels is qualitatively described in the table below.

	Narrowband Model	Wideband Model
Code-based ranging and positioning performance	No	Yes
Carrier-phase-based ranging and positioning performance	Yes	Yes
Time to first fix	Yes	Yes
Loss of lock, false lock, reacquisition	Yes	Yes

For the narrowband family of models, statistical models based on experimental data are convenient due to its ease of implementation and verification with reality. Several recent models characterize the variations of the direct signal using a Loo distribution function (Loo, 1985), which represents the large-scale fading as a lognormal distribution function (shadowing over a stationary channel region) and the small-scale fading modeled through a Rayleigh distribution function (multipath). In the Loo distribution function, the direct signal amplitude is lognormally distributed due to shadowing while the multipath component parameter, $2\sigma^2$, remains constant. The Loo probability distribution function is given by

$$f_{\text{Loo}}(r) = \frac{8.686r}{\sigma^2 \, \Sigma \, \sqrt{2\pi}} \int\limits_0^\infty \frac{1}{a} \exp\left[-\frac{r^2 + a^2}{2\sigma^2}\right] \exp\left[-\frac{(20\log a - M)^2}{2\Sigma^2}\right] I_0\left(\frac{ra}{\sigma^2}\right) da \qquad (5.41)$$

where a is the direct signal's amplitude and $\sqrt{2\sigma^2}$ describes the amount of diffuse multipath with M and Σ in dB, $MP = 10\log 2\sigma^2$ is the mean square value (of the associated Rayleigh distribution, that is, disregarding the direct signal) of the multipath component expressed in dB. M and Σ are respectively the mean and standard deviation of the associated normal distribution for the direct signal's amplitude ($A = 20\log(a)$) in dB. The Loo distribution includes as special cases the Rice distribution for large values of a, and the Rayleigh distribution for very small values of a. This makes it valid for a very wide range of conditions spanning from LOS to very shadowed (blockage) conditions. Additional information on the characteristics and the applicability of the Loo distribution function for land mobile satellite can be found in Fontan et al. (2008).

The three-state channel model described in Pérez-Fontán et al. (2001) characterizes the direct signal variations (narrowband channel) according to three scales: very large-scale (or very slow) fading, large-scale (or slow) fading, and small-scale (or fast) fading. The very large-scale fading represents strong changes of the shadowing/blockage conditions through the environment such as changes from a satellite blocked by a building to the same satellite unobstructed. This is modeled following a three-state Markov state chain where the states are described as LOS conditions, moderate shadowing, and deep shadowing. Together, the large- and small-scale variations are described through a Loo distribution function.

An evolution of the three-state channel model was proposed by Prieto-Cerdeira et al. (2010) incorporating the following improvements: a reduction in the number of states, the introduction of a random selection of statistical parameters describing each state providing more realistic variability over the same environment, and a state machine governed by a semi-Markov process. The new state classification does not necessarily correspond to intuitive physical definitions of the states as before (LOS, shadowing) but instead to channel variations that share similar statistical characteristics. The two states are termed for convenience, *good* and *bad* states, representing a range of LOS to moderate shadowing and moderate to deep shadowing, respectively. An example of the time series showing the two states is presented in Figure 5.19. This model has been adopted in ITU-R for the prediction of fading statistics over mixed propagation conditions and the generation of fading time series, including model parameters for frequencies in L/S, C, and Ku bands, from experimental measurements in France and Europe.

An extensive multi-satellite measurement campaign over the United States and Europe was performed in the MiLADY project, and the two-state model was extended to multisatellite in Arndt et al. (2012), allowing the generation of a correlated time series for two or more satellites simultaneously, making it very suitable for GNSS shadowing simulations.

For the characterization of channel effects in a wideband sense, the instantaneous multipath energy contained in the small-scale fading is spread or distributed along the delay axis characterized through instantaneous CIR, accounting for how scatterers in the environment contribute individually (or grouped into clusters). In satellite mobile communications, fully statistical wideband

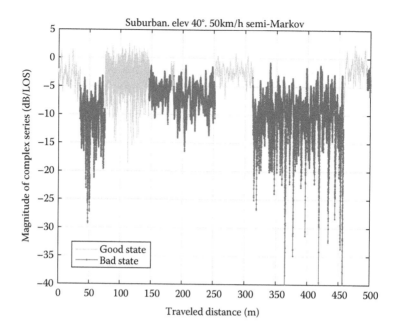

FIGURE 5.19 Direct signal power variation generated from statistical two-state narrowband model for 40° elevation angle in a suburban environment.

modeling exists assuming tap-delay lines at fixed delays grouping echoes that evolve following certain statistical conditions for allocation of power (for instance, multipath echoes excess delay following exponential distribution and exponential multipath power decay profile, and multipath echoes grouped into taps). The advantage of such models is of course their simplicity and flexibility and the main limitation is that tap delays are usually maintained at fixed delay locations, an assumption which strongly influences the results in terms of ranging errors for different delay values. Another limitation is related to the fact that they are valid for a single satellite only.

For satellite navigation, a complete physical modeling of the environment would be ideally preferred. In practice, the required level of detail of the characteristics of the environment and the different materials, the validity of ray-tracing mechanisms (geometrical optics plus universal theory of diffraction) versus other high-frequency approximations, and the computational complexity, makes such solutions often impractical for simulations in complex environments. A compromise is obtained with physical–statistical models or hybrid models, which combine some large-scale characteristics of the environment (e.g., location of objects that contribute to strong propagation effects, such as buildings or trees) with statistical variations usually aiming to account for diffuse multipath effects.

One of the models more commonly used for multiple satellite navigation purposes in land mobile satellite for various environments and including shadowing and multipath is the channel model for land mobile satellite navigation described in Lehner and Steingass (2005) and ITU-R (2009, 2015). The model is of physical–statistical nature and it is valid for a situation where a satellite is transmitting from a known position to a receiver on ground, where an elevation ε and an azimuth φ can be computed relative to the receiver heading and position. The model is based on deterministic and stochastic parameters and it is able to generate vectors that include complex envelope time series of direct signal and reflections, with corresponding path delay vectors. The parameters determining the stochastic behavior of the model have been derived from measurements obtained on a given scenario and parameters are available for urban and suburban vehicular and pedestrian cases in Munich, Germany and its surroundings. The geometry of the model is based on a synthetic environment representation. The channel model consists of a combination of the following parts (developed to support the simulation of realistic propagation behavior for many propagation scenarios of interest, and further validated with empirical analyses based on measured data):

- Shadowing of the direct signal:
  - House front module
  - Tree module
  - Light pole module
- Reflections module

It incorporates a large number of reflections generated stochastically based on statistical characteristics from channel measurements (number, distribution, duration, and mean power). For the simulation of multiple satellites, by construction, the shadowing of the direct signal is correlated as needed by the characteristics of the environment. The reflections module, through its statistical nature, does not necessarily reflect the characteristics of the environment between different satellites. An example of the output channel is presented in Figure 5.20.

For rural environments, where trees and alleys are the dominant effect on shadowing and multipath, a channel model was developed in Schubert et al. (2012) based on the synthesis approach, which composes complicated structures of less complex primitives. Trees are composed of volumes of individual effective point scatterers, not necessarily linked to a specific constituent of a treetop canopy such as a certain branch or leaf, but instead able to reproduce the physical behavior of strong scattering centers inside canopies in a simplified but realistic manner, that is, reproducing accurately the amount of power that is reradiated. Alleys are composed of individual trees, placed individually to resemble a given scenario, or generated stochastically according to certain typical characteristics of the environments such as typical tree dimensions, and forward and sideward distances between trees. An example is presented in Figure 5.21.

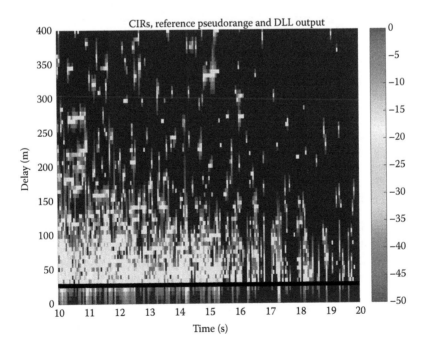

FIGURE 5.20 Example of channel output for channel model for land mobile satellite navigation described in Lehner and Steingass (2005).

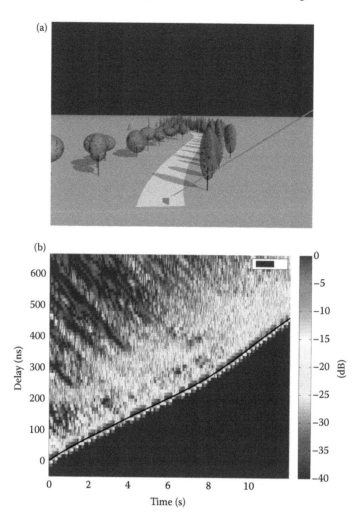

FIGURE 5.21 Example of scenery (a) and channel output (b) of rural channel model described in Schubert et al. (2012). (Courtesy of F. Schubert.)

In the frame of the ESA project PHYSTAT (Lemorton et al., 2013), a software simulator implementing a set of propagation channel models that combine deterministic and statistical approaches modeling physical principles (diffraction, specular or diffuse reflection, transmission) by means of stochastic parameters, was developed, including various satellite-to-user environments: aeronautical, maritime, and land mobile satellite. For the land mobile satellite case, the following elements are considered:

- The direct path from the satellite.
- The attenuated direct path through vegetation (tree), calculated using ray-tracing technique and a precalculated value of the linear attenuation through the canopy (which is calculated in advance using, for instance, multiple scattering technique [de Jong and Herben, 2004]).
- The tree scattered components evaluated with a cluster of point scatterers in the canopy (using theso-called geometry-based stochastic channel model[ing]—GSCM approach). The canopy is characterized by a volumetric scattering section, calculated in advance, using, for instance, also the multiple scattering technique (de Jong and Herben, 2004).
- The masking effect by the trunk is calculated through the uniform theory of diffraction (UTD) assuming three edges.

- The diffracted paths by
  - Buildings: calculated with edge-diffracted paths by the UTD
  - Poles: calculated through UTD with each pole considered as three edges
  - Static vehicles: calculated through UTD for edges in a cuboid, and physical optics (PO) for reflecting faces
- The scattered components from building facades are calculated by one of the following methods:
  - A three-component model (3CM) (Ait-Ighil, 2013), relying on the following components (see Figure 5.22): *The specular component* reproduces the forward reflection mechanism taking place on wide and smooth surfaces such as windows and flat walls. *The backscattering component* reproduces a double bounce reflection, or backward reflection, due to effects through windows or balconies. *The incoherent scattering component* reproduces the scattering phenomenon coming from all small features present on complex facades. The specular and backscattering components are calculated with a simplified version of the PO algorithm, whereas the incoherent scattering is evaluated from empirical models for rough surfaces.
  - A GSCM model using a cluster of scatterers located at the middle of each facade surface (Oestges et al., 2014). Each cluster is then characterized by the total scattered power density (per unit facade area), which enables the computation of the total scattered power (to be divided between the individual point scatterers) and the scattering cluster radius a, which is related to the angular spread of the facade as seen by a receiver in the street.
- The antenna pattern (combined angle-dependent amplitude/phase polarized pattern).

5.4.3 AERONAUTICAL, MARITIME, AND OTHER CHANNELS

Satellite navigation channel models for other dynamic users exist, such as for aircrafts and helicopters, for ships, or even for pedestrians in indoor environments.

The aeronautical satellite channel is characterized by a strong LOS component that is present most of the time (except for possible maneuvers blocking the satellite signal at low elevation angles). In addition, aircraft fuselage may cause specular and possibly diffuse scattering, introducing fading on the LOS signal. Depending on the type of terrain and the geometry, ground/sea reflections could arrive at the aircraft with a certain delay, phase, and attenuated power with respect to the LOS component.

The ITU-R satellite aeronautical wideband model (Steingass et al., 2008; ITU-R, 2012) assumes that the direct path is affected by a refractive, resulting in a Rician process with high direct-to-multipath ratio. The power spectrum of the refractive component can be approximated by an

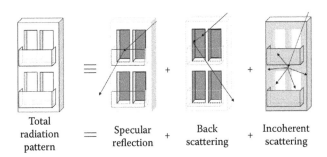

FIGURE 5.22 Components of the 3CM model (Ait-Ighil, 2013) representing the scattered components from buildings facades. (Courtesy of M. Ait-Ighil.)

exponential function, with an impulse at 0 Hz (DC component) that represents the direct path. The fuselage echo is modeled as a fixed delay of 1.5 ns and its power spectrum is characterized with a constant component (DC component) at −14.2 dB plus an exponential process. For code multipath, a bias is observed from the DC component depending on the relative phase with respect to the phase of the main LOS component, provided by the receiver discriminator function and the fixed delay. The ground reflection is modulated by terrain features, and implemented through a Markov state chain described for three altitude regimes.

For worst-case ground reflection, sea reflection may be assumed, and models for sea reflection are available on ITU-R (2012), based on Karasawa and Shiokawa (1988). More complex terrain scattering may be calculated through multiple-point scatterers distributed on the ground with a defined power and angular pattern with the power for each scatterer calculated through physical rough surface applied using digital terrain data, an approach followed on the aeronautical model in Lemorton et al. (2013).

For safety-of-life SBAS systems, airborne multipath error is characterized through a Gaussian distribution with a standard deviation $\sigma_{multipath}$ that is assumed to overbound the standard deviation of observed multipath errors. Such $\sigma_{multipath}$ depends on the elevation angle to the satellite according to the function (RTCA, 2006)

$$\sigma_{multipath}[i] = 0.13 + 0.53 e^{(-\theta[i]/10\,\text{deg})} \tag{5.42}$$

For helicopters, a periodic ON–OFF channel is often assumed with period depending on blade diameter, blade width, number of blades, and rotary angular velocity, and also on elevation and azimuth to the satellite. In Lavenant et al. (1997), various attenuation values depending on azimuth angle plus some small-scale fading probably due to diffraction are proposed.

The characteristics of the propagation channel for GNSS maritime users, analogously to the aeronautical case, incorporates direct signal from the satellite, reflection from the sea, and possibly interactions with the ship structure (depending on the antenna placement, ship size and structure, and geometry). In the case of rough sea, the ratio between diffuse and specular components increases, and for high waves, more than one coherent reflection may be observed. In Lemorton et al. (2013), sea surface reflection (specular and diffuse) is based on the models from Karasawa and Shiokawa (1988), where specular reflections exist when the sea conditions can be considered to be calm showing smooth characteristics; they are calculated through the classical Fresnel reflection coefficient, modified by sea roughness. For diffuse reflections, the sea surface affecting the maritime link is assumed to be made up of floor tiles, each contributing a given multipath power calculated through the fully polarimetric model for rough surface scattering in Barrick (1968). Also, the dynamics of the sea and the ship are included, with a statistical characterization of the sea surface and the movements of the ship (pitch, roll, and heave) depending on sea state, ship velocity, and its response.

The satellite-to-indoor channel is largely driven by two factors: the entry loss (or building attenuation), and the delay and angular spreading of the energy. The characterization of this channel is very complex due to multiple interactions. Channels suitable for the assessment of satellite navigation are available in the literature such as Pérez-Fontán et al. (2011) and Jost et al. (2014).

5.5 CONCLUSIONS AND FUTURE RESEARCH TOPICS

This chapter has reviewed the fundamentals of propagation effects relevant to GNSS, including state-of-the-art modeling and characterization for ionospheric, tropospheric, and environmental effects (shadowing and multipath). In general, the physical characterization of the effects is common to navigation, communications, or remote sensing; however, the specific parameters of interest in navigation are mainly the code and phase observable and therefore, channel models need to consider adequate representation suitable to represent the effects on those observables. In the future, the main challenges will be better accuracy in order to be able to properly discriminate propagation effects from antenna effects into a multiconstellation multifrequency GNSS environment, with different signal waveforms and modulations and different receiver architectures.

For ionospheric effects, a better understanding of small-scale plasma structures and strong scintillations will deserve further attention. For tropospheric effects, better mutual assimilation of tropospheric delay obtained from analysis of GNSS observations and NWP models, in order to obtain simultaneously high spatial and temporal resolution, and possibly slant estimation of tropospheric delay, are expected to be some of the future research areas.

The practical implementation of complex channel models into *radio frequency constellation simulator* (RFCS) allowing the performance evaluation of hardware receivers in a realistic laboratory setting is an important future challenge (for example, in terms of number of scatterers, diffuse energy, or update rate of the channel in the simulator).

An alternative approach often used for user receiver assessment is to use *record-and-replay* instrumentation that allows recording the channel in intermediate frequency and replaying it into any receiver configuration in a laboratory environment. The main limitation of such an approach is the fact that the antenna and RF front-end used for recording will have a significant influence in the assessment since the replay will be done at RF directly at the antenna port. In the future, antenna array solutions may be able to discriminate the angle of arrivals and techniques may be developed to radiate signals in the laboratory able to accurately reproduce the spatial distribution of the signals.

REFERENCES

Aarons, J., The longitudinal morphology of equatorial F-layer irregularities relevant to their occurrence, *Space Science Reviews*, 63(3–4), 209–243, 1993, doi:10.1007/BF00750769.

Ait-Ighil, M., *Enhanced Physical–Statistical Simulator of the Land Mobile Satellite Channel for Multipath Modelling Applied to Satellite Navigation Systems*, PhD Dissertation, University of Toulouse, Toulouse, France, 2013.

Arndt, D., T. Heyn, J. Konig, A. Ihlow, A. Heuberger, R. Prieto-Cerdeira, and E. Eberlein, Extended two-state narrowband LMS propagation model for S-band, in *2012 IEEE International Symposium on Broadband Multimedia Systems and Broadcasting (BMSB)*, Seoul, 1–6, 2012, doi:10.1109/BMSB.2012.6264301.

Barrick, D., Rough surface scattering based on the specular point theory, *IEEE Transactions on Antennas and Propagation*, 16(4), 449–454, 1968, doi:10.1109/TAP.1968.1139220.

Bassiri, S. and G.A. Hajj, Higher-order ionospheric effects on the GPS observables and means of modeling them, *NASA STI/Recon Technical Report A*, 95, 81411, 1993.

Béniguel, Y. and P. Hamel, A global ionosphere scintillation propagation model for equatorial regions, *Journal of Space Weather and Space Climate*, 1(1), A04, 2011, doi:10.1051/swsc/2011004.

Boehm, J., B. Werl, and H. Schuh, Troposphere mapping functions for GPS and very long baseline interferometry from European Centre for Medium-Range Weather Forecasts operational analysis data, *Journal of Geophysical Research: Solid Earth*, 111(B2), B02406, 2006, doi:10.1029/2005JB003629.

Cervera, M.A., R.M. Thomas, K.M. Groves, A.G. Ramli, and Effendy, Validation of WBMOD in the Southeast Asian region, *Radio Science*, 36(6), 1559–1572, 2001, doi:10.1029/2000RS002520.

Conker, R.S., M. Bakry El-Arini, C.J. Hegarty, and T. Hsiao, Modeling the effects of ionospheric scintillation on GPS/satellite-based augmentation system availability, *Radio Science*, 38(1), 1001, 2003, doi:10.1029/2000RS002604.

Davies, K., *Ionospheric Radio*, IET, London: Peter Peregrinus, 1990.

de Jong, Y.L.C. and M.H.A.J. Herben, A tree-scattering model for improved propagation prediction in urban microcells, *IEEE Transactions on Vehicular Technology*, 53(2), 503–513, 2004, doi:10.1109/TVT.2004.823493.

Deutsches Zentrum für Luft- und Raumfahrt e.V. (DLR), Space Weather Application Center—Ionosphere (SWACI), SWACI, 2015, http://swaciweb.dlr.de/, Accessed August 31, 2015.

Doherty, P.H., T. Dehel, J.A. Klobuchar, S.H. Delay, S. Datta-Barua, E.R. de Paula, and F.S. Rodrigues, Ionospheric effects on low-latitude space based augmentation systems, in *Proceedings of the 15th International Technical Meeting of the Satellite Division of the Institute of Navigation (ION GPS 2002)*, Portland, OR, pp. 1321–1329, 2002, http://www.ion.org/search/view_abstract.cfm?jp=p&idno=2141.

Doherty, P.H., S.H. Delay, C.E. Valladares, and J.A. Klobuchar, Ionospheric scintillation effects on GPS in the equatorial and auroral regions, *Navigation*, 50(4), 235–245, 2003, doi:10.1002/j.2161-4296.2003.tb00332.x.

European Space Agency (ESA), ESA's MONITOR Project (Ionospheric Monitoring Network), *MONITOR*, 2015, http://monitor.estec.esa.int/, Accessed August 31, 2015.

European Union, Ionospheric Correction Algorithm for Galileo Single Frequency Users, no. 1.1 (June), 2015.

Fontan, F.P., A. Mayo, D. Marote, R. Prieto-Cerdeira, P. Mariño, F. Machado, and N. Riera, Review of generative models for the narrowband land mobile satellite propagation channel, *International Journal of Satellite Communications and Networking*, 26(4), 291–316, 2008, doi:10.1002/sat.914.

Georgiadou, Y. and A. Kleusberg, On carrier signal multipath effects in relative GPS positioning, *Manuscripta Geodaetica*, 13(3), 172–179, 1988.

Hatch, R., The Synergism of GPS Code and Carrier Measurements. pp. 1213–1231, 1983, http://adsabs.harvard.edu/abs/1983igss.conf.1213H.

Hernández-Pajares, M., À. Aragón-Ángel, P. Defraigne, N. Bergeot, R. Prieto-Cerdeira, and A. García-Rigo, Distribution and mitigation of higher-order ionospheric effects on precise GNSS processing, *Journal of Geophysical Research: Solid Earth*, 119(4), 3823–3837, 2014, doi:10.1002/2013JB010568.

Hernández-Pajares, M., J.M. Juan, and J. Sanz, New approaches in global ionospheric determination using ground GPS data, *Journal of Atmospheric and Solar-Terrestrial Physics*, 61(16), 1237–1247, 1999, doi:10.1016/S1364-6826(99)00054-1.

Hernández-Pajares, M., J.M. Juan, J. Sanz, R. Orus, A. García-Rigo, J. Feltens, A. Komjathy, S.C. Schaer, and A. Krankowski, The IGS VTEC maps: A reliable source of ionospheric information since 1998, *Journal of Geodesy*, 83(3–4), 263–275, 2009, doi:10.1007/s00190-008-0266-1.

Hernández-Pajares, M., R. Prieto-Cerdeira, Y. Béniguel, A. García-Rigo, J. Kinrade, K. Kauristie, R. Orús-Pérez et al., MONITOR ionospheric monitoring system: Analysis of perturbed days affecting SBAS performance, in *Proceedings of the ION 2015 Pacific PNT Meeting*, pp. 970–978, Honolulu, Hawaii, 2015.

Hochegger, G., B. Nava, S. Radicella, and R. Leitinger, A family of ionospheric models for different uses, physics and chemistry of the Earth, Part C: Solar, *Terrestrial & Planetary Science*, 25(4), 307–310, 2000, doi:10.1016/S1464-1917(00)00022-2.

Humphreys, T.E., M.L. Psiaki, J.C. Hinks, B. O'Hanlon, and P.M. Kintner, Simulating ionosphere-induced scintillation for testing GPS receiver phase tracking loops, *IEEE Journal of Selected Topics in Signal Processing*, 3(4), 707–715, 2009, doi:10.1109/JSTSP.2009.2024130.

ITU-R, Model Parameters for an Urban Environment for the Physical–Statistical Wideband LMSS Model in Recommendation ITU-R P.681-6, Report P.2145. ITU, 2009.

ITU-R, Propagation Data Required for the Design of Earth–Space Aeronautical Mobile Telecommunication Systems, Recommendation ITU-R P. 682-3. P Series Radiowave Propagation. ITU, 2012.

ITU-R, Ionospheric Propagation Data and Prediction Methods Required for the Design of Satellite Services and Systems, Recommendation ITU-R P. 531-12. P Series Radiowave Propagation. ITU, 2013.

ITU-R, Propagation Data Required for the Design of Earth–Space Land Mobile Telecommunication Systems, Recommendation ITU-R P. 681-8. P Series Radiowave Propagation. ITU, 2015.

Jost, T., W. Wang, U.-C. Fiebig, and F. Pérez-Fontán, A wideband satellite-to-indoor channel model for navigation applications, *IEEE Transactions on Antennas and Propagation*, 62(10), 5307–5320, 2014, doi:10.1109/TAP.2014.2344097.

Kaplan, E.D. and C.J. Hegarty, *Understanding GPS: Principles and Applications*, Boston/London: Artech House, 2006.

Karasawa, Y. and T. Shiokawa, A simple prediction method for L-band multipath fading in rough sea conditions, *IEEE Transactions on Communications*, 36(10), 1098–1104, 1988, doi:10.1109/26.7526.

Kim, T., R.S. Conker, S.D. Ericson, C.J. Hegarty, M. Tran, and M.B. El-Arini, Preliminary evaluation of the effects of scintillation on L5 GPS and SBAS receivers using a frequency domain scintillation model and simulated and analytical receiver models, in *Proceedings of the 2003 National Technical Meeting of the Institute of Navigation*, pp. 833–847, Anaheim, CA, 2003, http://www.ion.org/search/view_abstract.cfm?jp=p&idno=3830.

Klobuchar, J.A., Ionospheric time-delay algorithm for single-frequency GPS users, *IEEE Transactions on Aerospace and Electronic Systems*, AES-23(3), 325–331, 1987, doi:10.1109/TAES.1987.310829.

Klobuchar, J.A., P.H. Doherty, M. Bakry El-Arini, R. Lejeune, T. Dehel, E.R. de Paula, and F.S. Rodrigues, Ionospheric issues for a SBAS in the equatorial region, in *Proceedings of the 10th International Ionospheric Effects Symposium*, Alexandria, VA, pp. 159–166, 2002.

Knight, M. and A. Finn, The effects of ionospheric scintillations on GPS, in *Proceedings of the 11th International Technical Meeting of the Satellite Division of the Institute of Navigation (ION GPS 1998)*, Nashville, TN, pp. 673–685, 1998, http://ww.ion.org/search/view_abstract.cfm?jp=p&idno=2999.

Lavenant, M.P., W.G. Cowley, J. McCarthy, C. Burnet, W. Zhang, S. Morris, and S. Cook, Digital communications for the satellite to helicopter channel, in *The Eleventh National Space Engineering Symposium: Proceedings*, Sydney, New South Wales, p. 83, 1997.

Lehner, A. and A. Steingass, A novel channel model for land mobile satellite navigation, in *Proceedings of the 18th International Technical Meeting of the Satellite Division of the Institute of Navigation (ION GNSS 2005)*, pp. 2132–2138, Long Beach, CA, 2005.

Lemorton, J., J. Israel, M. Ait-Ighil, G. Carrie, C. Oestges, T. Jost, and F. Pérez-Fontán, Physical–statistical models for mobile satellite communication systems below 10 GHz (PHYSTAT), Final Report, C4000104006/11/NL/ad, ESA, 2013.

Loo, C., A statistical model for a land mobile satellite link, *IEEE Transactions on Vehicular Technology*, 34(3), 122–127, 1985, doi:10.1109/T-VT.1985.24048.

Martellucci, A. and R. Prieto-Cerdeira, Review of tropospheric, ionospheric and multipath data and models for global navigation satellite systems, in *3rd European Conference on Antennas and Propagation, 2009. EuCAP 2009*, Berlin, pp. 3697–3702, 2009.

Mitchell, C.N. and P.S.J. Spencer, A three-dimensional time-dependent algorithm for ionospheric imaging using GPS, *Annals of Geophysics*, 46(4), 687–696, 2003, doi:10.4401/ag-4373.

Montenbruck, O., A. Hauschild, and P. Steigenberger, Differential code bias estimation using multi-GNSS observations and Global Ionosphere Maps. *Navigation*, 61(3), 191–201, 2014, doi:10.1002/navi.64.

Nakagami, M., The M-distribution—A general formula of intensity distribution of rapid fading, in W.C. Hoffman, *Statistical Methods in Radio Wave Propagation*, London: Pergamon Press Ltd., pp. 3–36, 1960, http://www.sciencedirect.com/science/article/pii/B9780080093062500054.

Niell, A.E., Global mapping functions for the atmosphere delay at radio wavelengths, *Journal of Geophysical Research*, 101(February), 3227–3246, 1996, doi:10.1029/95JB03048.

Oestges, C., N. Khan, F. Mani, and R. Prieto-Cerdeira, A geometry-based physical–statistical model of land mobile satellite channels in urban environments, in *2014 8th European Conference on Antennas and Propagation (EuCAP)*, The Hague, pp. 2261–2263, 2014, doi:10.1109/EuCAP.2014.6902264.

Parkinson, B.W., T. Stansell, R. Beard, and K. Gromov, A history of satellite navigation, *Navigation*, 42(1), 109–164, 1995, doi:10.1002/j.2161-4296.1995.tb02333.x.

Pérez-Fontán, F., V. Hovinen, M. Schönhuber, R. Prieto-Cerdeira, F. Teschl, J. Kyrolainen, and P. Valtr, A wideband, directional model for the satellite-to-indoor propagation channel at S-band, *International Journal of Satellite Communications and Networking*, 29(1), 23–45, 2011, doi:10.1002/sat.955.

Pérez-Fontán, F., M. Vázquez-Castro, C.E. Cabado, J.P. García, and E. Kubista, Statistical modeling of the LMS channel, *IEEE Transactions on Vehicular Technology*, 50(6), 1549–1567, 2001.

Petrie, E.J., M.A. King, P. Moore, and D.A. Lavallée, Higher-order ionospheric effects on the GPS reference frame and velocities, *Journal of Geophysical Research: Solid Earth*, 115(B3), B03417, 2010, doi:10.1029/2009JB006677.

Prieto-Cerdeira, R., R. Orús-Pérez, E. Breeuwer, R. Lucas-Rodríguez, and M. Falcone, Performance of the Galileo single-frequency ionospheric correction during in-orbit validation, *GPS World*, 25(6), 53–58, June 2014.

Prieto-Cerdeira, R., F. Pérez-Fontán, P. Burzigotti, A. Bolea-Alamañac, and I. Sanchez-Lago, Versatile two-state land mobile satellite channel model with first application to DVB-SH analysis, *International Journal of Satellite Communications and Networking*, 28(5–6), 291–315, 2010, doi:10.1002/sat.964.

Prikryl, P., P.T. Jayachandran, S.C. Mushini, D. Pokhotelov, J.W. MacDougall, E. Donovan, E. Spanswick, and J.-P. St.-Maurice, GPS TEC, scintillation and cycle slips observed at high latitudes during solar minimum, *Annals of Geophysics*, 28(6), 1307–1316, 2010, doi:10.5194/angeo-28-1307-2010.

Rino, C.L., A power law phase screen model for ionospheric scintillation: 1. Weak scatter, *Radio Science*, 14(6), 1135–1145, 1979, doi:10.1029/RS014i006p01135.

Rodrigues, F.S., M.H.O. Aquino, A. Dodson, T. Moore, and S. Waugh, Statistical analysis of GPS ionospheric scintillation and short-time TEC variations over Northern Europe, *Navigation*, 51(1), 59–75, 2004, doi:10.1002/j.2161-4296.2004.tb00341.x.

Rovira-Garcia, A., J.M. Juan, J. Sanz, and G. Gonzalez-Casado, A worldwide ionospheric model for fast precise point positioning, *IEEE Transactions on Geoscience and Remote Sensing*, 53(8), 4596–4604, 2015, doi:10.1109/TGRS.2015.2402598.

RTCA, Minimal operational performance standards for GPS/WAAS airborne equipment, Report P.2145. RTCA, 2006.

Saastamoinen, J., Atmospheric correction for the troposphere and stratosphere in radio ranging satellites, in S.W. Henriksen, Armandoncini, and B.H. Chovitz (eds.), *The Use of Artificial Satellites for Geodesy*, Washington DC: American Geophysical Union, pp. 247–251, 1972, http://onlinelibrary.wiley.com/doi/10.1029/GM015p0247/summary.

Sanz, J., J.M. Juan, G. González-Casado, R. Prieto-Cerdeira, S. Schlüter, R. Orús, Novel ionospheric activity indicator specifically tailored for GNSS users, *Proceedings of the 27th International Technical Meeting of The Satellite Division of the Institute of Navigation (ION GNSS+ 2014)*, Tampa, Florida, pp. 1173–1182, September 2014.

Sanz Subirana, J., J.M. Juan Zornoza, and M. Hernández-Pajares, *GNSS Data Processing Book, Vol. I: Fundamentals and Algorithms*, Vol. 1., 2 vols., TM-23. Noordwijk: European Space Agency, 2013.

SBAS Ionospheric Working Group, Ionospheric Research Issues for SBAS—A White Paper, 2003.

SBAS Ionospheric Working Group, Effect of Ionospheric Scintillations on GNSS—A White Paper, 2010.

Schaer, S.C., W. Gurtner, and J. Feltens, IONEX: The IONosphere Map Exchange Format Version 1, 1998, ftp://igscb.jpl.nasa.gov/igscb/data/format/ionex1.pdf.

Schubert, F.M., B.H. Fleury, R. Prieto-Cerdeira, A. Steingass, and A. Lehner, A rural channel model for satellite navigation applications, in *2012 6th European Conference on Antennas and Propagation (EUCAP)*, Prague, pp. 2431–2435, 2012, doi:10.1109/EuCAP.2012.6206522.

Secan, J.A., R.M. Bussey, E.J. Fremouw, and Sa. Basu, An improved model of equatorial scintillation, *Radio Science*, 30(3), 607–617, 1995, doi:10.1029/94RS03172.

Seo, J., T. Walter, and P. Enge, Correlation of GPS signal fades due to ionospheric scintillation for aviation applications, *Advances in Space Research*, GNSS Remote Sensing-2, 47(10), 1777–1788, 2011, doi:10.1016/j.asr.2010.07.014.

Sobel, D., A brief history of early navigation, *Johns Hopkins APL Technical Digest*, 19(1), 11–13, 1998.

Steingass, A., A. Lehner, F. Pérez-Fontán, E. Kubista, and B. Arbesser-Rastburg, Characterization of the aeronautical satellite navigation channel through high-resolution measurement and physical optics simulation, *International Journal of Satellite Communications and Networking*, 26(1), 1–30, 2008, doi:10.1002/sat.891.

Van Dierendonck, A., J. Pat Fenton, and T. Ford, Theory and performance of narrow correlator spacing in a GPS receiver, *Navigation*, 39(3), 265–283, 1992, doi:10.1002/j.2161-4296.1992.tb02276.x.

Walter, T., S. Datta-Barua, J. Blanch, and P. Enge, The effects of large ionospheric gradients on single frequency airborne smoothing filters for WAAS and LAAS, in *Proceedings of ION National Technical Meeting*, San Diego, CA, pp. 103–109, 2004.

Weber, E.J., J.A. Klobuchar, J. Buchau, H.C. Carlson, R.C. Livingston, O. de la Beaujardiere, M. McCready, J.G. Moore, and G.J. Bishop, Polar cap F layer patches: Structure and dynamics, *Journal of Geophysical Research: Space Physics*, 91(A11), 12121–12129, 1986, doi:10.1029/JA091iA11p12121.

Wheelon, A.D., *Electromagnetic Scintillation II. Weak Scattering*, Vol. 2, Cambridge: Cambridge University Press, 2003.

6 Tropospheric Attenuation Synthesizers

Charilaos Kourogiorgas and Athanasios D. Panagopoulos

CONTENTS

6.1   INTRODUCTION

In this chapter, models for tropospheric attenuation synthesizers are presented according to state-of-the-art research. First, the need and the motivation for developing such time series generators are briefly explained. Each individual attenuation factor and the total tropospheric attenuation is dealt with in each section of the chapter, with figures and equations.

Satellite and modern wireless terrestrial communications have started adopting frequencies above 10 GHz in order to provide high data rate services due to the high bandwidth that is available. They specifically employ the Ka band (20/30 GHz) for broadband satellite communication networks (Koudelka, 2011). More particularly, in satellite communications and services, Earth observation (EO) systems have started using a frequency close to 26 GHz for data downlink (Rosello et al., 2012; Toptsidis et al., 2012). Moreover, in deep space missions, scientific data will be delivered to ground using a carrier frequency at Ka band due to the high data volume that can be sent (EUCLID, 2014). Apart from scientific and observation missions, communications realized through satellites have started using Ka band, such as ASTRA. The plan of high-throughput satellites (HTS) and multibeam communications intend to use Ka and Q/V bands. In these high-frequency bands, atmospheric precipitation (mainly rain), clouds, water vapor, and oxygen molecules cause the attenuation of the signal power, while turbulence causes the scintillation of signal amplitude.

Owing to the very high values of attenuation for a small time percentage, static fade margins are not the most efficient and cost-effective technique for attenuation compensation and the reach of a certain availability or capacity threshold. Therefore, fade mitigation techniques have been developed, such as site diversity, time diversity, adaptive coding and modulation, power control, and data rate adjustment (Panagopoulos et al., 2004) for the reduction in cost of communications. These techniques require, apart from the knowledge of first-order statistics, that is, probability density functions, the dynamics of rain attenuation in order to be able to proceed to end-to-end simulations and evaluation methods of a system's performance.

Therefore, accurate time series synthesizers (generators) and channel models are of great significance for the satellite system design at high frequencies (Ku band and above). Moreover, time series synthesizers can be employed in the usual case that there are no measurements for various link characteristics and for all the regions across Earth. Since measurements of attenuation need at least a year (a very optimistic assumption in terms of time consumption estimation) and these can be obtained for a certain elevation angle, frequency, and ground position, synthesizers can be used (of course with precaution) for generating attenuation time series. Such synthesizers must be able to generate time series of attenuation that reproduce first-order statistics of attenuation and then the dynamics of attenuation. Here, it must be noted that the term "dynamics" in the literature has been translated in many ways. Dynamics can be considered the second-order statistics of attenuation, that is, temporal correlation, or the fade slope or fade duration statistics.

The propagation community has started drawing attention to the development of synthesizers for attenuation. In Ku and Ka bands, rain attenuation plays the most significant role and therefore, the number of synthesizers for generating rain attenuation time series is higher than other attenuation factors. However, as the Q/V bands are planned to be used in future satellite communication networks, time series generators are needed for other attenuation factors, since their effect on signal power and amplitude is increasing, in order to optimize the system's performance. It must be noted that very recently, the radio propagation sector of International Telecommunication Union (ITU-R.) has published the recommendation P. 1853-1 in which synthesizers of attenuation factors and total attenuation are presented.

In this chapter, tropospheric attenuation time series synthesizers found in the literature are presented and briefly explained. In Section 6.2, the most significant rain attenuation time series synthesizers are given, while Section 6.3 deals with integrated liquid water content (ILWC) and Section 6.4 deals with integrated water vapor content (IWVC) from which clouds and water vapor attenuation can be derived. In Section 6.5, scintillation time series generators are presented, and in Section 6.6, two methods for the total tropospheric impairments time series are presented.

At this point, it must be noted that the methodologies for generating time series of tropospheric attenuation based on space–time fields are excluded from this chapter since these are described analytically in Chapter 7 of this book.

6.2 RAIN ATTENUATION SYNTHESIZERS

Rain attenuation has the greatest impact for satellite links at Ku, Ka, and Q/V bands. Therefore, the number of synthesizers developed for the generation of time series of rain attenuation are numerous. A simple classification is the following:

1. Filter-based models
2. Models based on stochastic differential equations (SDEs)
3. Models based on conditional probabilities
4. Models based on Markov chains
5. Data-based models

Every category is briefly explained in the following subsections.

6.2.1 FILTER-BASED MODELS

In this category of models, the synthesizer depends on the derivation of a filter from which uncorrelated Gaussian noise is passing in order to create temporal correlated Gaussian samples which are then usually transformed to lognormal variables (since lognormal distribution is a well-accepted distribution with a very good behavior globally.

6.2.1.1 Maseng–Bakken Model

A well-accepted model of this category is the Maseng–Bakken (M–B) model (Maseng and Bakken, 1981). Although, in the M–B model, a SDE has been mainly introduced (which is explained later), the stochastic process of rain attenuation is being generated using a low-pass filter. The main assumptions of the model are

1. Rain attenuation follows the lognormal distribution.
2. The rate of change of rain attenuation is proportional to the instantaneous value of rain attenuation.
3. Rain attenuation can be described as a first-order Markov process.

In Figure 6.1, a block diagram for the generation of rain attenuation time series according to the M–B model is given. The input in the first block (n_t) are independent samples of a zero mean and unity variance Gaussian stochastic process. The first block is a low-pass filter with a cutoff frequency equal to the dynamic parameter of rain attenuation (β_A). The output of the filter is correlated samples of a Gaussian stochastic process, which then pass through the nonlinear transformation

$$A_t = A_m \exp(S_A X_t) \tag{6.1}$$

where A_m and S_A are the statistical parameters of the lognormal distribution of rain attenuation and then the rain attenuation time series are obtained. In the aforementioned model, it is considered that rain attenuation follows the unconditional exceedance probability, that is, the probability to rain on a specific point of the slant path is equal to one and therefore, the probability that rain attenuation exceeds the zero dB value is also equal to one. Figure 6.2 presents a snapshot of the time series of rain attenuation considering a ground station in Athens, with a frequency of 30 GHz and an elevation angle of 30°.

From the M–B model, the autocorrelation of the output of the low-pass filter is

$$\rho_X(t) = e^{-\beta_A t} \tag{6.2}$$

and its spectrum is

$$S_X(f) = \frac{1}{1 + (2\pi f/\beta_A)^2} \tag{6.3}$$

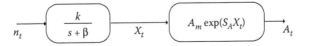

FIGURE 6.1   Block diagram of the M–B model.

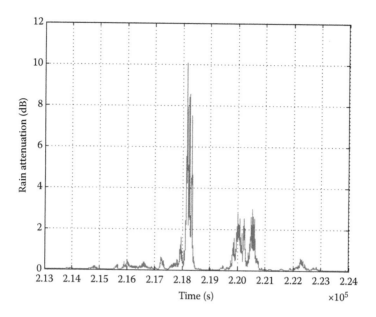

FIGURE 6.2   A snapshot of rain attenuation time series.

Lacoste et al. (2005) made the first attempt to consider the intermittency of rain attenuation by introducing an offset parameter. This offset parameter takes into account the exceedance probability of the zero-dB threshold for rain attenuation. The model relies on the M–B assumptions and uses the M–B model. Figure 6.3 presents the block diagram of the enhanced M–B (EMB) model.

The difference with the M–B model is that rain attenuation is obtained through the expression

$$A_t = \max[A_{1,t} - A_{offset}, 0] \tag{6.4}$$

with

$$A_{offset} = A_m \exp\left(S_A Q^{-1}\left[\frac{P_0}{100} \right] \right) \tag{6.5}$$

where $Q(\cdot)$ is the complementary cumulative distribution function (CCDF) of a zero mean unity variance Gaussian variable (Gaussian Q-function) and P_0 the probability to rain on a point and can be computed through the recommendation of ITU-R. P. 837 (ITU-R. P. 837-6, 2012). In Figure 6.4, a snapshot of rain attenuation is given using the EMB model. The above methodology has also been adopted in the first version of ITU-R. P. 1853-1 (2012).

However, Boulanger et al. (2013) explained that the synthesizer in Lacoste et al. (2005) does not give as an output a process following a lognormal distribution for the rain attenuation, given that it is raining somewhere on the slant path. Therefore, a new synthesizer is proposed as a result of a mixed law between a Dirac and lognormal distribution. Moreover, an algorithm is presented for

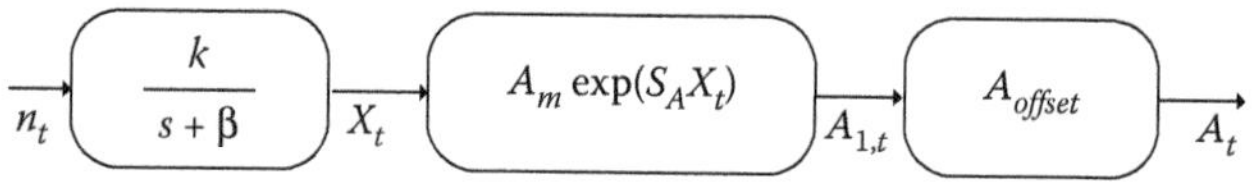

$$n_t \rightarrow \boxed{\frac{k}{s + \beta}} \xrightarrow{X_t} \boxed{A_m \exp(S_A X_t)} \xrightarrow{A_{1,t}} \boxed{A_{offset}} \xrightarrow{A_t}$$

FIGURE 6.3   Block diagram of the enhanced M–B model.

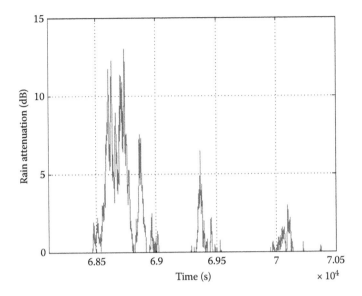

FIGURE 6.4   A snapshot of rain attenuation time series with EMB.

generating the rain attenuation time series for different correlation coefficient expressions and to not be restricted to the one of Equation 6.2.

The step-by-step algorithm of the synthesizer is the following:

1. Generate zero mean unity variance Gaussian random variables ($G(t)$) with arbitrary temporal correlation on Fourier domain (see details in Boulanger et al., 2013)
2. Calculate:

$$G_0 = \sqrt{2}\,\text{erfc}^{-1}\left(2\left(\frac{P_0}{100}\right)\right) \tag{6.6}$$

3. Calculate rain attenuation time series through the expression:

$$A(t) = \exp\left(\text{erfc}^{-1}\left(\text{erfc}\left(\frac{G(t)}{\sqrt{2}}\right)/(P_0/100)\right)\cdot\sqrt{2}\cdot\sigma + m\right) \quad \text{for} \quad G(t) > G_0 \tag{6.7}$$

$$A(t) = 0 \quad \text{for} \quad G(t) \leq G_0$$

6.2.2   Models Based on SDEs

The SDEs have been used in the propagation framework for the generation of rain attenuation. First, the M–B model introduced a first-order SDE for the description of rain attenuation (Maseng and Bakken, 1981). The SDE was of the form

$$da(t) = K_1(a(t))\cdot dt + \sqrt{K_2(a(t))}\,dW(t) \tag{6.8}$$

where K_1 is the drift coefficient, K_2 the diffusion coefficient, and $dW(t)$ the Brownian increments. Maseng and Bakken (1981) assumed that the rate of change of rain attenuation is proportional to the instantaneous value of rain attenuation. Therefore, by definition of the diffusion coefficient (Karlin and Taylor, 1981; Karatzas and Shreve, 2005), it is assumed that $K_2 \sim A_t^2$. So, given that

- The static probability density function (PDF) of a stochastic process described by Equation 6.8, is given by

$$p_{st}(a) = \frac{C}{K_2(a)} \exp\left\{2\int \frac{K_1(y)}{K_2(y)} dy\right\} \qquad (6.9)$$

C is the normalization constant
- Rain attenuation is assumed to follow lognormal distribution

The SDE that describes rain attenuation is

$$dA_t = \beta_A A_t \left[S_a^2 - \ln\left(\frac{A_t}{a_m}\right) \right] dt + \sqrt{2\beta_A} \, S_a A_t dW_t \qquad (6.10)$$

Now, using the following transformation:

$$X_t = M[A_t] = \ln\frac{(A_t/A_m)}{S_a} \qquad (6.11)$$

The SDE that describes the process X_t is

$$dX_t = -\beta_A X_t dt + \sqrt{2\beta_A} \, dW_t \qquad (6.12)$$

The analytical solution of Equation 6.12 is given in Kanellopoulos et al. (2007):

$$X_t = e^{\beta_A t}\left(X_0 + \sqrt{2\beta_A}\int_0^t e^{-\beta_A s} dW_s \right) \qquad (6.13)$$

where X_0 is the initial value of the process X_t. The process X_t is the Ornstein–Uhlenbeck process, which is a Gaussian zero mean unity variance stochastic process with an autocorrelation function given by Equation 6.2. Therefore, for the generation of rain attenuation using the above model, the following step-by-step algorithm can be used:

1. Generate time series of X_t process through Equation 6.13.
2. Calculate rain attenuation through the following transformation:

$$A_t = A_m \exp(S_A X_t) \qquad (6.14)$$

Apart from the lognormal distribution, the Weibull and gamma distributions have been identified for the description of rain attenuation. Kanellopoulos et al. (2014) proposed a synthesizer based on SDEs for gamma-distributed rain attenuation based on the assumption that $K_2 \sim A_t^2$, as in the M–B model. Given that the PDF of gamma distribution is

$$p_A(a) = \frac{w_a^{v_a}}{(\Gamma v_a)} \cdot a^{v_a - 1} \cdot \exp(-w_a \cdot a) \qquad (6.15)$$

where w_a, v_a are the two positive parameters of the gamma distribution, which can be calculated by regression fitting on experimental data or prediction models data and $\Gamma(.)$ is the gamma function. The proposed SDE is

$$dA_t = \beta_{AG} \cdot A_t \cdot \left(\frac{v_a}{w_a} + \frac{1}{w_a} - A_t \right) \cdot dt + \sqrt{(2/w_a) \cdot \beta_{AG} \cdot A_t^2} \, dW(t) \tag{6.16}$$

where β_{AG} is the dynamic parameter of rain attenuation. The solution of Equation 6.16 is

$$a(t) = \frac{a_0 \cdot e^{\sqrt{2\beta_{AG}/w_a} \cdot W(t) + \beta_{AG} \cdot (v_a/w_a) \cdot t}}{\left(1 + \beta_{AG} \cdot a_0 \int_0^t e^{\sqrt{2\beta_{AG}/w_a} \cdot W(s) + \beta_{AG} \cdot (v_a/w_a) \cdot s} \, ds \right)} \tag{6.17}$$

where a_0 is the initial value of rain attenuation and W the Brownian motion.

A block diagram for generating the rain attenuation time series for gamma-distributed rain attenuation is given in Figure 6.5. The first block is required in order to obtain the two statistical parameters of rain attenuation.

Kanellopoulos et al. (2013) proposed an SDE for Weibull-distributed rain attenuation. Similarly to Maseng and Bakken (1981) and Kanellopoulos et al. (2014) models, here it is assumed that $K_2 \sim A_t^2$. The PDF of a Weibull distribution is

$$p_X(x) = \frac{v}{w} x^{v-1} e^{-x^v/w}, \, x > 0 \tag{6.18}$$

where v and w are the two parameters of Weibull distribution that must be greater than 0. The proposed SDE is of the form of Equation 6.8 with drift coefficient of the proposed synthesizer:

$$K_1(A(t)) = \beta_A A(t) \left(w + \frac{w}{v} - A^v(t) \right) \tag{6.19}$$

where β_A is the dynamic parameter of rain attenuation. The diffusion coefficient is

$$K_2(A(t)) = \frac{2\beta_A w}{v} A^2(t) \tag{6.20}$$

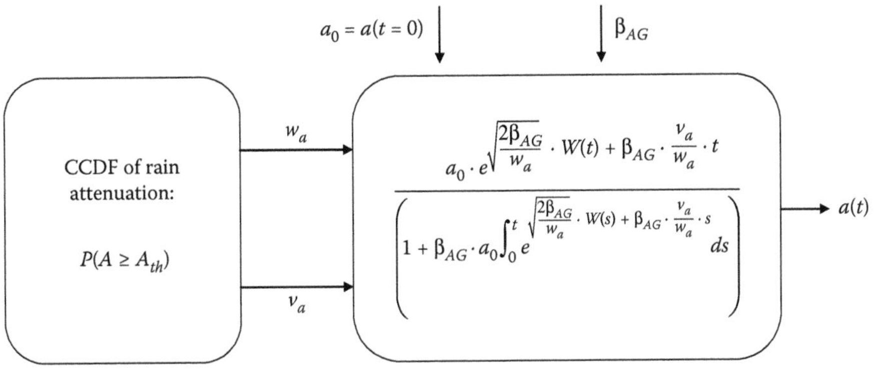

FIGURE 6.5 Block diagram of rain attenuation time series synthesizer for gamma-distributed rain attenuation.

The rain attenuation time series are then obtained through the solution of the SDE:

$$A(t) = \frac{A_0 \cdot e^{\sqrt{2\beta_A w/v} \cdot B(t) + \beta_A \cdot w \cdot t}}{\left(1 + v\beta_A \cdot A_0^v \int_0^t e^{v\sqrt{2\beta_A w/v} \cdot B(s) + v \cdot \beta_A \cdot w \cdot s} ds\right)^{1/v}} \tag{6.21}$$

Apart from the previous models, the model developed in Manning (1990) is based on SDEs, considering a second-order SDE for the description of rain attenuation. More particularly, it is considered that the rain rate is described by the following SDE:

$$dR_t = \beta_R R_t \left[S_R^2 - \ln\left(\frac{R_t}{R_m}\right) \right] dt + \sqrt{2\beta_R} S_R R_t dW_t \tag{6.22}$$

where β_R is the dynamic parameter of the rain rate, R_m the median value of rain rate, and S_R the standard deviation of $\ln R$. The SDE of Equation 6.22 is the same as the M–B for rain attenuation. The assumptions of the SDE of Equation 6.22 for the rain rate have also been tested in Burgueno et al. (1990) and Kourogiorgas et al. (2013) showing that the rain rate fulfills the same assumptions. Therefore, from Equation 6.22, the SDE that describes the underlined Gaussian process $x_R = (\ln R - \ln R_m)/S_R$ is

$$dx_{R,t} = -\beta_R X_{R,t} dt + \sqrt{2\beta_R} dW_t \tag{6.23}$$

According to Manning (1990), the underlined Gaussian process of the specific rain attenuation (Γ) is

$$dx_{\Gamma,t} = -\beta_\Gamma X_{\Gamma,t} dt + \sqrt{2\beta_\Gamma} dW_t \tag{6.24}$$

Given a very short path, the underlined Gaussian process also follows the expressions of Equations 6.23 and 6.24:

$$dX_{A,t} = -\beta_A X_{A,t} dt + \sqrt{2\beta_A} dW_t \tag{6.25}$$

However, as stated in Manning (1990), in the case that the slant path increases, the above expression (6.25) cannot be used, first, because the rain attenuation is not any more proportional to the specific rain attenuation, and also due to the increased path through rain, the process dW_t is smoothed. Therefore, a smoothed process ($\zeta(t)$) is considered:

$$\zeta(t) = \frac{1}{T_S} \int_{-\infty}^t \exp\left(-\frac{t'-t}{T_S}\right) \xi'(t') dt' \tag{6.26}$$

where $\xi'(t')$ is white Gaussian noise. So, Equation 6.26 is described through

$$d\zeta_t = -\gamma_S \zeta_t dt + \gamma_S dW_t \tag{6.27}$$

with $\gamma_S = (T_S)^{-1}$. The smoothed process is now used for driving rain attenuation dynamics, that is, the SDE, which describes the underlined Gaussian process for rain attenuation is

$$dX_t = -\beta X_t dt + \sqrt{2\beta}d\zeta_t \tag{6.28}$$

Using Equations 6.27 and 6.28, the following second-order SDE for the description of rain attenuation is used:

$$\frac{d^2X_A}{dt^2} = -\gamma\left(1 + \frac{\gamma_s}{\gamma}\right)\frac{dX_A}{dt} - \gamma\gamma_s X_A + \gamma_s\sqrt{2\gamma}\xi(t) \tag{6.29}$$

with $\xi(t)$ zero mean unity variance Gaussian noise.

Apart from the geostationary Earth orbit (GEO) slant paths, in Kourogiorgas and Panagopoulos (2015), a rain attenuation synthesizer based on the M–B model is proposed for low Earth orbit (LEO) and medium Earth orbit (MEO) slant paths. The characteristic of a LEO slant path is that the elevation angle changes over time and therefore the parameters of the SDE proposed in the M–B model change over time. The proposed synthesizer is an extension of the M–B model and the following SDE is proposed for the description of the reduced Gaussian process:

$$dX_t = -\beta_A(\theta)X_t dt + \sqrt{2\beta_A(\theta)}dW_t \tag{6.30}$$

where β_A is the dynamic parameter of rain attenuation that depends on the elevation angle of the link. The solution of Equation 6.30 is

$$X_t^{LEO} = e^{-\int_0^t \beta_A(\theta_s)ds}\left(X_0^{LEO} + \int_0^t \sqrt{2\beta_A(\theta_s)}\exp\left(\int_0^s \beta_A(\theta_{s'})ds'\right)dW_s\right) \tag{6.31}$$

and the rain attenuation time series can be obtained through

$$A_t^{LEO} = A_m(\theta_t)\exp[X_t^{LEO}S_A(\theta_t)] \tag{6.32}$$

In the same reference (Kourogiorgas and Panagopoulos, 2015), an expression for the dynamic parameter is proposed as a function of the dynamic parameter of the rain rate and link characteristics:

$$\beta_A(\theta_t, |\Delta t_A|) = -\frac{1}{2|\Delta t_A|}\ln\left\{1 - \frac{1}{S_A^2}\ln\left\{1 + \frac{H_1(\theta_t)}{L^2(\theta_t)}[\exp(b^2 S_R^2(1 - \exp(-2\beta_R(|\Delta t_R|)\cdot|\Delta t_R|))) - 1]\right\}\right\} \tag{6.33}$$

with

$$H_1(\theta_t) = 2L(\theta_t)g\sinh^{-1}\left(\frac{L(\theta_t)}{g}\right) + 2g^2\left[1 - \sqrt{\left(\frac{L(\theta_t)}{g}\right)^2 + 1}\right] \tag{6.34}$$

where g is a constant taking values from 0.75 km up to 3 km, L the projection of slant path on ground, b the coefficient at the power of the rain rate from ITU-R. P. 838-3 (2001), and β_R the dynamic parameter of the point rainfall rate.

A step-by-step algorithm has been presented in Kourogiorgas and Panagopoulos (2015) and repeated here

1. Calculate the time series of the elevation angles for the link between an Earth station and a LEO satellite. This can be achieved via the following two methods: first, either directly from the AGI's STK software (AGI STK) or from calculating the geographical coordinates of the subsatellite points and then calculating the elevation angle from the expressions given in Liu and Michelson (2012).
2. Afterward, for every time sample, compute the solution of Equation 6.30, using Equation 6.31 and by calculating the dynamic parameter for every elevation angle and for a given time resolution (6.33). Consequently, one can generate the time series of the random process X_t^{LEO}.
3. Using the values of the random process X_t^{LEO} and calculating the values $A_m(\theta_t)$ and $S_A(\theta_t)$, the time series of rain attenuation for a LEO slant path are generated.

The flow diagram of the algorithm is given in Figure 6.6.

The same synthesizer has been used in Kourogiorgas et al. (2014a) for MEO.

Moreover, the SDEs have been used for generating rain attenuation time series at links with mobile receivers (Arapoglou et al., 2012). More particularly, in Arapoglou et al. (2012), a rain attenuation synthesizer is developed and combined with the attenuation caused by power arches in order to have a channel model for railways. This model is based on the theoretical modeling of exceedance probability of rain attenuation for mobile receivers presented in Matricciani (1995). In the latter reference, it is found that the CCDF of rain attenuation for a mobile receiver ($P_M(A)$) is connected to that of a fixed receiver ($P_F(A)$) through

$$P_M(A) = \xi P_F(A) \tag{6.35}$$

where

$$\xi = \frac{\upsilon_R}{|\upsilon_M - \upsilon_R \cos\varphi|} \tag{6.36}$$

where υ_R is the wind speed, υ_M the speed of the mobile receiver, and φ the angle between the storm's velocity and the receiver's velocity.

Now, Arapoglou et al. (2012) assumed that rain attenuation induced in a mobile receiver follows lognormal distribution and that rain attenuation temporal variations are proportional to the instantaneous value of rain attenuation. Therefore, the M–B model can describe the rain attenuation time series. So, the following SDE describes the rain attenuation induced on a slant path with a mobile ground station:

$$dA_{M,t} = \beta_{A,M} A_{M,t} \left[S_{M,a}^2 - \ln\left(\frac{A_{M,t}}{a_{M,m}}\right) \right] dt + \sqrt{2\beta_{A,M}} S_{M,a} A_{M,t} dW_t \tag{6.37}$$

where $S_{M,a}$ and $a_{M,m}$ are the lognormal parameters of rain attenuation and the dynamic parameter of the mobile case is related to that of the fixed case through

$$\beta_M = \frac{1}{\xi} \beta_F \tag{6.38}$$

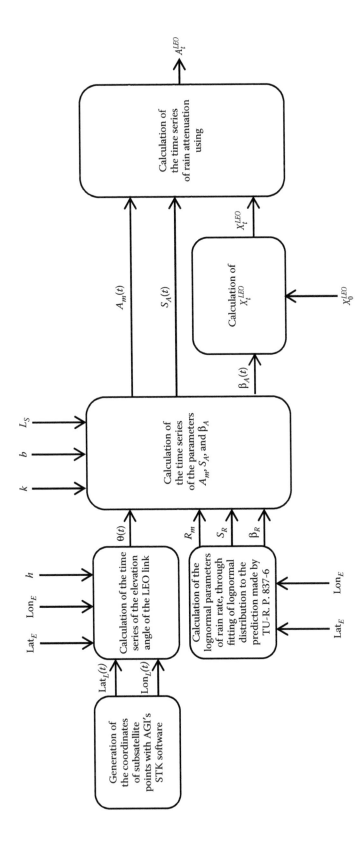

FIGURE 6.6 Block diagram for the generation of rain attenuation time series at LEO slant paths.

6.2.3 Synthetic Storm Technique

The synthetic storm technique (SST), given in Matricciani (1996), is a methodology for generating the rain attenuation time series given as input apart from the electrical characteristics, rain rate time series. In SST, the Taylor's hypothesis is used in order to translate the time series of rain rate to the spatial domain and calculate the rain attenuation along the path. Therefore, the SST depends on and uses as input parameter the mean wind speed of the location in which the time series are obtained. In Matricciani (1996), for the vertical structure of the rain rate, the two-layer model is taken into account (Matricciani, 1991). Matricciani (1996) presents the two-layer model in a graphical representation, also shown in Figure 6.7.

The final expression for the generation of the rain attenuation time series is given by the inverse Fourier transform of the following spectrum:

$$S_A(f) = S_{Y,A}(f)L_A \sin c[fL_A/v(\theta)]\exp[-j2\pi f_s \Delta x_0/v(\theta)]$$
$$+ r^{a_B} S_{Y,B}(f)\Delta L \sin c[f\Delta L/v(\theta)] \tag{6.39}$$

where

$$L_A = \frac{H_A - H_S}{\sin \theta} \tag{6.40}$$

is the slant path along Layer A and

$$\Delta L = L_B - L_A \tag{6.41}$$

is the slant path along Layer B with $L_B = (H_B - H_S)/\sin \theta$, H_S is the height of the ground station, and

$$H_B = H_A + 0.4 \tag{6.42}$$

which is the rain height considering the height of the melting layer.

6.2.4 Models Based on Data

Bertorelli et al. (2008) presented a methodology for generating the time series of rain attenuation based on rain attenuation events from large datasets with obtained measurements. Given that the rain attenuation events have been extracted by a given database for a ground station and specific link, rain attenuation is generated through the appropriate scaling for another link with different

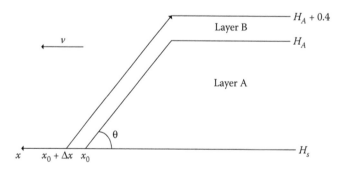

FIGURE 6.7 Vertical structure of rain rate in the two-layer model.

climatic and/or electrical, geometrical characteristics. The first version of the model has been developed with ITALSAT data in order to reproduce fade duration, fade slope, and the first-order statistics of rain attenuation.

The database used as input to the model needs to include only rain attenuation. Therefore, the effect of atmospheric gases must have been removed, while with rain gauges the true rain events can be derived. First, according to the model, the attenuation events of the databases are classified into 10 classes according to the attenuation peak. As noted in Bertorelli et al. (2008), it is important that each class contains a significant number of events. Then, for each class, the CCDF normalized to the total duration is calculated. The later are called "base functions."

After the calculation of "base functions," the unconditional CCDF of rain attenuation ("objective function") is calculated by prediction models from the literature. For the objective function, although the CCDF must take into account the climatology of the site in which time series are required through the CCDF of the rain rate, the elevation angle, frequency, polarization, and rain height must remain the same with these of the reference database. However, in case that for the site under investigation the rain attenuation CCDF is available but not the time series, then the measured CCDF is scaled to that of the reference site for which time series database exists through the following scaling factor:

$$F = \frac{H_{r1} - h_1}{H_{r2} - h_2} \frac{\sin(\theta_2)}{\sin(\theta_1)} \left(\frac{f_1}{f_2} \right)^{1.72} \tag{6.43}$$

where H_r is the rain height, h the altitude amsl of the station, θ elevation angle, and f the operating frequency of links 1 and 2.

Now, given that the objective function ($P(A)$) and the base functions ($P_j(A), j = 1, \ldots, 10$) are known, the weights (W_j) of each base function are needed such that

$$P(A) = \sum_{j=1}^{10} W_j P_j(A) \tag{6.44}$$

Multiplying the weights with the total observation period, the duration of each class is calculated. Then, from the classes, a subset of the rain events is chosen in a manner that the duration of these events is the same with the one calculated from the weights. Finally, the time series of rain attenuation is scaled using Equation 6.43.

6.2.5 Models Based on Conditional Probabilities

In this category of synthesizers, the conditional probabilities of rain attenuation are used. More particularly, Carrie et al. (2011) used the assumptions of the EMB model presented in Section 6.2.1.1 in order to synthesize rain events on-demand for the evaluation of fade mitigation techniques (FMTs). The on-demand characterization of the synthesizer occurs due to the main input parameters of the model, which are the max value of rain attenuation for the specific event, that is, A_{max}, the duration of the event (D), and the time instance when the maximum occurs, that is, D_{peak}. These parameters are needed in order to have the on-demand generation of rain attenuation for a rain event.

The methodology and framework of the "on-demand" synthesizer relies on the M–B assumptions and the EMB model. From these latter models, the conditional probability of rain attenuation given the values of rain attenuation at a past time instance, and a future time instance is needed. The latter conditional probability, assuming that rain attenuation is a first-order Markov process, is given by

$$p(A(t) \mid A(t - \Delta t_1), A(t + \Delta t_2)) = \frac{p(A(t + \Delta t_2) \mid A(t)) \times p(A(t) \mid A(t - \Delta t_1))}{p(A(t + \Delta t_2) \mid A(t - \Delta t_1))} \tag{6.45}$$

Based on the M–B model, the transitional probability density function of rain attenuation follows a lognormal distribution with a PDF

$$p(A(t+\Delta t)\,|\,A(t)) = \frac{1}{A(t+\Delta t)\sigma_{|A(t)}\sqrt{2\pi}}\exp\left(-\left[\frac{\ln\left(A(t+\Delta t)-m_{|A(t)}\right)}{\sigma_{|A(t)}\sqrt{2}}\right]^2\right) \quad (6.46)$$

with

$$m_{|A(t)}(\Delta(t)) = m(1-\exp(-\beta\,|\Delta t|) + \ln(A(t))\exp(-\beta\,|\Delta t|)$$

$$\sigma^2_{|A(t)}(\Delta(t)) = \sigma^2(1-\exp(-2\beta\,|\Delta t|)) \quad (6.47)$$

Since the TPDF of rain attenuation is a lognormal distribution, it follows that Equation 6.45 follows a lognormal distribution with parameters

$$m_{\Delta t_1,\Delta t_2} = \begin{cases} (1-\exp(-2\beta\Delta t_1))\times\exp(-2\beta\Delta t_2)\times\ln(A(t+\Delta t_2)) \\ +(1-\exp(-2\beta\Delta t_2))\times\exp(-2\beta\Delta t_1)\times\ln(A(t-\Delta t_1)) \\ +m\times(1-\exp(-2\beta\Delta t_1))\times(1-\exp(-2\beta\Delta t_2)) \\ \times(1-\exp(-2\beta(\Delta t_1+\Delta t_2))) \end{cases} / (1-\exp(-2\beta(\Delta t_1+\Delta t_2))) \quad (6.48)$$

$$\sigma^2_{\Delta t_1,\Delta t_2} = \sigma^2\times\frac{(1-\exp(-2\beta\Delta t_1))\times(1-\exp(-2\beta\Delta t_2))}{1-\exp(-2\beta(\Delta t_1+\Delta t_2))} \quad (6.49)$$

In order to take into account the offset parameter for the rain attenuation intermittency, the following expression is used:

$$P_{L-C}(A(t)\,|\,A(t-\Delta t_1),A(t+\Delta t_2))$$

$$= P(A(t)+A_{offset}\,|\,A(t-\Delta t_1)+A_{offset},A(t+\Delta t_2)+A_{offset}) \quad (6.50)$$

For the generation of rain attenuation events, the duration D of the rain event is needed, and the maximum attenuation (A_{max}) and the temporal position D_{peak} of the maximum attenuation are also needed. So, the algorithm starts with the initial values of attenuation $[0, A_{max}, 0]$:

- Add the A_{offset} to the initial values of attenuation.
- The interpolation starts from the middle and at Equations 6.48 and 6.49 uses the value of A_{offset} at time t_0 and $A_{max}+A_{offset}$ at time t_0+D_{peak}. If T_s is the sampling period of the desired time series, then in Equations 6.48 and 6.49, $\Delta t_1 = T_s$ and $\Delta t_2 = D_{peak}-T_s$.
- Generate zero mean and unity variance Gaussian noise.
- Transform exponentially using the lognormal parameters at Equations 6.48 and 6.49.

Another model based on conditional probabilities is the ONERA-Van de Kamp two-state model (Kamp, 2003). The model is based on the use of conditional probability of rain attenuation given the knowledge of two previous samples of rain attenuation. The PDF of rain attenuation, given the two previous samples, is given by

$$p(A \mid A_0, A_{-1}) = \frac{m_A}{2A\sigma_A} \sec h \left[\frac{\pi m_A \ln(A/m_A)}{2A\sigma_A} \right] \tag{6.51}$$

where m_A and σ_A are given by

$$m_A = A_0 \left(\frac{A_0}{A_{-1}} \right)^{a_2} \tag{6.52}$$

$$\sigma_A = A_0 \sqrt{\beta_2 \delta t} + A_0 \gamma_2 \left(1 - e^{-|\ln(A_0/A_{-1})|} \right)$$

The values of the above parameters have been extracted from measurements (Kamp, 2003).

The generation of rain attenuation time series occurs from the conditional PDF of rain attenuation and in Kamp (2003), it is recommended that the synthesizer be used for the generation of separate events. In case a specific maximum attenuation is needed, the time series are scaled with a multiplication factor such that the peak of the event is equal to the desired value.

6.2.6 Markov Chain Models

Markov chains have been used for the generation of rain attenuation time series. The first model presented here is the N-state Markov ONERA model (Castanet et al., 2003). This model is used to generate either event-on-demand or long time series. The method contains three parts. The first one is the "macroscopic model" and it is a Markov chain with two states of rain and no rain. As explained in Castanet et al. (2003), two probabilities are needed: the probability to rain (p_1) and the probability of transition from the state of rain to the state of no rain (p_{10}). The other probabilities of the transition matrix of the two-state Markov chain can be found through

$$p_{01} = \frac{p_1}{1 - p_1} p_{10} \tag{6.53}$$

where p_{01} is the probability from the state of no rain to the state of rain. The probability of no rain is equal to $1 - p_{01}$. The probability from rain state to no rain state can be obtained through fade duration models, as explained in Castanet et al. (2003).

The second part of the model is the microscopic model in which propagation events are generated. First, a maximum rain attenuation is assumed, that is, $A_{max} = X$, with X an integer and a minimum value of rain attenuation of 0 dB. The interval between minimum and maximum rain attenuation is discretized into the number of the states of the Markov chain:

$$N = \frac{A_{max} - A_{min}}{da} + 1 \tag{6.54}$$

where da is the distance in dBs between the two states of attenuation. So, the following transition matrix has to be defined and its elements to be calculated:

$$P = \begin{bmatrix} p_{11} & p_{12} & \cdots & & p_{1N} \\ p_{21} & \ddots & & & \vdots \\ \vdots & & p_{ij} & & \\ & & & \ddots & p_{N-1N} \\ p_{N1} & & & p_{NN-1} & p_{NN} \end{bmatrix} \tag{6.55}$$

where p_{ij} is the probability of transition from state i to state j. The computation of transition probabilities is based on the fade slope (ζ) statistics. Therefore, given a time series resolution of δt, it holds that

$$p_{ij} = p\left(s(t+\delta t) = s_j \mid s(t) = s_i\right) = p\left(\zeta = \frac{s_j - s_i}{\delta t} \mid s(t) = s_i\right) \tag{6.56}$$

So

- To stay in the same state, this must hold: $|\zeta\delta t| < (\delta a/2)$
- To go from state i to state $i \pm k$: $|\zeta\delta t| < (\delta a/2) \pm k\delta a$

From Equation 6.56 and the rules above, it holds that

$$p_{i,i\pm k} = \int_{k\delta a - (\delta a/2)}^{k\delta a + (\delta a/2)} p(\zeta \mid A_i)d\zeta \tag{6.57}$$

where $p(\zeta \mid A_i)$ is the PDF of fade slope. The fade slope statistics can be derived from the ITU-R. P. recommendations (ITU-R. P. 1623-1, 2005).

Another model based on Markov chains is the Deutschland für Luft- und Raumfahrt (DLR—German Aerospace Research Center) channel model (Fiebig, 2002). The DLR channel model or second-order Markov chain model is described in Fiebig (2002) in which it is assumed that rain attenuation is a second-order Markov process. The model can be discrete valued or continuous-time valued. In this model, the attenuation values are quantized and the value of rain attenuation at time t depends on the segment of the previous two samples, that is, at constant segment, the previous two values are equal; at up-segment, the rain attenuation is increasing, and at down-segment, the rain attenuation is decreasing. For every segment, a Gaussian PDF is formulated from which the new value is derived.

Therefore, although a second-order Markov chain is described by the N^3 transition probabilities, using the segmentation and the fact that the transition probabilities can be described by a Gaussian distribution, the states fall into 3N transition probabilities. Now, depending on whether rain attenuation increased, decreased, or remained equal in the previous two samples, three different Gaussian distributions are used for the description of the transition probabilities at the final state.

6.2.7 Copula-Based Time Series Synthesizer

Joint statistics of rain attenuation on temporal domain, that is, $P(A_t \geq A_{th}, A_t + \Delta_t \geq A_{th})$ are used mostly in time diversity systems. Kourogiorgas et al. (2014b) conducted the first analysis for describing these joint statistics with copulas. Here, the Gaussian copula (Nelsen, 2006) is examined. The general expression of the Gaussian copula for n random variables is given by

$$C(\mathbf{u}) = \Phi_R^n(\Phi^{-1}(u_1),...,\Phi^{-1}(u_n)) \tag{6.58}$$

where Φ_R^n is the normal multivariate cumulative distribution function (CDF) of Gaussian distribution with zero mean variables and correlation matrix $\mathbf{R}$. The function Φ^{-1} is the inverse CDF of the standard Gaussian distribution.

The Gaussian copula is characterized by the parameter ρ. Using joint exceedance probability of rain attenuation for various time delays and link characteristics, the parameter θ can be modeled as a function of time delay and the link characteristics. As an example, the measured joint distribution of rain attenuation for Spinno d'Adda derived from Fabbro et al. (2009) are used and the results are shown in Figure 6.8.

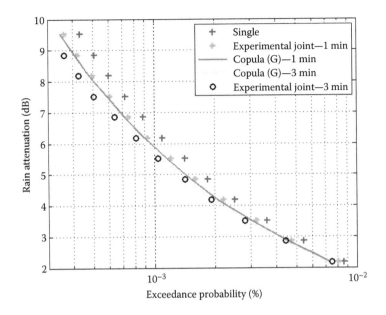

FIGURE 6.8 Testing Gaussian copula for modeling of joint statistics of rain attenuation.

For this specific copula, the copula parameter can be related to the Kendall's tau rank correlation (τ) through the expression

$$\tau = \frac{2}{\pi}\sin^{-1}(\rho) \tag{6.59}$$

The goal of this method of synthesis is to reproduce the CCDF that was given as input and the desired Kendall's tau correlation coefficient.

Since we want to reproduce the Kendall's tau coefficient as second-order statistics of the random variable, the Gaussian copula may be used. In case τ of Equation 6.59 is known, with the inverse of Equation 6.59, one can obtain the Gaussian copula parameter ρ. With the knowledge of the correlation matrix ($\mathbf{R}$) of the multivariate Gaussian copula, one can proceed to the generation of zero-mean Gaussian random variables correlated with matrix $\mathbf{R}$. Then, taking the normal CDF of the above generated values, the CDF of each sample of the desired random variable is calculated. Finally, taking the inverse CDF of the desired distribution, that is, gamma, lognormal, Weibull, etc., the random variables are generated.

In Figure 6.9, a snapshot of the time series is shown. It is considered that rain attenuation follows the Weibull distribution and the parameter ρ of Gaussian copula is extracted from measured CCDFs for various time delays. For this numerical example, the exceedance probability of rain attenuation is derived from the data in Fabbro et al. (2009) for Spinno d'Adda at 19 GHz.

6.3 CLOUDS ATTENUATION TIME SERIES

Although at frequencies close to 20 GHz, rain attenuation is the dominant fading mechanism, clouds also affect the signal and the attenuation caused by clouds is increasing as the operating frequency becomes higher and higher. One synthesizer for generating time series of clouds attenuation is the one proposed in ITU-R. P. 1853-1 (2012). More particularly, the synthesizer generates time series of ILWC, which then are transformed to clouds attenuation time series using the methodology proposed in ITU-R. P. 840-6 (2012).

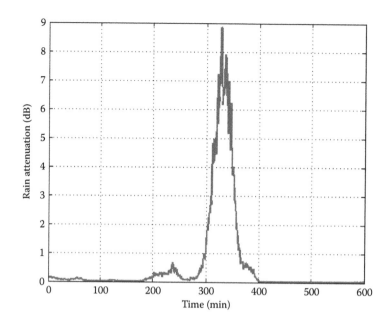

FIGURE 6.9 Time series of rain attenuation produced with the proposed synthesizer.

The main assumption of the synthesizer is that the exceedance probability of ILWC (L) is given by

$$P(L \geq L_{th}) = P(L > 0) \cdot P(L \geq L_{th} \mid L > 0) \tag{6.60}$$

The probability that ILWC exceeds the 0 mm value is taken from ITU-R. P. 840-6. Moreover, the conditional probability $P(L \geq L_{th} \mid L > 0)$ is considered to follow the lognormal distribution (Jeannin et al., 2008) with statistical parameters m, s again derived from the database ITU-R. P. 840-6. The methodology for generating ILWC time series is similar to the one proposed by Boulanger et al. (2013) for rain attenuation.

First, a Gaussian white noise $n(t)$ is generated and then the following Gaussian stochastic processes are calculated at the time instants kT_s for the sampling period T_s:

$$X_1(kT_S) = \rho_1 X_1((k-1)T_S) + \sqrt{1 - \rho_1^2}\, n(kT_S)$$
$$X_2(kT_S) = \rho_2 X_2((k-1)T_S) + \sqrt{1 - \rho_2^2}\, n(kT_S) \tag{6.61}$$

with

$$\rho_i = \exp(-\beta_i T_S), i = 1, 2 \tag{6.62}$$

Then the following Gaussian time series are calculated:

$$G_C(kT_S) = \gamma_1 X_1(kT_S) + \gamma_2 X_2(kT_S) \tag{6.63}$$

that is, the weighted sum of the processes X_1 and X_2. Since ILWC follows a conditional distribution, a truncation threshold is calculated:

$$a = Q^{-1}(P(L > 0)) \tag{6.64}$$

where Q is the Gaussian exceedance probability. The time series of ILWC are generated through

$$L(t) = \exp\left(\mathrm{erfc}^{-1}\left(\mathrm{erfc}\left(\frac{G_C(t)}{\sqrt{2}} \right) / (P(L > 0)/100) \right) \cdot \sqrt{2} \cdot s + m \right) \quad \text{for} \quad G(t) > a \tag{6.65}$$

$$L(t) = 0 \quad \text{for} \quad G(t) \leq a$$

The synthesizer as aforementioned generates ILWC time series assuming lognormal distribution for the conditional probability. Considering a ground station in Athens, Greece, the theoretical CCDF as well as the CCDF resulted from time series are shown in Figure 6.10a. The time series of ILWC are derived and shown in Figure 6.10b. Moreover, the underlined Gaussian processes are considered to have an autocorrelation function that is exponentially decaying with time. Therefore, the autocorrelation function of the underlined Gaussian process of ILWC is given by (Boulanger et al., 2011)

$$R_{G_C}(\tau) = A \exp(-\beta_1 \tau) + (1 - A) \exp(-\beta_2 \tau) \tag{6.66}$$

It must be noted that in ITU-R. P. 1853-1, the noise used for the filtering methods of rain attenuation and ILWC time series is the same, thus introducing a correlation for the two phenomena.

As aforementioned from the ILWC time series, the clouds attenuation time series can be derived from the recommendation of ITU-R. P. 840-6. In Figure 6.10b and c a snapshot of ILWC time series is shown for Athens, Greece and the clouds attenuation time series for an operating frequency of 40 GHz and an elevation angle of 30°.

Boulanger et al. (2011) proposed a simpler synthesizer in order to have a unified synthesizer for clouds and rain attenuation. The approach is motivated by the fact that the separation of rain attenuation in the presence of clouds from clouds attenuation is difficult and therefore a synthesizer that generates rain plus clouds attenuation is required. Therefore, time series of rain plus clouds attenuation are modeled. The first-order statistics are conditional with the conditioned PDF following a lognormal distribution:

$$P(A_{R+C} \geq A_{th}) = P(A_{R+C} > 0) \cdot \frac{1}{2} \mathrm{erfc}\left(\frac{\ln(A_{th}) - m_{R,C}}{\sqrt{2} s_{R,C}} \right) \tag{6.67}$$

Then, starting from Gaussian noise again, two parallel low-pass filters are used and then their output is summed in order to obtain a Gaussian process with the autocorrelation function being a decaying double-exponential function, similar to the ILWC time series generator methodology.

Another methodology for generating ILWC time series has been presented in Resteghini (2014). The methodology is based on the ILWC time series derived from databases. The ILWC daily time series are classified into frames of 24 h and then into classes based on the maximum value of ILWC for every day. The methodology can be extended to sites without any measurements through objective function and using the following scaling factor:

$$SF_L = \frac{P(L)_{OBF}(p)}{P(L)_{DB}(p)} \tag{6.68}$$

where $P(L)_{OBF}(p)$ is the CCDF of ILWC derived, for example, ITU-R. P. 840-6, for a desired site, and $P(L)_{DB}(p)$ the CCDF from the database for a probability level $p\%$, for example, 0.1%. The scaling factor is then applied to the time series of the database. Now, using the time series of every

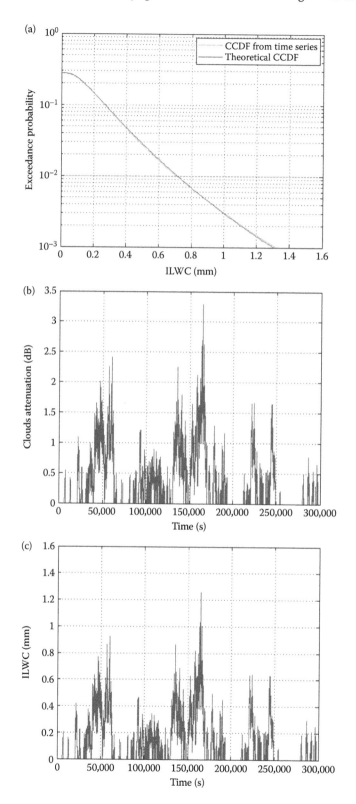

FIGURE 6.10 (a) CCDF of ILWC for Athens from time series and theoretical. (b) Snapshot of ILWC time series. (c) Snapshot of clouds attenuation.

class, a base function is calculated as the CCDF of ILWC for every class. Every base function has a weight such that their sum gives the CCDF for the desired site and this weight multiplied by the observation period (in number of days) will give the number of days that a certain class has been observed in the database. For the generation of the time series of ILWC, since there is a seasonal variation, the database is classified into months and the time series of this month are ordered randomly.

Resteghini (2014) introduced an algorithm for the classification of the cloud type according to the vertical extent of the cloud and for every cloud type, the probability to rain is derived from the database. Therefore, a dependence between ILWC and rain attenuation is introduced and can be used for the generation of ILWC and rain attenuation time series.

Very recently, an engineering model has been presented for the prediction of long-term cloud attenuation statistics using a time series synthesizer (Lyras et al., 2015) based on SDEs for the ILWC. It has been employed to evaluate optical communication systems but similarly, it can be used for radio frequencies (RFs).

6.4 TIME SERIES OF ATTENUATION DUE TO ATMOSPHERIC GASES

A synthesizer proposed for atmospheric gases is presented in Boulanger et al. (2011) and ITU-R. P. 1853-1. According to the methodology, it is assumed that oxygen attenuation is constant. Considering the attenuation due to water vapor, first the time series of IWVC are generated and then through the use of the simplified method of the calculation of water vapor attenuation, the attenuation time series are obtained. For IWVC, it is assumed that it is always present and that it follows the Weibull distribution (Jeannin et al., 2008):

$$P\left(\text{IWVC} \geq \text{IWVC}_{th}\right) = \exp\left(-\left(\frac{\text{IWVC}_{th}}{\lambda}\right)^{\kappa}\right) \tag{6.69}$$

As in the other time series generators of ITU-R. P. 1853-1, an underlined Gaussian process is assumed, which follows a decaying exponential autocorrelation function. After the generation through low-pass filtering of the later Gaussian process, a nonlinear device is used for obtaining the Weibull distribution of IWVC.

The block diagram for generating IWVC time series is shown in Figure 6.11. The last nonlinear device is actually transforming the Gaussian time series to time series that follow the Weibull distribution.

The correlation among ILWC, rain, and IWVC time series in ITU-R. P. 1853-1 is taken into account through the Gaussian noise used as input. More particularly, for clouds and rain, the same noise is used while for IWVC, the covariance between the input noise of IWLC and IWVC is equal to 0.8. Therefore, through the correlation on the input noise, a correlation is introduced in the IWVC time series. From the IWVC time series, the simplified model given in ITU-R. P. 676-9 (ITU-R. P. 676-9, 2012) is used for generating time series of attenuation due to water vapor.

Similar to the ILWC time series, the IWVC time series are generated using the methodology based on database (Resteghini, 2014). In the latter reference, a method is proposed for generating IWVC time series for different sites using a single database, as very briefly explained in the previous

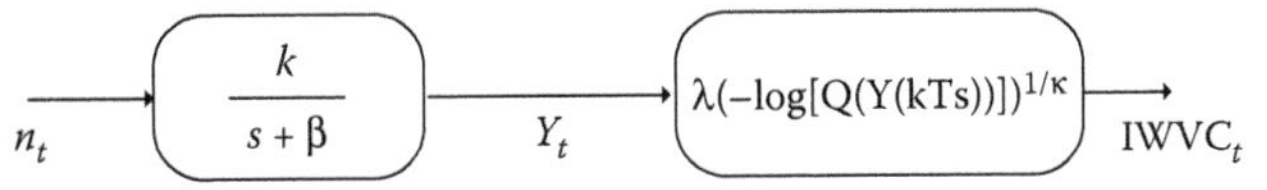

FIGURE 6.11 Block diagram of generating IWVC time series.

section. However, in Resteghini (2014), since the IWVC and ILWC are classified into different classes, the probability of simultaneous occurrence of the various classes between IWVC and ILWC is computed. Then a joint optimization procedure is used in order to have the weights of the joint base functions and preserve the number of the days when each joint class is observed equal to this given by the database. Then through random ordering based on seasonal variation, the joint time series of IWVC and ILWC are generated.

6.5 SCINTILLATION TIME SERIES

Apart from rain attenuation, the turbulence eddies that are shaped into the troposphere cause scintillation of the amplitude of the signal. Although the effect is not so severe in terms of fades as in rain attenuation, ARQ schemes and dynamic FMTs have been identified for the compensation of scintillation.

Therefore, time series of scintillation have been studied in the literature. More particularly, five synthesizers have been developed in the literature, and two filtering methods have been identified (Kassianides and Otung, 2003; ITU-R. P. 1853-1, 2012), one using the fractional Brownian motion (fBm) (Celadroni and Potorti, 1999) and one based on SDEs for scintillation induced on a GEO slant path (Kourogiorgas and Panagopoulos, 2013) while a synthesizer has been developed for LEO slant paths (Liu and Michelson, 2012).

The first filtering method proposed in ITU-R. P. 1853-1 (2012), for the generation of time series of scintillation is based on low-pass filtering of Gaussian zero mean and unity variance noise. The low-pass filter has a decreasing slope of −80/3 dB/decade, which is also the decreasing slope of the power spectrum of scintillation at high frequencies. The resulted time series follow Gaussian distribution with zero mean and unity variance and the power spectrum has the same shape of the low-pass filter. The cutoff frequency of the filter is considered equal to 0.1 Hz. For long-term statistics, the time series are multiplied by the variance of scintillation, which as it will be analyzed later depends on the kind of scintillation, that is, wet or dry.

In the second filtering method (Kassianides and Otung, 2003), the Gaussian noise passes through a low-pass filter and then through a nonlinear memoryless device as shown in Figure 6.12. The low-pass filter gives the appropriate shape at the power spectrum of scintillation, that is, low pass with a −80/3 roll-off factor and the nonlinear device the appropriate shape at the PDF of the generated time series. The low-pass filter has a characteristic function given in Equation 6.70. The $y(t)$ signal at Figure 6.12 is determined by the $x(t)$ signal through the expression in Equation 6.71. The parameters of Equation 6.71 are determined in Kassianides and Otung (2003).

$$T(z) = \frac{0.1810 + 0.0791z^{-1} + 0.0371z^{-2}}{1 - 0.4574z^{-1} + 0.1990z^{-2}} \tag{6.70}$$

$$y(t) = a_1 x(t) + a_3 x^3(t) + a_5 x^5(t) - a_6 x^6(t) + a_7 x^7(t) \tag{6.71}$$

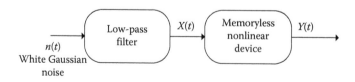

FIGURE 6.12   Block diagram of scintillation time series synthesizer.

Since the next two models are based on the fBm, few theoretical aspects are given. fBm, $B^H = \{B_t^H, t \in \mathbb{R}\}$ is characterized by the Hurst index (H, with $H \in (0,1)$) and it is a Gaussian process with the following properties (Mishura, 2008):

$$B_0^H = 0$$

$$E[B_t^H] = 0, t \in \mathbb{R}$$

$$E[B_t^H B_s^H] = \frac{1}{2}\left(|t|^{2H} + |s|^{2H} - |t-s|^{2H}\right), s,t \in \mathbb{R}$$

From the properties given above, it can be remarked that

$$E[(B_t^H - B_s^H)^2] = |t-s|^{2H}, s,t \in \mathbb{R}$$

The well-known standard Brownian motion has a Hurst index $H = 1/2$. For $H < 1/2$, the fBm is negatively correlated, while for $H > 1/2$, it is considered as positively correlated.

Considering the spectrum properties of the fBm, the following expression holds for the energy spectrum of fBm (Shao, 1995):

$$F_{B^H}(f) \sim f^{-(2H+1)} \tag{6.72}$$

In Figure 6.13, time series of fBm are shown for different values of the Hurst index.

Celadroni and Potorti (1999) proposed the fBm for the modeling of the scintillation time series. According to the analysis of the physical characterization of scintillation, a difference process is defined $W(t_1, t_2) = A(t_2) - A(t_1)$, with A being the attenuation values and for sufficiently small $t = t_2 - t_1$, is the difference process of scintillation. This difference process is assumed to be a function of rain attenuation, since scintillation variance depends on scintillation. It is stated in Celadroni and Potorti (1999) that the process $W(t,A)$ is assumed to be normal with zero mean and variance of σ_W^2, which is also validated through chi-square tests. Therefore, scintillation is modeled as an fBm process with variance:

$$\sigma_W^2 = Vt^{2H} \tag{6.73}$$

with H (the Hurst parameter) and V depending on the rain attenuation values. For the spectrum analysis, the power spectrum indicated in Karasawa and Matsudo (1991) is assumed, which gives that in log–log scale the power spectrum of scintillation follows an f^{-1} slope followed by $f^{-8/3}$. The corner frequency (f_c) is found to depend on the Hurst index and so to rain attenuation. The dependence of the Hurst index to the corner frequency is $H = 0.83 - 1.7f_c$.

Finally, an SDE is proposed based on the fBm for the generation of time series of scintillation. In this section, a time series synthesizer of the log-amplitude χ is proposed based on SDEs driven by fBm. The time series of log-amplitude will be described by

$$d\chi_t = f(t,\chi_t)dt + g(t,\chi_t)dB_H \tag{6.74}$$

where dB_H is the increment of the fBm with Hurst index H. Since for a given variance, χ_t can be considered as a Gaussian process with zero mean, we use the Langevin equation with fBm (fractional

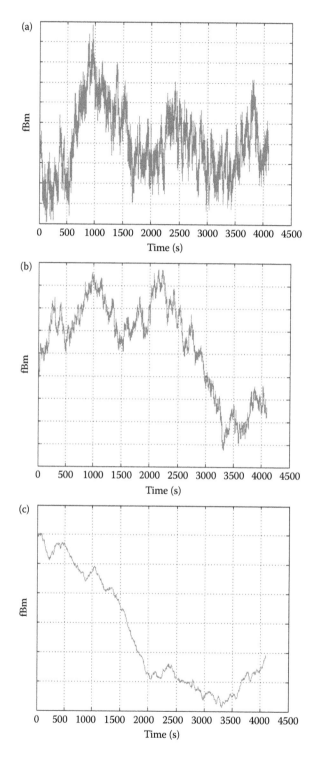

FIGURE 6.13 Time series of fractional Brownian motion with Hurst index (a) 0.25, (b) 0.5, and (c) 0.75.

Langevin equation) to model the time series of log-amplitude, giving that the variance is equal to 1 ($\chi_{t,1}$) (Shao, 1995):

$$d\chi_{t,1} = -\lambda\chi_{t,1}dt + \sigma dB_H \tag{6.75}$$

The parameters λ and σ depend on the dynamic parameters of the stochastic process and its long-term statistics. The solution of the above equation is (Shao, 1995; Cheridito et al., 2003)

$$\chi_{t,1} = e^{-\lambda t}\left(\chi_0 + \sigma\int_0^t e^{\lambda u}dB_u^H\right) \tag{6.76}$$

where χ_0 is the initial value of the log-amplitude. Here we consider it as equal to 0 dB. Therefore, Equation 6.76 becomes

$$\chi_{t,1} = e^{-\lambda t}\sigma\int_0^t e^{\lambda u}dB_u^H \tag{6.77}$$

The above process is also called the fractional Ornstein–Uhlenbeck process and it is a Gaussian process with zero mean and the variance of χ_t and χ_{t+s} is (Cheridito et al., 2003)

$$\mathrm{Cov}(\chi_{t,1},\chi_{t+s,1}) = \sigma^2\frac{\Gamma(2H+1)\sin(\pi H)}{2\pi}\int_{-\infty}^{+\infty}e^{isy}\frac{|y|^{1-2H}}{\lambda^2+y^2}dy \tag{6.78}$$

Therefore, the variance of the Ornstein–Uhlenbeck process is ($s = 0$)

$$\sigma_{\chi_{t,1}}^2 = \sigma^2\frac{\Gamma(2H+1)\sin(\pi H)}{2\pi}\int_{-\infty}^{+\infty}\frac{|y|^{1-2H}}{\lambda^2+y^2}dy \tag{6.79}$$

Since, first, we want to have a unitary variance process, we set

$$\sigma = \frac{1}{\sqrt{(\Gamma(2H+1)\sin(\pi H)/(2\pi))\int_{-\infty}^{+\infty}(|y|^{1-2H})/(\lambda^2+y^2)dy}} \tag{6.80}$$

For the energy spectrum, $F_{\chi_{t,1}}(f)$ of the fractional Ornstein–Uhlenbeck process holds that (Shao, 1995)

$$F_{\chi_{t,1}}(f) = \sigma^2\frac{F_{\omega_b}(f)}{\lambda^2+(2\pi f)^2} \tag{6.81}$$

where $F_{\omega_b}(f)$ is the energy spectrum of the fractional Gaussian noise and it holds that

$$F_{\omega_b}(f) \sim f^{-(2H-1)} \tag{6.82}$$

From Equations 6.81 and 6.82, it holds for high frequencies that

$$F_{\chi_{t,1}}(f) \sim f^{-(2H+1)} \tag{6.83}$$

with corner frequency close to $\omega_c = \lambda^{1/(H+1/2)}$.

Finally, in order to obtain the scintillation time series, we multiply the time series of $\chi_{t,1}$ derived from Equation 6.77 with the scintillation standard deviation σ_χ. Consequently, the time series of log-amplitude are generated from

$$\chi_t = \sigma_\chi \chi_{t,1} \tag{6.84}$$

From Equation 6.84, considering that σ_χ is constant with time, the scintillation log-amplitude is a Gaussian random variable with zero mean, standard deviation σ_χ, and a spectrum slope of $-(2H + 1)$.

However, scintillation variance follows a lognormal distribution and the question that can arise is for the time series of the scintillation variance. It has been shown that the instantaneous value of scintillation standard deviation is proportional to the instantaneous value of rain attenuation (A) with a power law of 5/12 under the presence of rain (Matricciani and Riva, 2008):

$$\sigma_\chi = CA^{5/12} \tag{6.85}$$

From Equation 6.85, and given that we take the unconditional distribution of rain attenuation and the M–B model (Maseng and Bakken, 1981) for the generation of time series of rain attenuation, we can further assume that the scintillation standard deviation follows the M–B model as well, with the same dynamic parameter with the rain attenuation process equal to 0.0002 s⁻¹ (Maseng and Bakken, 1981).

Liu and Michelson (2012) proposed a synthesizer based on low-pass filters with time-varying parameters for generating scintillation time series for different LEO satellite passes. In the proposed model, the spectrum of scintillation time series is the one proposed by Tatarskii, that is, constant at low frequencies and a slope of −80/3 dB²/Hz for high-frequency components, as already described in this chapter. However, Liu and Michelson (2012), considered the corner frequency of the power spectrum time dependent since it depends on the turbulent layer height and the satellite velocity, for a constant operating frequency. More particularly, the corner frequency (f_c) is equal to $1.43f_0$, where

$$f_0 = \frac{\upsilon_t}{\sqrt{2\pi\lambda z}} \tag{6.86}$$

where υ_t is the transverse velocity perpendicular to the Earth–space path, λ the wavelength, and z the length of the slant path between the Earth station and the turbulent layer. The height of the turbulent layer is considered to be Rician-distributed and its thickness lognormally distributed (Vasseur and Vanhoenacker, 2008). The short-term log-amplitude of the signal is considered as a Gaussian random variable with zero mean and the standard deviation of scintillation on the long-term as lognormally distributed or gamma distributed. The wet and dry scintillation is considered through the correlation of the standard deviation of scintillation with rain attenuation, as will be explained later on this subsection. For the generation of the short-term scintillation, time-varying parameters are used in low-pass filter. The low-pass filter is of fourth order and uses as input white Gaussian noise. However, its parameters are calculated for every sample.

One factor that has been studied in the literature is the correlation between rain and scintillation, or the wet (in the presence of rain) and dry scintillation. Matricciani and Riva (2008) found that in the presence of rain, the standard deviation of scintillation is equal to

$$\sigma_\chi = CA^{5/12} \tag{6.87}$$

when rain attenuation (A) is greater than 1 dB. However, Kamp et al. (1997) proposed another formula in which the standard deviation of scintillation is considered as lognormally distributed and the mean value and standard deviation are given by

$$m_\sigma = 0.23 + 0.024A$$
$$\sigma_\sigma = 0.52m_\sigma - 0.025 \tag{6.88}$$

However, in ITU-R. P. 1853 (2012), the scintillation is modeled considering not only dependence to rain attenuation but also to clouds and water vapor.

6.6 TOTAL TROPOSPHERIC ATTENUATION TIME SERIES

Two main methodologies have been developed for generating time series of tropospheric attenuation. One is based on ITU-R. P. 1853-1 methodology and the other is proposed in Resteghini (2014).

Starting from ITU-R. P. 1853-1, the block diagram is shown in Figure 6.14a directly derived from the recommendation. In the previous subsections, the generators for every attenuation factor and their interdependencies have been explained. The recommended synthesizer adds the time series of all the effects in order to calculate the time series of total attenuation, as

$$A(t) = A_R(t) + A_C(t) + A_V(t) + Sci(t) + A_O \tag{6.89}$$

where A_R, A_C, A_V, Sci, and A_O are the rain attenuation, clouds attenuation, water vapor attenuation, scintillation, and oxygen attenuation time series. The oxygen attenuation is considered constant.

The second synthesizer proposed in Resteghini (2014) is based on the generation of time series from database, including the appropriate scaling to take into account the difference in the geometric and electrical characteristics of the link. Although the synthesizers of each attenuation factor for this specific synthesizer have been briefly presented in the previous sections, the generation of total tropospheric attenuation time series is based on the block diagram of Figure 6.14b. First, since the very slow variations of oxygen attenuation, attenuation due to oxygen is considered constant. Then, as briefly explained in Section 6.4, the time series of ILWC and IWVC are derived, after the appropriate scaling, jointly from the input database in order to have correlated/dependent ILWC and IWVC time series. Using ITU-R. P. 840 and ITU-R. P. 676, clouds attenuation and attenuation due to water vapor is calculated for every sample of ILWC and IWVC time series, respectively.

Moreover, using the cloud type algorithm (CTA) in Resteghini (2014), the cloud types, associated with the ILWC time series, with the higher probability to rain are identified. Then, identifying the clouds intervals, that is, time period in which ILWC is greater than zero, a rainy mask is considered equal to 50% of the clouds interval during which is considered that it is raining, giving priority of the presence of rain to the cloud types with the highest probability of rain. Then, in the rain mask, a rain attenuation event is superimposed with equal length. In this way, the rain attenuation and clouds attenuation are presented in interdependent manner.

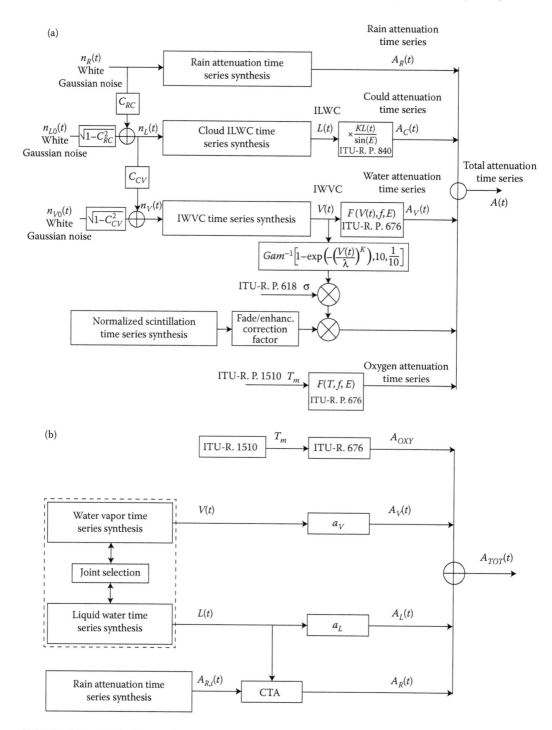

FIGURE 6.14 Block diagram for generating total tropospheric attenuation time series. (a) ITU-R. P. 1853 and (b) Resteghini method derived from ITU-R. P. 1853 (2012) and Resteghini (2014), respectively.

6.7 RESEARCH ISSUES AND CONCLUDING REMARKS

The existing time series synthesizers in the literature and in ITU-R recommendation have not been tested with attenuation measurement since they are not available experimentally, especially for frequencies above 30 GHz. The tropospheric attenuation channel models should be further refined considering new data for the temporal and spectral properties of the tropospheric phenomena. These new space–time models will be applied for the optimum design of FMTs.

Rain attenuation channel modeling and all the channel models for the rest of the tropospheric components of the total should be carefully reexamined considering the experimental data from the new experiments. This is an important issue in order to design the next-generation satellite networks in terms of the chosen air interfaces, including the number of available antennas, modulation/demodulation complexity, level of robustness to transceiver chain nonlinearities, latency, frequency band selectivity, required transmission directivity, and available degrees of freedom (time, frequency, space, etc.).

As further research direction on this subject is the development of a stochastic radio channel modeling framework for the radio intersystem interference between satellite and terrestrial networks in order to optimize the use of radio spectrum.

REFERENCES

AGI's STK, https://www.agi.com/products/by-product-type/applications/stk/.

Arapoglou, P-D.M., K.P. Liolis, and A.D. Panagopoulos, Railway satellite channel at Ku band and above: Composite dynamic modeling for the design of fade mitigation techniques, *International Journal of Satellite Communications and Networking*, 30, 1–17, 2012.

Bertorelli, S., C. Riva, and L. Valbonesi, Generation of attenuation time series for simulation purposes starting from ITALSAT measurements, *IEEE Transactions on Antennas and Propagation*, 56(4), 1094–1102, 2008.

Boulanger, X., G. Carrie, L. Castanet, and L. Feral, Overview of a more simplified new channel model to synthesize total attenuation time series for satellite communication systems at Ka and Q/V band, *ESA Workshop on Radiowave Propagation*, Noordwijk, The Netherlands, 2011.

Boulanger, X., L. Feral, L. Castanet, N. Jeannin, G. Carrie, and F. Lacoste, A rain attenuation time-series synthesizer based on a Dirac and lognormal distribution, *IEEE Transactions on Antennas and Propagation*, 61(3), 1396–1406, 2013.

Burgueno, A., E. Vilar, and M. Puigcerevr, Spectral analysis of 49 years of rainfall rate and relation to fade dynamics, *IEEE Transactions on Communications*, 38(9), September 1990.

Carrie, G., F. Lacoste, and L. Castanet, A new "event-on-demand" synthesizer of rain attenuation time series at Ku-, Ka- and Q/V-bands, *International Journal of Satellite Communications and Networking*, 29, 47–60, 2011.

Castanet, L., T. Deloues, and J. Lemorton, Channel modeling based on N-state Markov chains for SatCom system simulation, *ICAP 2003*, Exeter UK, April 2003.

Celadroni, N. and F. Potorti, Modeling Ka-band scintillation as a fractal process, *IEEE Journal on Selected Areas in Communications*, 17(2), 164–172, 1999.

Cheridito, P., H. Kawaguchi, and M. Maejima, Fractional Ornstein–Uhlenbeck processes, *Electronic Journal of Probability*, 8, 2003.

EUCLID Scientific Mission, http://sci.esa.int/euclid/46661-mission-operations/, Accessed May 2014.

Fabbro, V., L. Castanet, S. Croce, and C. Riva, Characterization and modelling of time diversity statistics for satellite communications from 12 to 50 GHz, *International Journal of Satellite Communications*, 27, 87–101, 2009.

Fiebig, U.C., A time-series generator modelling rain fading, in *Open Symposium on Propagation and Remote Sensing, URSI Commission F*, Garmisch-Partenkirchen, Germany, 2002.

ITU-R. P. 676-9, *Attenuation by Atmospheric Gases*, International Telecommunication Union, Geneva, 2012.

ITU-R. P. 837-6, *Characteristics of Precipitation for Propagation Modelling*, International Telecommunication Union, Geneva, 2012.

ITU-R. P. 838-3, *Specific Attenuation Model for Rain for Use in Prediction Models*, International Telecommunication Union, Geneva, 2001.

ITU-R. P. 840-6, *Attenuation due to Clouds and Fog*, International Telecommunication Union, Geneva, 2013.

ITU-R. P. 1623-1, *Prediction Method of Fade Dynamics on Earth–Space Paths*, International Telecommunication Union, Geneva, 2005.

ITU-R. P. 1853-1, *Tropospheric Attenuation Time Series Synthesis*, International Telecommunication Union, Geneva, 2012.

Jeannin, N., L. Feral, H. Sauvageot, and L. Castanet, Statistical distribution of integrated liquid water and water vapor content from meteorological reanalysis, *IEEE Transactions on Antennas and Propagation*, 56(10), 3350–3355, 2008.

Kamp, M.M.J.L., J.K. Tervonen, E.T. Salonen. Tropospheric scintillation measurements and modelling in Finland. *10th International Conference on Antennas and Propagation*. Edinburgh, Scotland, April 14–17, 1997.

Kanellopoulos, S.A., C. Kourogiorgas, A.D. Panagopoulos, S.N. Livieratos, and G.E. Chatzarakis, Channel model for satellite communication links above 10 GHz based on Weibull distribution, *IEEE Communication Letter*, 18(4), 568–571, 2014.

Kanellopoulos, S.A., A.D. Panagopoulos, and J.D. Kanellopoulos, Calculation of the dynamic input parameter for a stochastic model simulating rain attenuation: A novel mathematical approach, *IEEE Transactions on Antennas and Propagation*, 55(11), 3257–3264, 2007.

Kanellopoulos, S.A., A.D. Panagopoulos, C. Kourogiorgas, and J.D. Kanellopoulos, Slant path and terrestrial links rain attenuation time series generator for heavy rain climatic regions, *IEEE Transactions on Antennas and Propagation*, 61(6), 3396–3399, 2013.

Karatzas, I. and S.E. Shreve, *Brownian Motion and Stochastic Calculus*, New York: Springer-Verlag, 2005.

Karlin, S. and H. Taylor, *A Second Course in Stochastic Processes*, New York: Academic, 1981.

Karasawa, Y. and T. Matsudo, Characteristics of fading on low-elevation angle earth–space paths with concurrent rain fade and scintillation, *IEEE Transactions on Antennas and Propagation*, 39, 657–661, 1991.

Kassianides, C.N. and I.E. Otung, Dynamic model of tropospheric scintillation on earth–space paths, *IEE Proceedings*, 150(2), 97–104, 2003.

Koudelka, O., Q/V-band communications and propagation experiments using ALPHASAT, *Acta Astronautica*, 69(11–12), 1029–1037, 2011, doi: 10.1016/j.actaastro.2011.07.008. Alphasat.

Kourogiorgas, C. and A.D. Panagopoulos, A tropospheric scintillation time series synthesizer based on stochastic differential equations, in *2013 Joint Conference: 19th Ka and Broadband Communications, Navigation and Earth Observation Conference and 31st AIAA ICSSC*, Florence, Italy, October 14–17, 2013.

Kourogiorgas, C. and A.D. Panagopoulos, A rain attenuation stochastic dynamic model for LEO satellite systems operating at frequencies above 10 GHz, *IEEE Transactions on Vehicular Technology*, 64(2), 829–834, 2015.

Kourogiorgas, C., A.D. Panagopoulos, and P-D.M. Arapoglou, Rain attenuation time series generator for medium Earth orbit links operating at Ka band and above, *EuCAP 2014*, the Hague, Netherlands, April 6–11, 2014a.

Kourogiorgas, C., A.D. Panagopoulos, J.D. Kanellopoulos, S.N. Livieratos, and G.E. Chatzarakis, Investigation of rain fade dynamic properties using simulated rain attenuation data with synthetic storm technique, in *2013 7th European Conference on Antennas and Propagation (EuCAP)*, Florence, Italy, pp. 2277–2281, April 8–12, 2013.

Kourogiorgas, C., A.D. Panagopoulos, S.N. Livieratos, and G.E. Chatzarakis, Time diversity prediction modeling using copula functions for satellite communication systems operating above 10 GHz, in *General Assembly and Scientific Symposium (URSI GASS), 2014 XXXIth URSI*, Beijing, China, pp. 1–4, August 16–23, 2014b.

Lacoste, F., M. Bousquet, L. Castanet, F. Cornet, and J. Lemorton, Improvement of the ONERA-CNES rain attenuation time series synthesiser and validation of the dynamic characteristics of the generated fade events, *Space Communications*, 20, 45–59, 2005.

Liu, W. and D.G. Michelson, Effect of turbulence layer height and satellite altitude on tropospheric scintillation on Ka-band Earth-LEO satellite links, *IEEE Transactions on Vehicular Technology*, 59(7), 3181–3192, 2012.

Lyras, N.K., C. Kourogiorgas, and A.D. Panagopoulos, Cloud attenuation time series synthesizer for Earth-space links operating at optical frequencies, *ICEAA-AWPC 2015*, Torino, Italy, September 2015.

Manning, R.M., A unified statistical rain-attenuation model for communication link fade predictions and optimal stochastic fade control design using a location-dependent rain-statistics database, *International Journal of Satellite Communications and Networking*, 8, 11–30, 1990.

Maseng, T. and P. Bakken, A stochastic dynamic model of rain attenuation, *IEEE Transactions on Communications*, COM-29(5), 660–669, 1981.

Matricciani, E., Rain attenuation predicted with two-layer rain model, *European Transactions on Telecommunications*, 2(6), 715–727, 1991.

Matricciani, E., Transformation of rain attenuation statistics from fixed to mobile satellite communication systems, *IEEE Transactions on Vehicular Technology*, 44(2), 565–569, 1995.

Matricciani, E., Physical-mathematical model of the dynamics of rain attenuation based on rain rate time series and a two-layer vertical structure of precipitation, *Radio Science*, 31(2), 281–295, 1996.

Matricciani, E. and C. Riva, 18.7 GHz tropospheric scintillation and simultaneous rain attenuation measured at Spino d'Adda and Darmstadt with Italsat, *Radio Science*, 43, 2008.

Mishura, Y., *Stochastic Calculus for Fractional Brownian Motion and Related Processes*, Berlin, Heidelberg, Germany: Springer, 2008.

Nelsen, R., *An Introduction to Copulas*, New York: Springer, 2006.

Panagopoulos, A.D., P-D.M. Arapoglou, and P.G. Cottis, Satellite communications at Ku, Ka and V bands: Propagation impairments and mitigation techniques, *IEEE Communication Surveys and Tutorials*, 6(3), 2–14, Third Quarter, 2004.

Resteghini, L. A time series synthesizer of tropospheric impairments affecting satellite links developed in the framework of the ALPHASAT experiment, Doctoral Dissertation, Politecnico di Milano, 2014.

Rosello, J., A. Martellucci, R. Acosta, J. Nessel, L.E. Braten, and C. Riva, 26-GHz data downlink for LEO satellites, in *6th European Conference on Antennas and Propagation (EUCAP)*, Prague, Czech Republic, March 2012.

Shao, Y., The fractional Ornstein–Uhlenbeck process as a representation of homogeneous fulerian velocity turbulence, *Physica D*, 83, 461–477, 1995.

Toptsidis, N., P-D.M. Arapoglou, and M. Bertinelli, Link adaptation for Ka band low Earth orbit Earth observation systems: A realistic performance assessment, *International Journal of Satellite Communications and Networking*, 30(3), 131–146, 2012.

van de Kamp, M., Rain attenuation as a Markov process: How to make an event, *COST280 Second International Workshop*, Noordwijk, Netherlands, May 2003.

Vasseur, H. and D. Vanhoenacker, Characterisation of tropospheric turbulent layers from radiosonde data, *Electronics Letters*, 34(4), 318–319, 1998.

7 Review of Space–Time Tropospheric Propagation Models

Nicolas Jeannin and Laurent Castanet

CONTENTS

7.1 INTRODUCTION

The increase of bandwidth needs fostered by growing data rate demand has triggered a shift of fixed-satellite links toward frequency higher than 20 GHz. A counterpart to the large bandwidth available at those frequency bands are strong propagation impairments in case of adverse weather conditions. Static power margins are inefficient to cope with the severe attenuation level that can be experienced in case of rain on the link. Thus, to cope with those strong propagation impairments, and maintain a satisfying level of link availability, fade mitigation techniques (FMTs) have been developed for satellite links as for instance: up-link power control, site diversity, onboard power adjustment, reconfigurable antennas, or adaptive coding and modulation (Panagopoulos et al., 2005).

- Up-link power control consists of increasing the emitted power in case of adverse propagation conditions to maintain the link budget. This technique is usually bounded by the limitation of the amplifier and does not allow coping with the complete range of propagation impairments.
- Site diversity or smart gateways (Jeannin et al., 2014b) can be used to mitigate the impact of propagation impairments on the feeder link by using distant-redundant sites. In case of impairments on one site, a redundant site is used, decreasing the probability of outage. It relies on the partial decorrelation of the impairments on the redundant sites to increase the availability. The assessment of the availability and the optimization of the control loop require the use of space–time-correlated attenuation time series over the various sites.
- Reconfigurable antennas or onboard power reallocation in a multibeam satellite system aims at allocating more power to areas affected by adverse propagation conditions. The performance assessment of those disposals should be made considering space–time-correlated link budgets at the scale of satellite coverage to optimize the areas in which the additional power should be focused. Their control requires knowledge of the propagation conditions over the various users or anticipation through weather forecast.
- The use of adaptive modulation and coding standardized in DVB-S2 (Morello and Reimers, 2004), DVB-S2X, and DVB-RCS2 combined or not with the adjustment of link bandwidth enables to adjust the link budget of a link in real time to adapt to the propagation conditions. The change of coding and modulation induces a change of data rate or of bandwidth usage of the link. As all the users are not experiencing the same propagation conditions simultaneously, the data rate and bandwidth demand can be adjusted through radio resource management algorithms. To assess the efficiency of those algorithms, space–time-correlated link budget considering all the users of the system are required.

Considering those various disposals, the optimization and the planning of systems using those FMTs require the simulation space–time-correlated link budgets, to assess the performance of the system at the scale of the satellite as the optimization of the bandwidth and power utilization should be made considering all or a significant fraction of the links of the system. To perform those space–time link budgets, space–time-correlated propagation data are needed. Those data need to be statistically representative of the climate, in terms of correlation and probability distribution to derive the proper figure from system optimization. Those data need to cover an area in-line with satellite coverage (for instance a continent) and to have a sufficient temporal resolution to design the control loop of the FMT.

There are no data directly available that fulfill those constraints to simulate those space–time-correlated link budgets on a generic basis. Indeed, for now, even the most ambitious propagation experiment (Koudelka, 2011; Castanet et al., 2014) measuring the state of the channel with a beacon does not collect data over more than 10 sites. This is not sufficient considering the thousands of links to be possibly considered in a satellite system and collected data are not directly scalable to a different radio–electric configuration (frequency, elevation, and polarization).

Alternatives to those propagation data could be sought among meteorological data that could be converted into propagation models using appropriate physical models. For instance, data from weather radar networks provide with time intervals of some minutes a cartography of the rain rate over a wide area with a spatial resolution around 1 km as illustrated in Figure 7.1. Those rain rate or reflectivity values can be converted into attenuation on an Earth–space link assuming a particular shape of drop size distribution and a uniform vertical structure of the precipitation medium.

Those data could be useful for system sizing; however, they lack homogeneity and are not available on a global basis as some areas in the world are lacking coverage. Therefore, they cannot be used for system-sizing purposes on a generic manner. The same kinds of drawbacks can be expected from the use of rain rate time series from rain gauges. If those weather data cannot be used directly to model the propagation channel in system simulations, it will be shown that they prove to be extremely valuable to parameterize the propagation models.

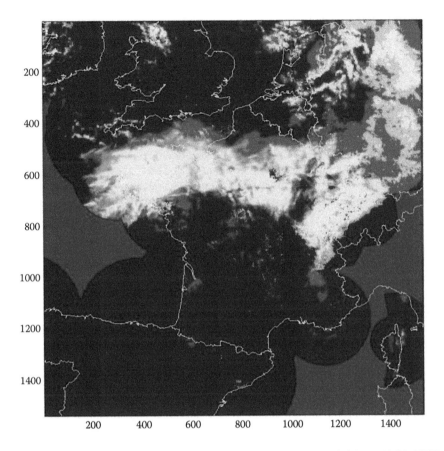

FIGURE 7.1 Radar composite image over Western Europe on 03/04/2004 at 12:00 UTC. (From Meteo-France.)

Data from meteorological models usually do not meet the requirements to be used directly in space–time-correlated link budgets. Indeed, data from global models such as the ones from ECMWF (European Center for Medium-Range Weather Forecast) of NCEP (National Center for Environmental Prediction) give a representation of the state of the atmosphere that includes the various variables of interest to depict the propagation channel (water vapor, liquid vapor, and precipitation) but with an insufficient resolution with regard to the variability of the phenomena as the spatial resolution is of some tens of kilometers and the temporal one is of 3–6 hours.

Numerical weather forecast models with a higher resolution exist and could be used to compute the propagation parameters directly on the outputs as the resolution can be lower than 1 km with time steps of the order of 1 minute. However, they have to be applied to reduced areas for computing reasons and the statistical soundness of the output has to be checked. Their prospective use is discussed in Section 7.4.3.

As there are no data that can be used directly to simulate space–time-correlated link budgets, the use of space–time tropospheric propagation model is required. The simulation of the various propagation impairments could be of interest (attenuation by clouds, gas, rain, sky brightness temperature, scintillation, depolarization…) but up to now, most of the efforts have been made to model rain attenuation as it constitutes the main effect to be tackled by adaptive FMTs. Indeed it is the one that has the highest magnitude and that is the less often encountered. Nevertheless, with the migration to frequency bands higher than Ka band, the significance of the other impairments becomes much higher.

As the simulation of rain fields could be of interest for different purposes (remote sensing, hydrology...), a wide variety of concepts to model them can be found in the literature. Those concepts have often been reused and adapted to fit the requirements of propagation simulation. It explains the large variety of proposed models that are available. An overview of the various methodologies proposed to generate space–time correlated propagation fields is given in this chapter. Each subsection corresponds to a different class of models. Cell-based models, random fields, fractal-based models, and models based on the downscaling of meteorological data are discussed in the following sections. Only a brief description of the various models and of the general principle is proposed and the reader is invited to go through the reference for further details. In addition, most of the attention is dedicated to the simulation of the meteorological variable. The conversion of those meteorological parameters in attenuation that can be handled by various models, that are for some of them standardized by ITU-R is not tackled into detail in this chapter.

7.2 CELL-BASED MODELS TO MODEL RAIN ATTENUATION

7.2.1 RATIONALE

Cellular approaches split up the rain field into individual areas: the rain cells. These cellular approaches have been developed to reproduce the structure of individual storms that can be observed from weather radar observations (Misme and Waldteufel, 1980). They allow describing the inner structure of the cells from a small number of parameters. Whatever the cellular model considered, the two-dimensional generation of a rain rate field from modeled cells requires the determination of a key parameter: the spatial density of the cells, expressed as a function of the model parameters. The generation generally targets the reproduction of the local rainfall rate complementary cumulative distribution function (CCDF) on a given area. In addition, to correlate the field, the spatial organization of the cells needs to be described through random walk, whose parameters are derived from radar data. To include the temporal evolution, the displacement birth, growth, decay, and death of the various cells need to be parameterized. The basics of different approaches namely ExCell and HyCell using those cellular models are described in the following sections.

7.2.2 EXCELL AND ITS DERIVATIVE

7.2.2.1 ExCell

The ExCell (exponential cell) model, developed at Politecnico di Milano (Capsoni et al., 1987, 2006; Ferrari, 1997), is a proposal for the modeling of rain cells based on the analysis of radar images from the Spino d'Adda station. This review of radar scans indicated that a rain cell could be modeled via an exponential expression

$$R(x,y) = R_M \exp\left(-k\sqrt{\frac{\rho^2}{\rho_0^2}}\right) \quad \text{or for elliptic cells,} \quad R(x,y) = R_M \exp\left(-k\sqrt{\frac{\rho^2}{(a^2+b^2)}}\right)$$

where ρ is the distance to the center of cell ($\rho^2 = x^2 + y^2$), R_M is the maximum rain rate in the cell (assumed to occur in the center of the cell, and the cell located in the origin of the coordinate system), and ρ_0 is the distance between the center of the cell and the point in which the rain rate has fallen by a factor $1/e$ from the maximum.

The exponential profile starts at a threshold of 5 mm h^{-1}. Below this threshold, a plateau of 4 mm h^{-1} is introduced, to avoid a subestimation of the CDF.

Depending on the values of k, the cell profile changes. $k = 1$ renders a Gaussian profile, $k = 2$ an exponential profile, and $k > 3$ a hyperexponential profile. Figure 7.2 illustrates the impact of k.

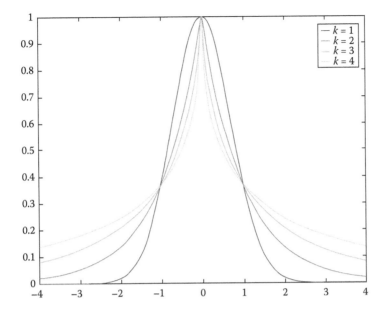

FIGURE 7.2 Impact of the shape constant k on the profile of the modeled cell.

The most important difference between the model and the real cell are

- The real rain cells have on occasion more than one rain rate peak whereas the model proposes a cell with a single peak
- On real cells, the rain rate peak is not always located at the mass center of the cell
- The shape of the cell has a noninteger (fractal) dimension whereas the model proposes cells with a profile always circular or elliptical
- The rain rate of the real cells goes down to zero, whereas in the model, it is constant at the plateau level

A modification of the model, called lowered ExCell, attempts to render the model more "real" with a better approximation to the actual CDF of the rain. The model modifies the expression of the cell into

$$R(\rho) = (R_M + R_{Low})\exp\left(-\frac{\rho}{\rho_0}\right) - R_{Low} \qquad \rho \le \rho_{max}$$

$$R(\rho) = 0 \qquad\qquad\qquad\qquad\qquad \rho > \rho_{max}$$

where R_M is the peak rain rate in the cell, R_{Low} is the rain cell lowering factor, and ρ_{MAX} is the maximum rain cell radius (for values greater than this, the cell will have a rain rate of zero):

$$\rho_{MAX} = \rho_0 \ln\left(1 + \frac{R_M}{R_{Low}}\right)$$

For the derivation of the rain cell spatial density, the following law is proposed:

$$N(\ln(R_M + R_{Low}),\rho_0) = -\frac{1}{4\pi\rho_0^{-2}}\frac{\partial^3 P(R + R_{Low})}{\partial\ln(R + R_{Low})^3}\bigg|_{R=R_M} p(\rho_0 \mid R_M), \quad P(R) = P_0 \ln^n\left(\frac{R_{a\sin t} + R_{Low}}{R + R_{Low}}\right)$$

As one of its many applications, the ExCell model was combined with "frontal" model (Barbaliscia et al., 1992). A population of ExCells is injected into fronts (described succinctly by the "fraction of territory covered by fronts"—the median value for Italy being 0.2—and by the average front depth, assumed to be 300 km) resulting in the operative prediction algorithm (Barbaliscia and Paraboni 2002) for real-time handling of onboard power margin. It turns out to offer a very realistic description of a horizontal rain structure and goes toward the category of models allowing the simulation of rainfall for large regions and long time periods (many years).

7.2.2.2 Goldhirsh Approach

Goldhirsh (2000) proposed a method for simulating typical two-dimensional rain rate fields, at any particular geographic location, from the knowledge of the local CDF of the rain rate (ITU-R Recommendations P 837-3). The method relies on the rain cell modeling by ExCell (Capsoni et al., 1987) and allows filling in the observation area A_o with a population of cells with an exponential profile and rotational symmetry.

To determine the rain cell spatial density over A_o, Goldhirsh uses the analytic formulation of Capsoni et al. (1987), slightly corrected by Awaka (1989), which derives from radar measurements in the region of Milan (Italy) (Figure 7.3).

7.2.2.3 MultiEXCELL

MultiEXCELL (Luini and Capsoni, 2011) allows generating complete rain fields (an example of which is shown in Figure 7.4), where the rainfall spatial distribution is realistically reproduced. Such a goal is achieved by simulating the natural rain cells' aggregative process that has been observed in the real rain fields derived by the weather radar of Spino d'Adda. Considering the strong impact of the rainy coverage on the rainfall spatial correlation, the model also includes a methodology that allows estimating, on a global basis, the distribution of the fractional area of a map covered by rain, based on the ECMWF ERA-40 database.

Radar data collected both at Spino d'Adda (Italy) and Bordeaux (France) are employed to validate the MultiEXCELL model in terms of its accuracy in reproducing the local rainfall statistics, as well as the correct intercellular distance and rainfall spatial distribution.

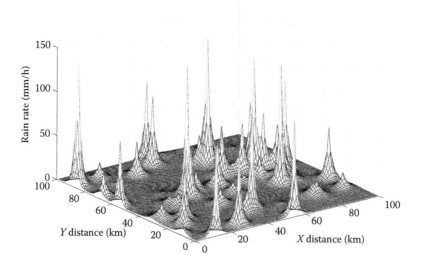

FIGURE 7.3 Simulation of a rain field composed of rain cells thrown at random on a medium-scale area.

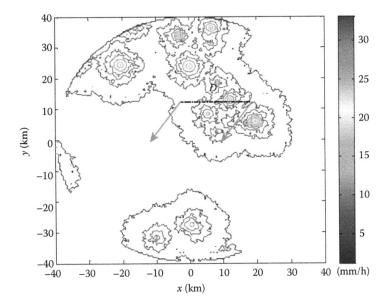

FIGURE 7.4   Example of the rain fields generated by the MultiEXCELL model affecting a two-station site diversity system: the arrows indicate two Earth–space telecommunication links at distance D.

7.2.3 HyCell

The HyCell rain cell model combines the concepts exposed by the ExCell model for the modeling of the horizontal profile of rain within the cell (the event interior, according to the hydrology terminology) with observations from radar networks indicating that the variation follows a shape with Gaussian form.

This model therefore uses a combination of an exponential and Gaussian functions to model rain cells: the exponential part used to model the stratiform rain area surrounding a convective event and the Gaussian portion modeling the more intense part, avoiding the high rain rate peaks seen in the ExCell model cells. The hybrid approach also helps to model high rain rates occurring over nonnegligible areas, such as storms in tropical regions.

The cells of elliptic shape are modeled as follows (Féral, 2003a,b):

$$R(x,y) = \begin{cases} R_G \exp\left[-\left(\dfrac{x^2}{a_G^2}+\dfrac{y^2}{b_G^2}\right)\right] & R \geq R_1 \\[4mm] R_E \exp\left[-\left(\dfrac{x^2}{a_E^2}+\dfrac{y^2}{b_E^2}\right)^{\frac{1}{2}}\right] & R_2 < R < R_1 \end{cases}$$

where R_G, a_G, and b_G are the peak rain rate and the distances along the axes (O_x) and (O_y), for which the rain rate decreases by a factor $1/e$ with respect to R_G, respectively, thus defining the Gaussian portion of the cell. R_E, a_E, and b_E define the exponential component with a similar meaning. R_1 separates the Gaussian and exponential components.

The HyCell model of the cell provides a framework for the analysis of small areas (<5 km). When larger areas are involved, it is necessary to simulate a rain field rather than a rain cell.

The methodology consists of a conglomeration of rain cells modeled by HyCell and of two analytical expressions of the rain cell spatial density, both derived from the statistical distribution of the rain cell size. The scene generation requires, as input parameters, the local CDF of the rain rate and the area S_r over which it is raining. The rain rate field is then generated numerically, according to an iterative scheme, under the constraint of accurately reproducing the local CDF intrinsic to the simulation area S_o, while rigorously following the rain cell spatial density. At this stage, the rain cell location is uniformly distributed over the mid-scale simulation area S_o (Figure 7.5a) (Féral, 2003b).

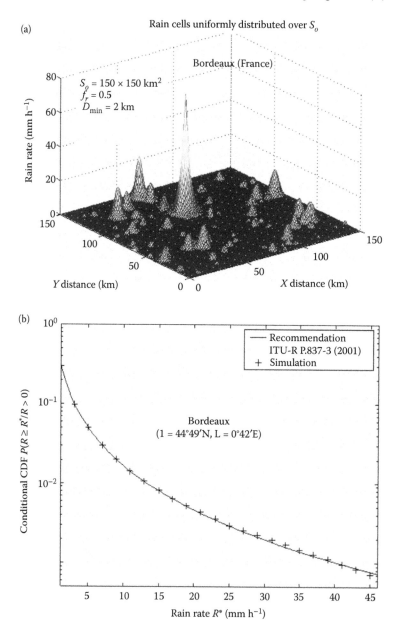

FIGURE 7.5 Generation of rain fields at mid-scale with the HyCell-extended approach. (From Féral, L. et al. *Radio Science*, 41, 2006.) (a) Statistical modeling of a rain field in the region of Bordeaux, France ($S_o = 150 \times 150 \text{ km}^2$, $S_r/S_o = f_r = 0.5$). (b) Conditional CDFs $P(R \geq R*/R > 0)$ for Bordeaux derived from Recommendation ITU-R (solid line) and computed from the generated scene shown in Figure 7.2 (crosses).

(Continued)

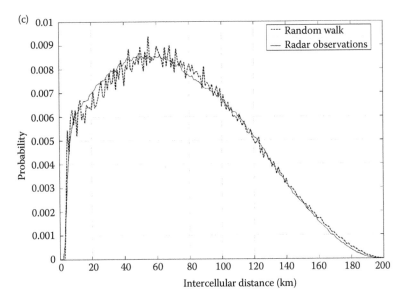

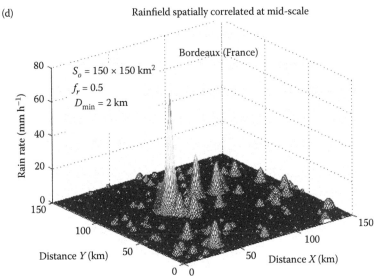

FIGURE 7.5 (Continued) Generation of rain fields at mid-scale with the HyCell-extended approach. (From Féral, L. et al. *Radio Science*, 41, 2006.) (c) ICDD obtained from simulations for an optimal parameterization of the random walk ($\alpha = 4$ and IADD = IADD 15 km) and ICDD derived from radar observations at mid-scale. (d) Same as (a) but now the rain cells are located over S_o by random walk, defining a rain field spatially correlated at mid-scale.

From radar observations at mid-scale, the rain field is modeled by a doubly aggregative isotropic random walk. Starting from an initiatory point randomly located in S_o, the walk constructs small-scale groups from which a cell number results from a draw in a uniform discrete law. The amplitude of the walk within a small-scale aggregate results from a random draw in the distribution of the nearest-neighbour distances (intercellular distance distribution: ICDD) derived from radar observations (Figure 7.5c). The amplitude of the walk between two points from distinct aggregates results from a random draw in the distribution of the interaggregate distances (interaggregate distance distributions: IADDs) also derived from radar observations (Féral et al., 2006). Coupled with the HyCell modeling of rain fields at mid-scale, the random walk allows locating the rain cells within the simulation area S_o. The rain fields generated that way are spatially correlated at mid-scale while accounting for the local climatological characteristics, intrinsic to the simulation area S_o (see Figure 7.5d).

7.2.3.1 Inclusion of the Temporal Fluctuations

The framework of COST 280 (Bousquet et al., 2003) combines the cellular approach and a time-series synthesizer.

The multichannel model generates a short-duration (1–2 hours) space-correlated rain attenuation time series of total impairment (rain attenuation + scintillation) over each selected spot beam (one time series per active Earth station). The generation is carried out in three steps (Bousquet et al., 2003):

- Generation of a mid-scale rain attenuation field (corresponding to a spot beam) with the HyCell model (Féral et al., 2003a,b), or from radar images such as radar data of the French network ARAMIS
- Synthesis of a dedicated time series with the two-samples model for each cell generated in the coverage (Van de Kamp, 2002)
- Weighting of the generated time series according to the space variation inside each rain cell (following the HyCell model)

Radar images are used under some simplifying assumptions. It is assumed above all that the birth and death dynamic process of the rain cells is not to be modeled that results in the fact that all rain cells survive and that their structure (i.e., their HyCell model parameters) remains constant during the whole observation period. All N rain cells are assumed to have an elliptical shape and move uniformly in the same average direction and with the same advection speed v.

Figure 7.6 presents a result with this simulation procedure where $N = 3$ UESs are impinged and the simulation duration is 2 hours.

7.3 RANDOM FIELD OR CORRELATED TIME-SERIES-BASED MODELS

Rather than a description of the rain medium by an aggregation of cells of a predefined shape with a known probability density, another widely used methodology to model rain rate or rain attenuation fields is to consider rain or other meteorological fields as a multidimensional random process whose characteristics can be derived from observations. The tropospheric propagation fields are then obtained through the simulation random processes with similar characteristics. Different techniques have been used to generate space–time-correlated inputs: the transformation of a stationary Gaussian random field, the simulation of fractal fields, or the use of a spatially correlated time series. The main features of the various methods are described in the following sections. These models generally enable an easier inclusion of the temporal fluctuations as it is not required to include the individual evolution of the fields.

7.3.1 Simulation by Transformation of Gaussian Random Fields

The rationale of this methodology is to simulate random processes with a marginal probability distribution and correlation function characteristics representative of the ones of rain rate or rain attenuation fields observed mainly from weather radar data. One of the main difficulty lies in the non-Gaussian distribution of the rain rate or rain attenuation fields. As the most efficient algorithm for the simulation of random fields is the simulation of Gaussian field, the idea is to generate Gaussian fields whose values are mapped to values corresponding to rain or rain attenuation

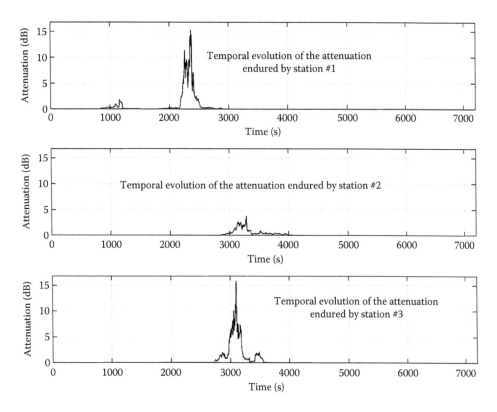

FIGURE 7.6 Evolution with the time of rain attenuation as seen by three stations.

distributions. One of the difficulties of these algorithms is to predistort the correlation functions of the Gaussian fields to obtain the correlation function observed on rain rate or rain attenuation data after the transformation. It has to be noticed that one of the mandatory assumption with this framework is to use a stationary process that can bound in space and in time the domain of applicability of the methodology.

Bell (1987) proposed a stochastic space–time model of rain fields based on rainfall space–time covariance. The model allows the generation, over an observation area A_o, of a two-dimensional rain rate field of which the spatial correlation is Cr(X) and that the probability distribution of R is lognormal, with parameters $\mu_{\ln R}$ and $\sigma_{\ln R}$. It relies on the generation of random fields $g(X)$ that have Gaussian statistics and are spatially correlated and represented as a Fourier series

$$g(X) = \sum_K a_K \exp(i K^T X)$$

where superscript T denotes matrix transpose.

This approach enables the overall advection (mean horizontal moving) of the rain field to be modeled by using the time-shift property of the Fourier transform

$$g(X - Vt) = \sum_K a_K \exp(i K^T X) \exp(-i K^T V t)$$

where V is the advection velocity of the field.

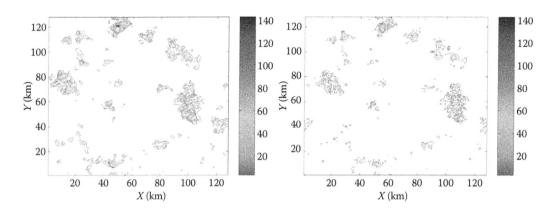

FIGURE 7.7 Example of two rain fields generated from [k] at t_o (left) and $t_o + 12$ minutes (right). The advection velocity of the field is 20 km h^{-1} along the (O_x) axis.

On the other hand, rain field time development is introduced by constraining the coefficients a_K to evolve in time, so that their correlation should decrease exponentially with time lag Δt

$$\mathrm{corr}[a_K(t + \Delta t), a_K(t)] = \exp(-\Delta t / \tau_K)$$

where τ_K is a time constant accounting for the time correlation of the field at spatial frequency k.

This is accomplished by letting the coefficients $a_K(t)$ satisfy Markov equations in time (Bell, 1987). This approach can be extended to the generation of attenuation fields whenever the attenuation distribution is supposed to be lognormal. Figure 7.7 shows modeling of the rain field at t_o ($\mu_{\mathrm{lnR}} = 1.14$ mm h^{-1}, $\sigma_{\mathrm{lnR}} = 1.1$ mm h^{-1}) and at $t_o + \Delta t$, where $\Delta t = 12$ minutes. The overall advection velocity of the field is supposed to be 20 km h^{-1} along the (O_x) axis.

It can be mentioned that this initial work on the space–time variability of rainfall has been extended for mesoscale rainfall characterization in recent activities in the remote-sensing domain (Kundu and Bell, 2003).

Gremont and Filip (2004) presented a spatiotemporal rain attenuation model for application to FMTs. Their approach (Gremont, 2002) is directly based on the Bell approach, except that they act directly on the rain attenuation distribution. In Karagiannis et al. (2012), a methodology to generate a multidimensional field using a stochastic differential equation is employed to extend the formalism and could allow the generation of nonstationary processes.

7.3.2 STOCHASTIC MODELS USED TO GENERATE RAIN RATE FIELDS IN HYDROLOGY

7.3.2.1 String-of-Beads Models

The string-of-beads model is a stochastic rainfall model based on the combined observations of a large network of daily rain gauges and an S-band weather radar in South Africa.

The model is designed to perform two tasks: first, the simulation of long sequences (of a year-long duration) to obtain variables such as average areal rainfall, and wet–dry area ratio (WAR); the second task of the model is to adapt it to perform *nowcasts* of rain for use in real-time flood management (Pegram, 2002).

The main characteristics a model must capture are the arrival of storms, the duration and intensity of rain, and a realistic movement of cells in direction and velocity. The basic proposal of the model is to identify the location of wet and dry events over the time axis. Once the position of a rain event is determined in time, it is modeled in space as a convective or stratiform event, with length constrained by the length of the event determined before, and also described by a velocity and direction. The forecasting component is included by making an explicit use of previous images to create

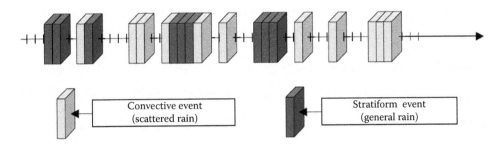

FIGURE 7.8 Diagram of the 1D process in time, the "string."

the next rain image within an event, using a linear combination plus noise. This approach renders usable forecasts up to an hour.

In the "string" part of the model, the climate sequence of rain can be seen as an intercalation of generally wet periods and longer dry periods. Considering this, modeling the process starts by properly addressing the presence of a wet or dry period, thus modeling the time structure.

This one-dimensional wet/dry process can be seen as a string on which "beads" representing the rain event over an area are threaded. The string is therefore the alternation of wet and dry periods. Furthermore, a wet period is subdivided into two: a scattered rain period related to top convective events, and a general rainy period associated to stratiform events. This is shown in Figure 7.8. The process is modeled by a three-state Markov chain. The description of the model is given in Pegram and Clotier (2001).

In the "bead" part of the model, each radar image is treated as a sample from an exponentiated and spatially correlated Gaussian field. The pixel-scale intensity (PSI) of each image is considered to be modeled by a lognormal function, described by two parameters, which vary from image to image.

In image analysis mode, a rain rate estimate averaged over a pixel site u with (x_u, y_u) is a realization $x(u)$ of a random variable $X(u)$. The data $x(u)$ are assumed to be drawn from a lognormal distribution with estimated parameters m and s (corresponding to the real μ and σ). Each pixel is normalized and standardized as $y(u) = [\text{Ln}(x(u) - m]/s$, and $y(u) = -3$ if $x(u) = 0$.

The images $y(u)$ are then transformed into $Y(f)$ via fast Fourier transform (FFT) and the resulting spectrum fitted by a function $p(f) \sim f^\beta$. The exponent is obtained after radially averaging the power spectrum, assuming isotropy. Its typical range is between 1.9 and 2.6. After determining this value, a power law filter is constructed and the field $Y(f)$ is divided by it, resulting in a field $z(u)$, which if the assumption of lognormality holds, should be Gaussian but still with a temporal structure. In addition to this and according to the authors, due to the nature of the original data (integers), before the temporal correlation can be studied, the data must be transformed again into normality by associating cumulative probabilities to each element and drawing them from a uniform distribution.

7.3.2.2 Modified Turning Bands Model

The modified turning bands model (MTB) (Mellor, 1996; Mellor and Metcalfe, 1996; Mellor and O'Connell, 1996), is a stochastic representation of space–time rain fields that allows modeling of the fields with parameters directly obtained from radar observations. The model is designed to represent the primary features of frontal rainfall such as bands, clusters and cells, and principal sources of the inhomogeneity on the fields. It is based on a technique developed by G. Matheron in 1973 to generate Gaussian fields based on the sum of several stochastic processes and called the turning bands method. The structure of the model is shown in Figures 7.9 and 7.10.

In this structure, three lines (one horizontal, the others at prescribed angles to the horizontal) are projected from a determined point in space. Over these lines, a stochastic process is superposed. The characteristic features of these processes are perpendicularly projected into the area over which the storm is to be synthesized. The projections are called "bands." Over the horizontal

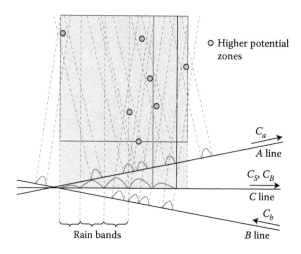

FIGURE 7.9   Structure of a rain field using the MTB model. Event at time t.

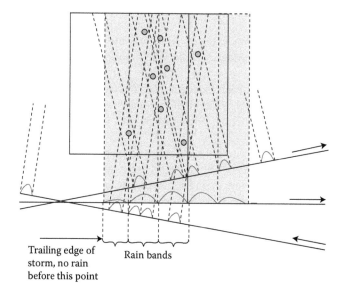

FIGURE 7.10   Structure of the event at time $t + \Delta t$.

line, a function (superposition of sinusoids) is placed to modulate the details of the field so that it decays on the edges and contains inside a banded structure. The function changes in time so that the storm is seen to move along the c-line.

The dynamic behavior of the line c can be modeled as

$$c(x,t) = \begin{cases} A\sin(\dfrac{\pi}{L}(x - c_S t))\left\{ \sin\left[\dfrac{2\pi}{W}(x - c_B t) + \phi\right] + 1 \right\}, & 0 \le x - c_S t \le L \\ 0 \quad other\ cases \end{cases}$$

where A is the amplitude of the modulating function, W is the width of the rain bands, L is the length of the storm, ϕ is the phase offset (arbitrary), and c_S and c_B are the velocities of the storm and rain bands, respectively. Removing the $c_S t$ and $c_B t$ factors yields the static modulating function.

FIGURE 7.11 Steps of the formation of the rain field with the MTB model (ordered from low to high).

The stochastic processes along the inclined lines are Poisson point processes, with inverted parabolas centered on the points. These processes slide along the lines with predetermined speeds c_a and c_b. The inverted parabolas are projected into the region as bands and summed, the result of which is then multiplied by the modulating function projected from the c-line.

The field generated by this procedure is taken to represent the time-varying spatial potential function of rain cells in the region, and is used as the rate function of an inhomogeneous Poisson process that controls the birth of the cells. That is, where the field takes a high value, there is a large probability of rain cells occurring there, and where the field is low, there is a smaller probability of rain cell occurrence. The rain cells themselves are described as parabolas of revolution whose maximum height is the peak cell rate. Figures 7.11 and 7.12 are examples of the results of the model.

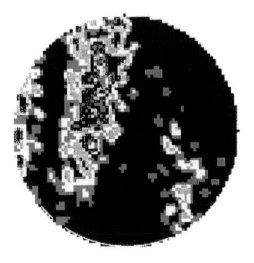

FIGURE 7.12 A simulated rain field.

Parameter estimation is achieved by analyzing the dynamics of rain cells and rain bands in radar images. Rain cell estimation algorithms have been developed that are based on the technique of full correlation analysis and the theoretical covariance properties of the model, while the overall storm velocity, orientation, and width are estimated from the storm/rain band profile.

The MTB-modeling system can be used either in the simulation mode or in real-time forecasting mode. In the latter case, the model is conditioned on the latest radar images, which, through an inverse technique, is used to infer the potential function for rain cells. Having estimated the velocity and curvature of the storm, an ensemble of forecasts can then be generated, with the spread of the ensemble representing uncertainty about the possible future evolution of the storm.

7.3.3 FRACTAL FIELDS

A large number of studies (Lovejoy, 1982; Rys and Waldvogel, 1986; Féral and Sauvageot, 2002; Callaghan and Vilar, 2003) suggest that fractal methods may be of use in characterizing the shapes of rain cells.

In general, the fractal dimension D characterizes any self-similar system; if the linear dimension of a fractal observable is changed by a scale factor f, then, for any value of f, the values of the fractal observable will be changed by the factor f^D. For surfaces, the value of the surface dimension, D_S, lies in the range of $2 \leq D_S \leq 3$. A smooth surface has $D_S = 2$. Similarly, for a contour line, the dimension of the line D_L satisfies $1 \leq D_L \leq 2$, and $D_L = 1$ for smooth lines. The more twisted and "wriggly" the contour line is, the higher the value of D_L. If pathological cases are disregarded (Voss, 1985), a planar section of a fractal surface has

$$D_L = D_S - 1$$

The fractal nature of rain has been studied for many years, and its characterization as a fractal and multifractal field is well documented (Lovejoy and Schertzer, 1990; Olsson and Niemczynowicz, 1996). Unfortunately, there is little consensus on the exact form of the fractal field, due to differing methods of calculating the fractal dimension and/or characteristic multifractal function. The majority of the published works use multifractal methods to deal with the intermittency and anisotropy of the rain field (Lovejoy and Schertzer, 1990; Olsson and Niemczynowicz, 1996; Deidda, 1999).

Callaghan developed a procedure to simulate rain fields that produces simulated fields that are monofractal fields. This is justified by multifractal analysis of meteorological radar data recorded in the south of England (Callaghan, 2004), which shows that log rain rate fields may be accurately characterized as monofractal fields, as their $K(q)$ functions are straight lines. The transformation of the variables from rain rate to log rain rate allows us to linearize the problem, showing that rain rate fields can be characterized as "meta-Gaussian." This is in agreement with other works published recently (Ferraris et al., 2003).

The rain field simulator presented by Callaghan is based on the Voss successive random additions algorithm for generating fractional Brownian motion in multiple dimensions (Voss, 1985). This method of simulating rain fields uses a monofractal, additive (in the logarithmic domain) discrete cascade model for simulating rain fields in two spatial dimensions. The model produces events on demand, customized to an input rain rate parameter and the desired rain event type (stratiform or convective).

The details of the procedure have been presented in Callaghan (2006). The rain field simulator produces two-dimensional snapshots of simulated rain fields. The resulting simulated rain field has an appropriate spectral density exponent, fractal dimension, and behavior that is visually consistent with the experimentally observed convective or stratiform type of events (according to what is desired) (Figure 7.13).

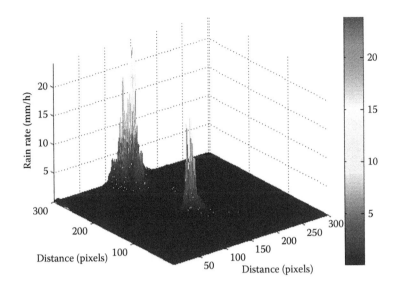

FIGURE 7.13   Simulated convective fractal rain field with $R_{0.01} = 26$ mm h^{-1}.

The time variation of the two-dimensional-simulated rain field was achieved through the use of Taylor's frozen storm hypothesis (Taylor, 1938). This hypothesis postulates the equivalence between the spatial autocorrelation at a fixed point in time and the temporal autocorrelation at a fixed position in space. However, for this to hold, the spatial argument of the former must be interpreted as a time lag of the latter and the spatiotemporal field must be a fixed spatial field moving with a constant velocity. It has been shown (Zawadski, 1973) that this holds approximately for time lags under about 40 minutes.

To implement this time variation, each simulated rain field can be cut down into a number of smaller "snapshots" of appropriate size. The variation in time can then be simulated by moving the position of the snapshots in the full-size-simulated array by a small amount Δx and Δy, and then saving the resulting new snapshot to give the rain field at time $t = 1$. The process is repeated to give snapshots at time $t = 1...n$. Δx and Δy are chosen to give similar wind velocities as those experienced during measured rain events in the climate of interest.

This method of introducing time variation produces rain fields that advect, but do not evolve in time. Figure 7.14 gives a schematic example of this process using a successive random additions algorithm for three-dimensional arrays. The successive random addition algorithm can also be implemented in three dimensions to give a temporal variation provided by the third dimension of the array. This assumes an equivalency of time and space, and provides simulated rain fields that evolve in time, but do not advect.

It is then possible to use a combination of multiple events to form a simulated database covering a specific time period. Each realization of the simulated rain fields is independent of each other; hence, simulating multiple arrays and taking the simulated rain rate value from the same location in multiple arrays will not produce a realistic simulated time series. However, it is possible to produce a large number of simulated arrays and taking them as a group, scale them to $R_{0.01}$. These simulated databases are then capable of reproducing long-term statistics.

7.3.4   CORRELATED TIME SERIES

7.3.4.1   Correlated Time Series from Measurements

This technique proposed in Bertorelli and Paraboni (2005) allows generating a large set of experimentally derived time series of rain attenuation to be employed in the channel simulation of satellite-based resource-sharing networks.

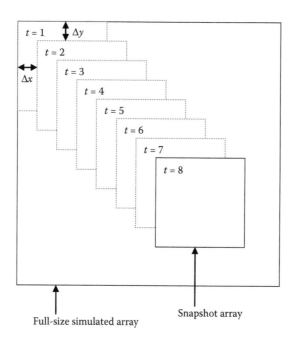

FIGURE 7.14   Schematic diagram of the process used to simulate the variation in time of the rain field using the frozen storm hypothesis.

The time series derived from the ones collected with the Italian satellite (ITALSAT) at 18.7, 39.6, and 49.5 GHz in Spino d'Adda near Milan according to the procedure outlined in Bertorelli et al. (2008) are allocated to the various links with an algorithm that, on the long run, preserves both the statistical behavior of all the sites and the spatial correlation between the couples of sites.

The procedure can be divided into two distinct processes: the first one, called a *conditioning process*, expresses the probability that two or more sites are in rainy conditions at the same time (common rainy time), whereas the second one, called a *conditioned process*, expresses the probability of the attenuations conditioned to be within the common rainy time.

Being rain attenuation by far, the most important tropospheric impairment, especially at high frequencies, the evaluation of the correlation between rain attenuations in different locations as a function of distance is particularly useful for time-series generators to be employed in the simulation of satellite networks, where the proper degree of spatial correlation must be preserved to obtain reliable results in the performance assessment. Moreover, the knowledge of this parameter is essential for communication systems that make use of the site diversity technique, where the performance of the system improves with increasing the separation between the primary station and the secondary station because the attenuation becomes more and more decorrelated with distance.

Data relative to the spatial distribution of rain attenuation used during the European Space Agency (ESA) contract have been derived from the meteorological radar data of Spino d'Adda; these data have been analyzed to determine the correlation function of both the conditioning and conditioned processes governing the rain log attenuation. All the correlations have been obtained with a procedure that makes use of a parameter called "*statistical dependence index*" χ, expressing the probability that two or more variables (rain log attenuation in this case) exceed a selected threshold at the same time, normalized to the same probability in the case of statistical independence. The full procedure and the algorithm that permits to obtain the correlation coefficient ρ from χ are described in Bertorelli and Paraboni (2005).

The statistical dependence index χ has been directly evaluated from radar data for distances smaller than 50 km and then integrated with data obtained from rain gauges with higher site separations. The correlation coefficient (integrated with the one for large distances taken from Barbaliscia and Paraboni, 2002) is well represented by the following expression:

$$\rho = w_1 \exp\left(-\frac{d}{d_1}\right) + w_2 \exp\left(-\frac{d}{d_2}\right)^2$$

where the parameters are different if we take into consideration the conditioning or the conditioned process. For the conditioning process we have

$$w_1 = 0.7$$
$$d_1 = 86 \text{ km}$$
$$w_2 = 1 - w_1 = 0.3$$
$$d_2 = 700 \text{ km}$$

whereas for the conditioned one the parameters are

$$w_1 = 0.94$$
$$d_1 = 25 \text{ km}$$
$$w_2 = 1 - w_1 = 0.06$$
$$d_2 = 500 \text{ km}$$

The equations above can be used to calculate the covariance matrix of rain attenuation for both the processes (the conditioning and the conditioned one), the only data to be used as input being the distances for every couple of sites. In fact, the process is an N-dimensional Gaussian one (N being the number of Earth stations constituting the satellite network; as known in the literature, rain attenuation is normally distributed); the whole distribution is fully defined by the "single-site" parameters (marginal averages and standard deviations of rain attenuation) together with the correlation coefficients relative to couples of variables. For this reason, there is no need to classify occurrences relative to triplets or combinations of a higher order of sites.

This technique is based on the following mechanism: first, a pseudorandom binary process assigns a descriptor that characterizes the state (rainy/nonrainy) of any point of the served region in a basic time interval of a given length T (e.g., 1 hour); then, a second descriptor assigns the median log attenuation. Even though the first descriptor is binary, it is assimilated to a Gaussian process in which the generation of the samples is driven by a threshold calibrated so as the "rainy" state occurs for a wanted percentage of time. The second descriptor is normal by nature (average of Gaussian variables) and is driven by the marginal probability characterizing the point in question. A set of ITALSAT-derived time series of duration T is then assigned to the points after applying the proper attenuation scaling in frequency, polarization, and elevation. It is important to outline that the generators of the Gaussian processes can be trained to preserve the local marginal statistics and the correlations expressed by the covariance matrix.

7.4 DOWNSCALING OF METEOROLOGICAL DATA

The models described in the previous sections have an intrinsic limitation in terms of coverage. In fact, the statistical assumptions made to describe the fields are usually accurate on limited areas and over a limited duration. For instance, cellular models realistically depict the position of rain cells on an area of some tens of kilometers due to the assumptions on the random walk and stochastic models that have a limited range of validity due to the stationarity assumption used in the modeling. A possible way to overcome these limitations is to delegate the description of the scale larger than the scales of validity of the modeling to meteorological data from observation or from meteorological models. These data do not have those intrinsic limitations due to the statistical assumption and it may be a solution to extend the domain of applicability of the techniques mentioned in the previous sections. This process is often referred to as downscaling. The various methodologies presented in the previous sections could be used in conjunction with coarse-resolution data. It requires the use of a model to enable the inclusion of the coarse meteorological information in the high-resolution locally applicable models. These coarse-resolution data can be remote-sensing data or outputs of numerical weather forecast models. Various techniques using these concepts developed for propagation applications are described in this section.

7.4.1 Downscaling of Data from Numerical Weather Forecast Models

7.4.1.1 Local Area Model

In Hodge et al. (2006) and Hodge and Watson (2007), a method to generate attenuation time series is presented and is based on the use of proven numerical weather prediction models (i.e., UM or MM5) in conjunction with a propagation model. This approach has two unique aspects. First, the spatial correlation and dynamic behavior of the attenuation fields are inherited from the meteorological environment at some kilometers of resolution. Second, the model can provide forecasts of attenuation.

The approach takes historical estimates of the meteorological environment from NWP systems and estimates the link attenuation. The conversion from the meteorological environment into a radio-wave propagation environment uses a set of physical models. Each model provides an estimate of a propagation phenomenon based on the physics of the problem rather than a statistical data-analysis approach.

A schematic diagram of the model is given in Figure 7.15. The system is capable of generating time-coincident multiple-site historical attenuation time series for simulation and applications of these time series include investigations into the effects of site and time diversity. Examples of outputs from this model are given below.

Time series have been simulated on a 3-month period (April 1 to July 1, 2004), using UM NWP archived data with 1-hour interval and UK weather radar network archived with a 15-minute

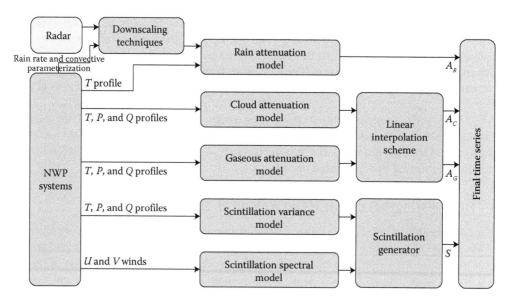

FIGURE 7.15 Simplified box diagram of the time-series generator.

interval, and with a Voss downscaling algorithm (Paulson, 2004) for getting a 10-second time step. The applicability of the methodology is limited by the domain covered by the meteorological data but could be applicable to other meteorological products if available over the considered areas.

7.4.1.2 Reanalysis Models

Another option that enables the use of globally available inputs spanning a long duration is the use of reanalysis data as base data from the downscaling process. As a counterpart to the global data availability, the scales range in which statistical models are wider. An example of such a methodology is presented in Jeannin et al. (2012).

The simulation procedure relies on the generation of space–time-correlated Gaussian random fields, converted into rain rate fields constrained by the local rain rate distribution using the algorithm of Bell (1987). The local average at each time step of the rain rate fields is constrained by data from ECMWF ERA-40 database through the analysis of the fraction of the resolution cell covered by rain. As a consequence, high-resolution-simulated fields reproduce, on average, the rain amount given by the reanalysis database over any area and also reproduce the average evolution of rain event over long durations. The advection of the field is matched to the wind vector at the 700-hPa pressure level, derived from reanalysis data. An example of a simulated rain rate field is illustrated in Figure 7.16.

The evolution of the field during its motion is simulated using a Markov process on the components of the spatial Fourier transform of the field. The correlation parameters of the fields have been derived from different weather radar datasets in temperate and tropical climates by converting the correlation of rain rate fields into an equivalent correlation of an underlying Gaussian field. Scales lower than the resolution of the radar data are not simulated as the correlation under this resolution is not known. Examples of time series generated with this methodology are illustrated in Figure 7.17.

This methodology has also been extended to include gaseous and cloud attenuation in Jeannin et al. (2011).

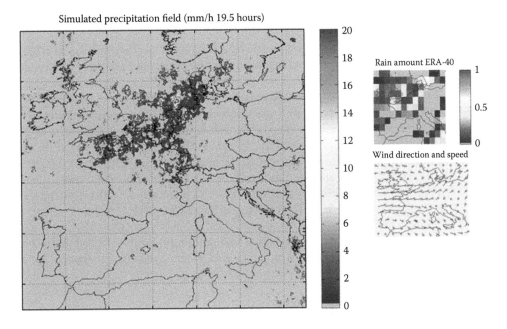

FIGURE 7.16 Example of a rain rate field generated over Europe. The thumbnails on the right are corresponding to the concurrent ERA-40 inputs.

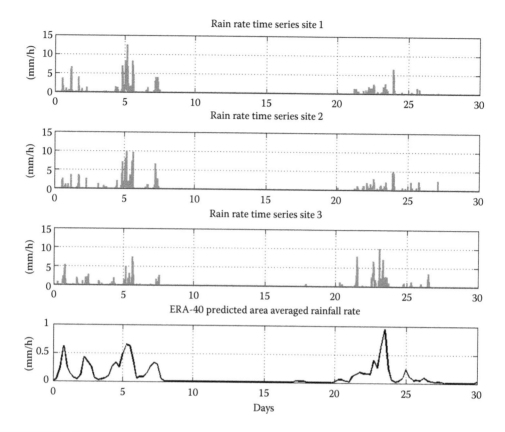

FIGURE 7.17 Simulation outputs for January 2000 in three sites and concurrent ERA-40 rain outputs. Site 1 and 2 are 10 km away. Site 3 is 100 km away from the two others.

7.4.2 Downscaling from Radar Data

As what is done with data from a numerical weather forecast model, a convenient way to reduce the scales covered by the statistical algorithms may be to use data already at a relatively high resolution such as weather radar data.

The spatial and temporal resolutions of radar data may not be sufficient to account for the complete depiction of the space–time-correlated rain attenuation fields. The pixel size is reduced using multifractal disaggregation based on a multiplicative cascade technique (Paulson and Zhang, 2007), and constrained by the extrapolation of multifractal scaling exponents measured on the same data. Finer-scale temporal sampling was introduced by fractal interpolation using the local average subdivision (LAS) algorithm of Fenton and Vanmarcke (1990). The resulting fine-scale, spatial–temporal rain fields were used to simulate arbitrary terrestrial networks of microwave links spanning squares of size 50 km (Paulson and Zhang, 2009; Paulson and Basarudin, 2011). The initial coarse-scale data used in Paulson and Basarudin (2011) are historical databases of Opera rain radar data as shown in Figure 7.18. These data are composites of various European networks and this technique can thus be applied to any kind of link within this area. It could also be extended where other weather radar data do exist.

7.4.3 Direct Use of Meteorological Products

With the increase of computing capabilities and the improvement of a numerical model, it could be foreseen to directly convert the outputs from nonhydrostatic high-resolution numerical weather

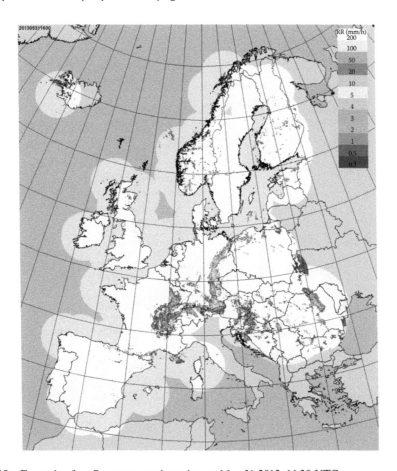

FIGURE 7.18   Example of an Opera composite: rain rate May 31 2013, 16:30 UTC.

simulation models. In fact, these models such as WRF (weather research and forecast model), Meso-NH, and COSMO are capable of describing the state of the atmosphere with a high resolution. Giving at the input of these models, coarse resolution reanalysis of the past weather enables to perform reanalysis of the state of the atmosphere at a high resolution. In fact, subkilometric resolution can be reached by those models and the resolution can be sufficient to compute the propagation effects directly on the raw output of the data that comprises all the necessary outputs to accurately compute the various propagation effects. In fact profiles of pressure, temperature, humidity, and of the specific content various states of water (Morrison, 2010) (rain, ice, cloud, snow…). The general workflow to simulate the propagation fields, further described in Jeannin et al. (2014a,b) is illustrated in Figure 7.19.

An example of simulation is illustrated in Figure 7.20.

Still, the simulation over large areas and long durations is extremely demanding in terms of computation power and require a large storage capability. In addition, as it constitutes a completely independent way of simulating the propagation effects that have their own bias, the coherence with measured statistics is not ensured and has to be confirmed.

It is however a promising alternative to models based on a statistical analysis of various sources of data as the modeling of the underlying physical processes ensures the soundness of the results. For instance, it guarantees the consistency between the various effects generated due to the explicit resolution of the various microphysics quantities. It is also worth noticing that this kind of methodology could also be used to forecast the propagation impairments and could thus be used to control the FMTs for which no feedback from the user terminal on the quality of the link is available (i.e.,

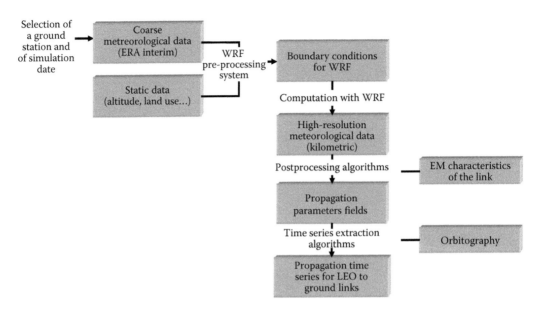

FIGURE 7.19   Workflow to obtain space–time propagation parameter fields from mesoscale weather forecast models.

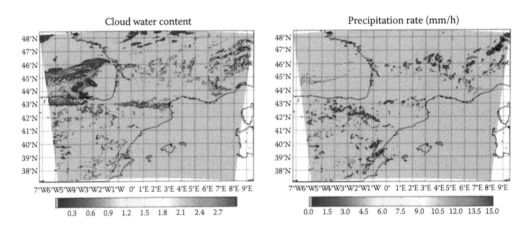

FIGURE 7.20   Example of an integrated cloud liquid water vapor and precipitation rate generated by WRF over south of France and north of Spain with a resolution of 1 km with WRF.

for instance in broadcasting systems). In this case, the initialization of the model has to be made using not only reanalysis data that correspond to past weather conditions but using forecasts from large-scale meteorological models.

7.5   SUMMARY AND CONCLUSION

The use of space–time-correlated propagation impairment fields is of prime importance for the optimization of radio frequency (RF) systems using an adaptive fade countermeasure to cope with atmospheric attenuation.

As there are no data available that are directly suitable to act as input for system needs, the use of models is required. A large variety of algorithms and techniques have been developed to model these propagation fields and in particular rain attenuation, ranging from a cellular model to stochastic field models. Whatever the methodology, the aim of these models is to reproduce observed statistical features from available data such as weather radar data. Cell models aim at reproducing the shape and density of the rain cells over a radar coverage. Random field models are designed to reproduce the correlation properties observed on the data and fractal ones are designed to reproduce the scaling properties.

The domain of applicability of these models without support from external data is bounded to some hundreds of kilometers due to the assumptions inherent to the modeling. It can be extended by coupling the models with coarse data available over a large scale as data from remote sensing and from numerical weather forecast models.

Up to now, most of the effort has been devoted to rain that is the major source of impairment at Ka band but moving to higher frequencies such as Q/V band, the modeling of the other impairments should get additional attenuation.

REFERENCES

Awaka, J., A three-dimensional rain cell model for the study of interference due to hydrometeor scattering, *Journal of Communication Research Laboratory*, 36(147), 13–44, 1989.

Barbaliscia, F. and A. Paraboni, Multiple-site attenuation prediction models based on the rainfall structures (stratified and convective) for advanced TLC or broadcasting systems, in *XXVIIth General Assembly of the International Union of Radio Science*, Maastricht, The Netherlands, August 17–24, 2002.

Bell, T.L., A space–time stochastic model of rainfall for satellite remote-sensing studies, *Journal of Geophysics Research*, 92(D8), 9631–9643, 1987.

Bertorelli, S. and A. Paraboni, Simulation of joint statistics of rain attenuation in multiple sites across wide areas using ITALSAT data, *IEEE Transactions on Antennas and Propagation*, 53(8), 2611–2622, 2005.

Bertorelli, S., C. Riva, and L. Valbonesi, Generation of attenuation time series for simulation purposes starting from ITALSAT measurements, *IEEE Transactions on Antennas and Propagation*, 56(4), 1094–1102, 2008.

Bousquet, M., L. Castanet, L. Féral, J. Lemorton, and P. Pech, Application of a model of spatial correlated time series into a simulation platform of adaptive resource management for Ka-band OBP satellite systems, in *COST 272–280 International Workshop on Satellite Communications from Fade Mitigation to Service Provision*, Noordwijk, The Netherlands, May 2003.

Callaghan, S.A., Fractal analysis and synthesis of rain fields for radio communication systems, PhD thesis, University of Portsmouth, June 2004.

Callaghan, S.A., *Fractal Modelling of Rain Fields: From Event-on-Demand to Annual Statistics*, EuCAP 2006, Nice, November 2006.

Callaghan, S.A. and E. Vilar, Analysis of the fractal dimension of rain rate contours with reference to wide area coverage of satellite communications, *Foreign Radioelectronics. Successes of Modern Radioelectronics*, No. 6, Radiotechnika, Moscow, 2003.

Capsoni, C., F. Fedi, C. Magistroni, A. Pawlina, and A. Paraboni, Data and theory for a new model of the horizontal structure of rain cells for propagation applications, *Radio Science*, 22, 395–404, 1987.

Capsoni, C., L. Luini, A. Paraboni, and C. Riva, Stratiform and convective rain discrimination deduced from local P(R), *IEEE Transactions on Antennas and Propagation*, 54(11), 3566–3569, 2006.

Castanet, L., X. Boulanger, B. Casadebaig, B. Gabard, F. Carvalho, F. Lacoste, and F. Rousseau, Multiple-site diversity experiments at Ka and Q bands carried out by CNES and ONERA in south of France, in *20th Ka Band Utilization Conference*, Salerno, Italy, 2014.

Deidda, R., Multifractal analysis and simulation of rainfall fields in space, *Physics and Chemistry of the Earth Part B—Hydrology, Oceans and Atmosphere*, 24(1–2), 73–78, 1999.

Fenton, G.A. and E. Vanmarcke, Simulation of random fields via local average subdivision, *ASCE Journal of Engineering Mechanics*, 116(8), 1733–1749, 1990.

Féral, L. and H. Sauvageot, Fractal identification of supercell storms, *Geophysical Research Letters*, 29(14), 31-1-31-4, 2002.

Féral, L., H. Sauvageot, L. Castanet, and J. Lemorton, HYCELL: A new hybrid model of the rain horizontal distribution for propagation studies. Part 1: Modelling of the rain cell, *Radio Science*, 38(3), 1056, 2003a, doi: 10.1029/2002RS002802.

Féral, L., H. Sauvageot, L. Castanet, and J. Lemorton, HYCELL: A new hybrid model of the rain horizontal distribution for propagation studies. Part 2: Statistical modelling of the rain rate field, *Radio Science*, 38(3), 1057, 2003b, doi:10.1029/2002RS002803.

Féral, L., H. Sauvageot, L. Castanet, J. Lemorton, F. Cornet, and K. Leconte, Large-scale modelling of rain fields from a rain cell deterministic model, *Radio Science*, 41, 2006.

Ferraris, L., S. Gabellani, U. Parodi, N. Rebora, J. von Hardenberg, and A. Provnzale, Revisiting multifractality in rainfall fields, *Journal of Hydrometry*, 4, 544–551, 2003.

Goldhirsh, J. Two-dimension visualization of rain cell structures, *Radio Science*, 35(3), 713–729, 2000, doi:10.1029/1999RS002274.

Gremont, B., *Simulation of Rainfield Attenuation for Satellite Communication Networks*, Paper PM3014, First COST 280 Workshop, Malvern, UK, July 2002.

Gremont, B. and M. Filip, Spatio-temporal rain attenuation model for application to fade mitigation techniques, *IEEE Transactions on Antennas and Propagation*, 52(5), 1245–1256, 2004.

Hodge, D.D. and R.J. Watson, *Predicting the Effects of Weather on EHF and SHF Communications Links*, EUCAP, Edinburgh, 2007.

Hodge, D.D., R.J. Watson, and G. Wyman, Attenuation time series model for propagation forecasting, *IEEE Transactions on Antennas and Propagation*, 54(6), 1726–1733, 2006.

Jeannin, N., G. Carrié, L. Castanet, and F. Lacoste, Space–time propagation model of total attenuation for the dimensioning of high frequency multimedia SatCom systems, in *17th Ka and Broadband Communications, Navigation and Earth Observation Conference*, Palermo, Italy, 2011.

Jeannin, N., L. Castanet, J. Radzik, M. Bousquet, B. Evans, and P. Thompson, Smart gateways for terabit/s satellite, *International Journal of Satellite Communication Network*, 32, 93–106, 2014a, doi: 10.1002/sat.1065.

Jeannin, N., L. Féral, H. Sauvageot, L. Castanet, and F. Lacoste, A large-scale space–time stochastic simulation tool of rain attenuation for the design and optimization of adaptive satellite communication systems operating between 10 and 50 GHz, *International Journal of Antennas and Propagation*, 2012, 2012, doi: 10.1155/2012/749829.

Jeannin, N., M. Outeiral, L. Castanet, C. Pereira, D. Vanhoenacker-Janvier, C. Riva, C. Capsoni, L. Luini, M. Cossu, and A. Martellucci, Atmospheric channel simulator for the simulation of propagation impairments for Ka band data downlink, in *The 8th European Conference on Antennas and Propagation (EuCAP 2014)*, The Hague, The Netherlands, pp. 1–5, April 6–11, 2014b.

Karagiannis, G.A., A.D. Panagopoulos, and J.D. Kanellopoulos, Multidimensional rain attenuation stochastic dynamic modeling: Application to earth–space diversity systems, *IEEE Transactions on Antennas and Propagation*, 60(11), 5400–5411, 2012.

Koudelka, O., Q/V-band communications and propagation experiments using ALPHASAT, *Acta Astronautica*, 69(11–12), 1029–1037, 2011, doi: 10.1016/j.actaastro.2011.07.008.Alphasat.

Kundu, P.K. and T.L. Bell, A stochastic model of space–time variability of mesoscale rainfall: Statistics of spatial averages, *Water Resources Research*, 39(12), 1328, 2003.

Lovejoy, S., Area–perimeter relation for rain and cloud areas, *Science*, 216, 185–187, 1982.

Lovejoy, S. and D. Schertzer, Multifractals, universality classes and satellite and radar measurements of clouds and rain fields, *Journal of Geophysical Research*, 95, 2021–2034, 1990.

Luini, L. and C. Capsoni, MultiEXCELL: A new rain field model for propagation applications, *IEEE Transactions on Antennas and Propagation*, 59(11), 4286–4300, 2011, doi: 10.1109/TAP.2011.2164175.

Mellor, D., The modified turning bands (MTB) model for space–time rainfall. I. Model definition and properties, *Journal of Hydrology*, 175, 113–127, 1996.

Mellor, D. and A.V. Metcalfe, The modified turning bands (MTB) model for space–time rainfall. III. Estimation of the storm/rainband profile and a discussion of future model prospects, *Journal of Hydrology*, 175, 161–180, 1996.

Mellor, D. and P.E. O'Connell, The modified turning bands (MTB) model for space–time rainfall. II. Estimation of raincell parameters, *Journal of Hydrology*, 175, 129–159, 1996.

Misme, P. and P. Waldteufel, A model for attenuation by precipitation on a microwave earth–space link, *Radio Science*, 15, 655–665, 1980.

Morello, A. and U. Reimers, DVB-S2, the second generation standard for satellite broadcasting and unicasting, *International Journal of Satellite Communications and Networking*, 22, 249–268, 2004.

Morrison, H., An overview of cloud and precipitation microphysics and its parameterization in models, in *11th WRF Users' Workshop*, National Centre for Atmospheric Research (NCAR), Boulder, CO, June 2010.

Olsson, J. and J. Niemczynowicz, Multifractal analysis of daily spatial rainfall distributions, *Journal of Hydrology*, 187, 29–43, 1996.

Panagopoulos, A.D., P.-D. Arapoglou, and P.G. Cottis, Satellite communications at Ku, Ka and V bands: Propagation impairments and mitigation techniques, *IEEE Communication Surveys and Tutorials*, 6(3), 2–14, 2005.

Paulson, K., Fractal interpolation of rain rate time-series, *Journal of Geophysics Research—Atmosphere*, 109, 2004, doi:10.1029/2004JD004717.

Paulson, K.S. and H. Basarudin, Development of a heterogeneous microwave network, fade simulation tool applicable to networks that span Europe, *Radio Science*, 48(4), 2011, doi: 10.1029/2010RS004608.

Paulson, K.S. and X. Zhang, Estimating the scaling of rain rate moments from radar and rain gauge, *Journal of Geophysical Research*, 112, 2007, doi: 10.1029/2007JD008547.

Paulson, K.S. and X. Zhang, Simulation of rain fade on arbitrary microwave link networks by the downscaling and interpolation of rain radar data, *Radio Science*, 44, RS2013, 2009, doi: 10.1029/2008RS003935.

Pegram, G., Space time modeling of rainfall using the string of beads model: Integration of radar and rain-gauge data, Web site: http://www.fwr.org/wrcsa/1010102.htm.

Pegram, G.G.S. and A.N. Clotier, High resolution space–time modeling of rainfall: The string of beads model, *Journal of Hydrology*, 241, 26–41, 2001.

Rys, F.S. and A. Waldvogel, Fractal shape of hail clouds, *Physical Review Letters*, 56(7), 784–787, 1986.

Taylor, G.I., The spectrum of turbulence. *Proceedings of the Royal Society of London A*, 164, 476–490, 1938, doi: 10.1098/rspa.1938.0032.

Van de Kamp, M.M.J.L., Short-term prediction of rain attenuation using two samples, *Electronics Letters*, 38(23), 1476–1477, 2002.

Voss, R.F., Random fractal forgeries, in R.A. Earnshaw (ed.), *Fundamental Algorithms for Computer Graphics*, NATO ASI Series F, Computer and System Sciences, Berlin, Vol. 17, 1985.

Zawadzki, I.I., Statistical properties of precipitation patterns. *Journal of Applied Meteorology*, 12(3), 459–472, 1973.

8 Impact of Clouds from Ka Band to Optical Frequencies

Lorenzo Luini and Roberto Nebuloni

CONTENTS

8.1 INTRODUCTION

Near-Earth and deep-space telecommunications have progressively moved toward higher-frequency bands to satisfy the increasing demand for high data rates. Currently, the microwave bands used for satellite communications are mostly Ku (user links) and Ka bands (feeder links). Bandwidth congestion and the need for new broadband and interactive services are leading toward even higher carrier frequencies such as in the Q/V and W bands. Moreover, advances in laser system technology have made attractive free space optics (FSO) for Earth–space communications (Hemmati et al., 2011; NASA, 2014).

In such systems, the impact of propagation impairments produced by Earth's atmosphere is dramatic and must be quantified through suitable models. Furthermore, ad hoc fade mitigation techniques must be adopted, because simple countermeasures such as adding a fixed extra power margin to the link budget are inadequate due to the unfeasible margins required (for instance, W-band total attenuation can exceeds 40 dB during 0.1% of time in a temperate continental area, for a reference link elevation of 40° [Riva et al., 2014]). Though the impact and the relative importance of the impairments are highly frequency dependent, both microwave and optical propagation through the atmosphere produce signal attenuation due to gases, precipitation (snow, rain, and hail), clouds, and turbulence, as well as depolarization due to rain droplets and ice crystals. A review of radio propagation effects up to above 30 GHz is provided in Ippolito (2008), whereas the W band is specifically addressed in Riva et al. (2014); optical propagation effects are investigated by Andrews and Phillips (1998), Strickland et al. (1999), and Kim et al. (2001).

Clouds are usually the major source of signal fading in a nonrainy atmosphere at Q band and above. In the microwave region, up to about 100 GHz, cloud attenuation is proportional to the frequency and depends only on the profile of the liquid water content (i.e., the mass of liquid water present in the unit volume of air) across the propagation path. The optical properties of the atmosphere are deeply affected by the concentration, size, and phase of atmospheric particulates. Specifically, the occurrence of thick cloud layers across the propagation path usually results in beam blockage, hence making FSO impracticable.

This chapter reviews the impact of clouds on microwave and optical wave propagation for applications to satellite communications. The fundamentals of cloud microphysics and electromagnetic theory are summarized in Sections 8.2 through 8.4. Section 8.5 presents some models proposed so far in the literature to estimate cloud attenuation. The overview is far from being comprehensive, as we have selected only those models that do not simply define relationships typically between the specific attenuation and liquid water content, but also take advantage of input meteorological data to predict yearly cloud attenuation. Models of increasing complexity are addressed, starting from semiempirical methodologies receiving as input simple local measurements such as the ground temperature and water vapor concentration (Altshuler–Marr and Dintelmann–Ortgies) to much more complex models exploiting global atlases of cloud cover and water content (Dissanayake–Allnut–Haidara [DAH], Salonen and Uppala, the latter currently adopted in Recommendation ITU-R P.840-6, and stochastic model of clouds [SMOC]).

8.2 PHYSICAL BACKGROUND

Clouds are made of small water droplets or ice crystals suspended in the air, which form after water vapor condensation or sublimation when the air is cooled. The only difference between clouds and fog is that in the latter case the process occurs close to the ground. Fog and cloud particles modify the optical properties of the atmosphere, reducing its transparency. In meteorology, fog is present when the horizontal visibility drops below 1 km (Ahrens, 2013). Although to a much lesser extent, microwave propagation is affected as well by fog and clouds.

The effects on wave propagation through the troposphere depend on their frequency of occurrence and on their physical properties (basically, particle phase, shape, size, and number concentration). The following is a brief review of the cloud properties of interest for the calculation and the prediction of microwave and optical wave attenuation through cloud layers.

8.2.1 CLOUD OCCURRENCE

Figure 8.1 shows a global map of cloud occurrence drawn from the reanalyses of the numerical weather prediction (NWP) data produced by the European Centre for Medium-Range Weather Forecasts (ECMWF, 2014). The quantity shown is the total cloud cover, that is, the fraction of the area of every $0.75° \times 0.75°$ pixel that is covered by clouds. Four daily samples of the total cloud cover are available. The data are averaged over the period 1984–2013.

The distribution of cloud cover depends on latitude, climate, and environment (land or sea). Clouds are more frequent over marine pixels due to the availability of sea water: for instance, the average total cloud cover exceeds 0.5 in about 80% of marine pixels and in 60% of continental pixels, respectively. At mid-latitude (30–60°), about half of the pixels over land exhibit a fraction of cloudy area slightly larger than 0.5.

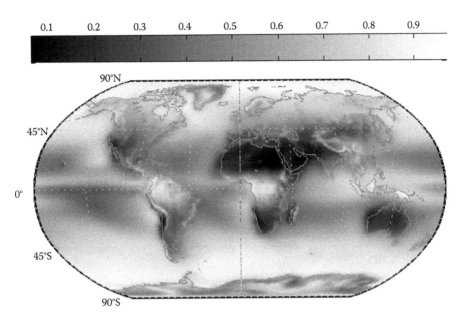

FIGURE 8.1 Global map of the fraction of sky covered by clouds (ECMWF data over the period 1984–2013).

8.2.2 CLOUD CLASSIFICATION

Following a well-established classification, clouds can be distinguished into four major groups according to the height of their base above Earth's surface (Ahrens, 2013):

1. *High-level clouds:* Cold bodies consisting of ice crystals whose height is approximately between 5 and 13 km at mid-latitudes.
2. *Mid-level clouds:* Primarily made of water droplets with ice crystals sometimes present (height 2–7 km at mid-latitudes).
3. *Low-level clouds:* Composed of water droplets and, if the air temperature is cold enough, of ice and snow (height <2 km at mid-latitudes).
4. *Vertically developed clouds:* Turbulent structures generated through either thermal convection or frontal lifting, and usually associated with precipitation.

The bounds of each group depend on latitude, the limiting base heights of clouds increasing when one moves from polar to tropical latitudes through temperate regions. Clouds belonging to a given group are further subdivided into types according to their appearance, as reported in Table 8.1.

Cloud type information is used by some statistical models to predict cloud attenuation, as we shall see later. In this respect, the products of actual high-resolution radiometers on board Earth observation satellites include cloud masks and cloud type identification (see, for instance, EUMETSAT, 2014).

8.2.3 MICROPHYSICAL PROPERTIES OF CLOUDS

Cloud microphysics deals with the properties of the elemental components of a cloud body, that is, liquid droplets and ice crystals. In this respect, the size of the particles (relative to the wavelength of the incident radiation in the present context) and their number concentration determine the amount of energy subtracted from the incoming wave. The particle size of every atmospheric population is spread over a spectrum, otherwise named particle size distribution (PSD). A flexible and convenient

TABLE 8.1

Cloud Classification and Approximate Height of Cloud Base According to Ahrens (2013)

Group	Name	Type	Tropical (km)	Mid-Latitude (km)	Polar (km)
1	High clouds	Cirrus (Ci)	6–18	5–13	3–8
		Cirrostratus (Cs)			
		Cirrocumulus (Cc)			
2	Middle clouds	Altostratus (As)	2–8	2–7	2–4
		Altocumulus (Ac)			
3	Low clouds	Stratus (St)	0–2	0–2	0–2
		Stratocumulus (Sc)			
		Nimbostratus (Ns)			
4	Vertically developed clouds	Cumulus (Cu)	–	–	–
		Cumulonimbus (Cb)			

mathematical representation for the PSD of a large variety of atmospheric particles from aerosols to clouds and precipitation is the four-parameter modified gamma function (Deirmedjian, 1969):

$$n(r) = ar^{\alpha} \exp(-br^{\gamma}) \tag{8.1}$$

where r is the particle radius and $n(r)\,dr$ is the number of particles per unit volume of air with radius comprised between r and $r + dr$. The parameters α, b, and γ determine the shape and the width of the distribution around its maximum, whereas a is proportional to the particle number concentration. Other widely used PSDs such as the three-parameter gamma distribution, the exponential distribution, and the power-law distribution are special cases of Equation 8.1 (Petty and Huang, 2011). A few authors use the three-parameter lognormal distribution to model the PSD, which is not considered here.

Alternative formulations of Equation 8.1 make use of different sets of independent parameters, corresponding to physical quantities. Commonly used quantities in cloud microphysics are the particle concentration N, the liquid water content w, the mode radius of the distribution r_c, and the effective radius r_e, which can be expressed in terms of a, b, α, and γ as follows:

$$N = \int_{r_1}^{r_2} n(r)dr \simeq \frac{a}{\gamma} \frac{\Gamma((\alpha+1)/\gamma)}{b^{(\alpha+1)/\gamma}} \tag{8.2}$$

$$w = \frac{4}{3}\pi\rho_w \int_{r_1}^{r_2} r^3 n(r)dr \simeq \frac{4}{3}\pi\rho_w \frac{a}{\gamma} \frac{\Gamma((\alpha+4)/\gamma)}{b^{(\alpha+4)/\gamma}} \tag{8.3}$$

$$r_c = \left(\frac{\alpha}{b\gamma}\right)^{1/\gamma} \tag{8.4}$$

$$r_e = \frac{\int_{r_1}^{r_2} r^3 n(r)dr}{\int_{r_1}^{r_2} r^2 n(r)dr} \simeq \frac{\Gamma((\alpha+4)/\gamma)}{\Gamma((\alpha+3)/\gamma)} \frac{1}{b^{1/\gamma}} \tag{8.5}$$

where $\Gamma(x)$ is the gamma function (equal to $x!$ for an integer x) and ρ_w is the water density.

Despite physical arguments requiring a nonzero lower bound (r_1) and a finite upper bound (r_2) on particle size, integration in Equations 8.2, 8.3, and 8.5 can be safely extended from 0 to infinity with negligible errors. In fact, from a well-known property of the modified gamma function, the nth-order moment of the distribution can be written in closed form as

$$M_n = \int_0^\infty r^n n(r) dr = \frac{a}{\gamma} \frac{\Gamma((\alpha+n+1)/\gamma)}{b^{(\alpha+n+1)/\gamma}} \tag{8.6}$$

As we shall see later, the liquid water content, that is, the mass of cloud liquid water per unit volume of air, is a key quantity in the computation of microwave attenuation by liquid clouds. A similar formula as Equation 8.3 for the ice water content is not as straightforward because ice crystals are a mixture of ice and air, hence some equivalent size should be used. The mode radius r_c is the size corresponding to the maximum of the PSD. The effective radius r_e is of widespread use in radiative transfer parameterizations and satellite retrieval algorithms (Slingo et al., 1982; King et al., 2003). A comprehensive review of the properties of the modified gamma distribution and of the conversion relationships between representations based on different sets of parameters is given in Petty and Huang (2011).

The microphysical properties of clouds are highly variable due to the complexity of the formation processes. The growth of droplets depends on several factors such as size, type, and concentration of air pollutants, which act as cloud condensation nuclei, temperature, turbulence scale and intensity, and cooling rate of particles in the atmosphere. Nonetheless, it is useful to give a few numbers, representative of the average characteristics of the different cloud types. According to ITU-R P.840–6, the liquid water content in fog is typically about 0.05 g/m³ for medium fog (visibility in the order of 300 m) and 0.5 g/m³ for thick fog (visibility in the order of 50 m). Maximum values for nonprecipitating clouds are comparable to the ones of thick fogs. As a term of comparison, the liquid water content of rain is 0.31 g/m³ at 5 mm/h (when one uses the classical exponential PSD derived by Marshall and Palmer [1948]).

The PSD of clouds has been measured by several investigators through the years. As a general rule, clouds growing in a strongly polluted atmosphere consist of a higher number of small particles than the ones developing in a natural atmosphere (Hess et al., 1998). Table 8.2 lists the values of r_c, α, and γ for a number of PSDs of liquid clouds modeled through modified gamma distributions (Deirmedjian, 1969; Tampieri and Tomasi, 1976; Shettle, 1989; Miles et al., 2000). As most of the PSDs were given in normalized form, w is reported for a nominal $N = 100$ particles/cm³, which occasionally produces anomalously high values of w. Scaling N to the actual value of the liquid water content is required when one needs to calculate signal attenuation from the PSDs through the physical approach described in Section 8.4.3. Shettle's models for clouds (see Table 8.2) are three-parameter gamma distributions that are used by the low-resolution propagation model LOWTRAN7 (Abreu and Anderson, 1996). Tampieri and Tomasi (1976) reduced a large dataset of empirical size spectra into a limited number of PSDs, which are representative of average spectra for different types of fogs and clouds. A total 13 Cu distributions are shown in the table ($\alpha = 1$–8, $\gamma = 0.48$–11.75, and $r_c = 0.64$–15.98 μm). Finally, mean values for the datasets of marine and continental St clouds are reported as classified by Miles et al. (2000). On the other side, ice clouds are composed of crystals with several different habits (i.e., shapes), each one with its own microphysical parameters. Hence, the identification of a single PSD for ice clouds is generally a complicated task. Two PSDs for thin cirrus and cirrus are given in Shettle's paper.

In using PSD data, one should be aware of the inner variability of the phenomenon rather than of measurement limitations. The numbers in Table 8.2 should be taken as the product of curve fitting on experimental size spectra averaged in time and space, and representative of average cases. The type of clouds and the associated size spectra can vary considerably depending on the environment and on local conditions. Moreover, the microphysical properties can change in space within a cloud

TABLE 8.2

PSD Parameters r_c, α, and γ and Liquid Water Content w Calculated Assuming a Nominal $N = 100$ Particle/cm³ Concentration for Several Cloud Types

Source	Cloud Type	r_c (μm)	α	γ	w (g/m³)	Comments
Shettle (1989)	Cu	6	3	1	0.401	
	St	3.33	2	1	0.116	
	St/Sc	2.67	2	1	0.060	
	As	4.5	5	1	0.102	
	Ns	4.7	2	1	0.326	
Deirmendjian	C1 (Cu)	4.0	6	1	0.062	C6 is the superior mode of a precipitating
(1975)	C5 (Ns)	6.0	4	1	0.296	cloud showing a bimodal PSD
	C6 (Ns)	20.0	2	1	25.083	
Tampieri and	Cu(1)	0.64	1	0.48	0.141	Data reduction and fit of experimental size
Tommasi (1976)	Cu(2)	3.19	3	1.04	0.056	spectra
	Cu(3)	3.53	8	2.15	0.023	A lot of Cu size spectra were reportedly
	Cu(4)	4.80	5	2.16	0.065	bimodal: they can be obtained as linear
	Cu(5)	5.28	3	1.07	0.243	combinations of the modes reported here
	Cu(6)	5.39	8	1.51	0.096	Not shown here are numbers for St, As, Ns,
	Cu(7)	6.92	5	6.6	0.132	and orographic clouds due to lack of
	Cu(8)	7.06	8	4.8	0.149	measured data
	Cu(9)	7.72	2	2.39	0.356	
	Cu(10)	10.26	8	7.41	0.429	
	Cu(11)	10.40	4	2.34	0.675	
	Cu(12)	15.6	9	11.75	1.450	
	Cu(13)	15.98	2	3.97	2.085	
Miles et al.	St (marine)	10.3	7.6	1	0.91	Average parameters from data:
(2000)	St (continental)	5.0	7.7	1	0.10	$w = 0.18$ g/m³ (marine)
						$w = 0.19$ g/m³ (continental)

Note: Cu = cumulus, St = stratus, St/Sc = stratus/stratocumulus, As = altostratus, Ns = nimbostratus.

(e.g., from base to top). Finally, the averaging scale (both in time and in space) of the measurements produces different estimates of the PSD parameters (Miles et al., 2000). For instance, significant differences have been reported between marine and continental low-level stratus (but also within each group). The former exhibits small number concentration and large mean and effective radius, and vice versa for the latter, due to the higher concentration of atmospheric condensation nuclei over land. Moreover, continental clouds evidence large variations in the vertical profiles of the number concentration (Miles et al., 2000). Deng et al. (2009) quantified the variations of N, w, and r_e, for warm clouds of different types by *in situ* measurements in a highly polluted area. Large variations of N are evidenced for a specific cloud type. Values exceeding 1000 particles/cm³ are reported in the case of Cu. w spreads over about one order of magnitude for Ns and As (0.01–0.1 g/m³), whereas Cu undergoes even higher variations, w values in excess of 0.1 g/m³ being frequent. The effective radius and its variability within each cloud type goes as the inverse of N. Cu exhibits the smallest values (r_e in the range 2.5–5 μm), whereas Ns has the largest particles (5–9 μm).

It follows that simple propagation models relying on the knowledge of the PSD (see Section 8.4.2) can only work on a statistical basis. A straightforward approach would be that of scaling the PSD according to the time series of the liquid water content (which are available from measurements or NWP models), while assuming an average PSD shape at least for a given cloud type. On the other hand, synthesizing instantaneous time series of the signal received after propagation through clouds would require simulation of the evolution of the atmospheric droplets. Alternatively, a stochastic approach based on reproducing the features evidenced by measurement signals may be used.

8.3 REFRACTIVE INDEX OF LIQUID WATER AND ICE

The refractive index n quantifies the effect of the chemical composition, thermodynamic phase, and temperature of the cloud droplets on their electromagnetic and optical characteristics. The double-Debye model (Manabe et al., 1987) provides an analytical expression for the real and imaginary parts of the permittivity of water ε' and ε'', respectively, for frequencies up to 300 GHz over a temperature range from $-40°C$ to $30°C$:

$$\varepsilon'(f) = \varepsilon_2 + \frac{(\varepsilon_0 - \varepsilon_1)}{1 + (f/f_D)^2} + \frac{(\varepsilon_1 - \varepsilon_2)}{1 + (f/f_S)^2} \tag{8.7}$$

$$\varepsilon''(f) = \frac{f(\varepsilon_0 - \varepsilon_1)}{f_D[1 + (f/f_D)^2]} + \frac{f(\varepsilon_1 - \varepsilon_2)}{f_S[1 + (f/f_S)^2]} \tag{8.8}$$

where

$\varepsilon = \varepsilon' + j\varepsilon'' = n^2$

$\varepsilon_0 = 77.66 + 103.3(\Theta - 1), \quad \varepsilon_1 = 5.48, \quad \text{and} \quad \varepsilon_2 = 3.51$

$f_D = 20.09 - 142.4(\Theta - 1) + 294(\Theta - 1)^2 \quad \text{and} \quad f_s = 590 - 1500(\Theta - 1) \quad \text{(GHz)}$

$\Theta = \dfrac{300}{T}$ (with T in kelvin)

Figure 8.2 shows the real and imaginary parts of the refractive index of liquid water and ice from the visible (400 nm, i.e., 750 THz) to the microwave band (10 cm, i.e., 3 GHz). As for liquid water, the data are the ones tabulated by Hale and Querry (1973) and Deirmendjian (1975) in the range 0.4–200 μm and 0.2–1 mm, respectively, whereas in the microwave range, the Debye equations above have been used. The refractive index is plotted at 25°C. As for ice, the data are taken from a recent compilation that encompasses measurements from the UV to the microwave range (Warren and Brandt, 2008), and for temperatures near the melting point (–7°C).

The real part of the refractive index for ice and water is similar in the range from the visible up to the third optical window (1550 nm) and nearly invariant with the wavelength. On the other side, the imaginary part, which is responsible of absorption, undergoes heavy variations, increasing by more than six orders of magnitudes from 400 nm to 2 μm.

8.4 PROPAGATION THROUGH CLOUDS

Wave attenuation through a population of atmospheric particles as clouds or precipitation is usually calculated according to one of the following two approaches:

- *Physical models:* Based on the microphysical and electromagnetic properties of individual particles. First, the interaction between the incoming radiation and a single particle is considered. The individual contributions are subsequently added up to find the amount of energy subtracted to the incoming wave. The task is rather demanding if multiple scattering is taken into account.
- *Empirical models:* Rely on empirical relationships linking wave attenuation to an easy-to-measure quantity such as the liquid water content in the case of liquid clouds, the ice water content for cirri, or, again, the precipitation rate in the case of rain.

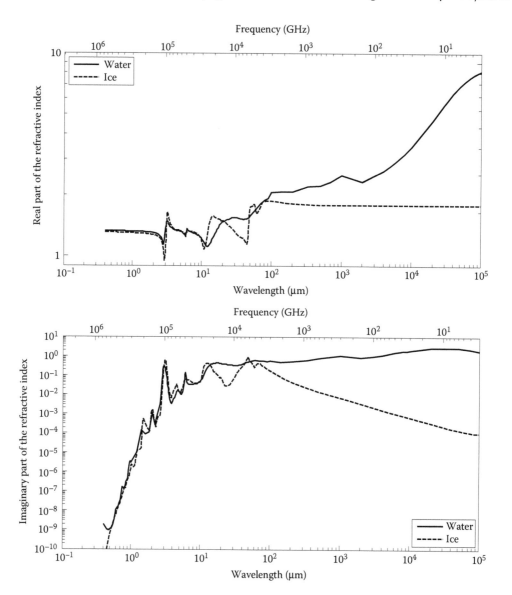

FIGURE 8.2 Real and imaginary part of the refractive index of water (25°C) and ice (−7°C).

Physical models provide accurate results but they require data such as the PSD, which are usually available only in a few sites (not mentioning the critical aspects discussed in Section 8.2.3). Empirical models are not as accurate but on the other side they can be in principle applied on a global scale.

The following is a brief review of the fundamental concepts that allow quantification of the interaction between an electromagnetic wave and a population of particles.

8.4.1 WAVE–PARTICLE INTERACTION

When an electromagnetic wave impinges on a particle, part of the incident energy is scattered in all directions and part of it is absorbed by the particle and dissipated into heat. On the whole, a certain amount of energy is subtracted to the incoming wave. Following Van de Hulst's reasoning

(Van de Hulst, 1957), the energy flux scattered by a generic particle within an infinitesimal solid angle $d\Omega$ in the direction (ϑ, φ) is constant and can be written as

$$dP_{sca} = Ir^2 d\Omega = \frac{I_0}{k^2} F(\vartheta, \varphi) d\Omega \tag{8.9}$$

where I is the scattered intensity (which decreases as $1/r^2$), I_0 is the intensity of the incoming wave (a scalar plane wave is assumed for the sake of simplicity), r is the distance from the particle, k is the propagation constant in the medium surrounding the particle (e.g., air), and $F(\vartheta, \varphi)$ is a dimensionless function that describes the angular distribution of the scattered intensity. Note that, while the energy of the incident radiation is conveyed along parallel rays, the spherical scattered wave irradiates in the form of a diverging pencil of rays.

By integrating Equation 8.9 over a spherical surface A with radius r, we obtain the total energy flux scattered in all directions:

$$P_{sca} = \oint_A Ir^2 d\Omega = \frac{I_0}{k^2} \int_{4\pi} F(\vartheta, \varphi) d\Omega \tag{8.10}$$

It follows from Equation 8.9 that the ratio between dP_{sca} and I_0 has the dimensions of a differential area

$$d\sigma_{sca} = \frac{Ir^2 d\Omega}{I_0} = \frac{1}{k^2} F(\vartheta, \varphi) d\Omega \tag{8.11}$$

Similarly, let us define an area σ_{sca} such that the total energy flux P_{sca} scattered in all directions equals the intensity of the incoming wave multiplied by σ_{sca}, that is,

$$P_{sca} = \sigma_{sca} I_0 \tag{8.12}$$

From Equations 8.10 and 8.12, we get

$$\sigma_{sca} = \frac{1}{k^2} \int_{4\pi} F(\vartheta, \varphi) d\Omega \tag{8.13}$$

In a similar way, if part of the incoming wave is absorbed by the particle, we define an area σ_{abs} such that

$$P_{abs} = \sigma_{abs} I_0 \tag{8.14}$$

Finally, if P_{ext} is the amount of energy flux subtracted to the incoming wave, then

$$P_{ext} = \sigma_{ext} I_0 \tag{8.15}$$

Energy conservation requires that

$$\sigma_{ext} = \sigma_{sca} + \sigma_{sca} \tag{8.16}$$

σ_{sca}, σ_{abs}, and σ_{ext} are named scattering cross section, absorption cross section, and extinction cross section, respectively. If σ_{ext} and I_0 are known, the energy flux subtracted to the wave is determined.

In general, σ_{ext} depends on the wavelength and on the polarization of the incident wave, on particle shape, size, and chemical composition, and on its orientation with respect to the direction of the incident wave.

8.4.2 PROPAGATION THROUGH A LAYER OF PARTICLES

We now turn to investigate wave interaction with a population of particles of different size and with identical optical properties and homogeneously distributed in space. If the particles behave as independent scatterers (see below), σ_{sca}, σ_{abs}, and σ_{ext} can be replaced by the corresponding cross sections per unit volume. For instance, in the case of extinction:

$$\beta_{ext} = \int_{r_1}^{r_2} \sigma_{ext}(r)\,n(r)\,dr \tag{8.17}$$

where $n(r)$ is the PSD. β_{ext} is usually referred to as volume extinction coefficient and is typically expressed in Np/km. In general, r is a certain size representative of particle shape. As droplets within (liquid) water clouds are well approximated by spheres, r is their radius. Moreover, Mie theory provides a closed-form expression for σ_{ext} in the case of homogeneous spheres at an arbitrary wavelength (Deirmedjian, 1969).

We are now in a position to calculate the attenuation experienced by the traveling wave recalling the energetic definition of the extinction cross section. Let us consider a plane wave propagating along the z-axis, which enters a slab of length L along z, uniformly filled with scatterers whose size distribution is described by $n(r)$. As the incident wave travels through an infinitesimal layer of width dz, the corresponding decrease of energy flux (assuming a unit cross section) is

$$dI = -\left[\int_{r_1}^{r_2} \sigma_{ext}(r)n(r)dr\right]I(z)dz \tag{8.18}$$

The energy flux after a propagation distance z is therefore

$$I(z) = I_0 \exp\left\{-\left[\int_{r_1}^{r_2} \sigma_{ext}(r)n(r)dr\right]z\right\} = I_0 \exp(-\beta_{ext}z) \tag{8.19}$$

where I_0 is the energy flux at $z = 0$. The above equation known as Beer–Lambert law, when evaluated at $z = L$, returns the energy flux of the outgoing wave. The fraction of energy transmitted is therefore

$$\tau(z) = \frac{I(z)}{I_0} = \exp(-\beta_{ext}z) \tag{8.20}$$

and it is called transmissivity. In microwave engineering, the above ratio is more commonly measured in dB, and named wave attenuation A, that is,

$$A(\lambda) = \beta_{ext}(\lambda)L \quad (\text{dB}) \tag{8.21}$$

where L is in km and β_{ext} is expressed in units of dB/km, and often called specific attenuation, and the wavelength dependence is made explicit. Finally, using physical units and in the case of water clouds, we can rewrite Equation 8.17 as follows:

$$\beta_{ext}(\lambda) = 4.343 \times 10^{-3} \int_{r_1}^{r_2} \sigma_{ext}(r,\lambda) n(r)\, dr \quad \text{(dB/km)} \tag{8.22}$$

where β_{ext} is in dB/km, $\sigma_{ext}(r,\lambda)$ is in μm^2, $n(r)$ is in $\mu m^{-1} cm^{-3}$, and r is in μm.

There are two fundamental assumptions behind the above reasoning:

- *Independent scattering:* Wave scattering by a given particle is independent of other particles. The above assumption holds if the particles are far enough from each other (with respect to their size). If the relative positions of the particles change randomly, independent scattering implies that the individual contributions can be added up in intensity rather than in amplitude and phase.
- *Multiple scattering* effects are negligible. In principle, the incident wave on a particle is attenuated by scattering and absorption on other particles and the particle is excited by the scattered waves as well. When these processes are negligible, the electromagnetic radiation incident on each particle is the original wave (though attenuated). Therefore, by superposition of effects, the total energy removed from the incoming wave by a cloud of particles is the sum of the ones subtracted by the single particles.

Maximum particle concentrations in very dense clouds reach 1000 per cm^3, which means a mutual distance on the order of 50 times their size (assuming a 10 μm radius): independent scattering is therefore fulfilled. Furthermore, the atmospheric particulate is continuously reshuffled by the action of the gravitational force, wind, and turbulence, which results in a random phase relationship between scattered waves.

A straightforward explanation of multiple scattering effects descends from the quantum theory of light. When a photon collides with a particle it is either scattered, with a probability equal to its albedo $\sigma_{sca}/\sigma_{ext}$, or absorbed with a probability $\sigma_{abs}/\sigma_{ext}$. The direction of a scattered photon depends on the scattering phase function of the particle. In the single-scattering approximation, photons that interact with a particle are considered lost. When first-order multiple scattering is taken into account, the photons scattered by a particle are collected by the receiver if the scattering direction is within the angle of view of the receiver. Higher-order scattering involves photons scattered several times before eventually reaching the receiver. Unless a scattered photon travels along the propagation direction of the original wave (forward scattering), it exhibits a propagation delay with respect to the unscattered photons; hence it can be considered as additive noise. To conclude, the forward scattered photons reduce attenuation whereas all the others scattered photons decrease the signal-to-noise ratio (SNR). The impact of multiple scattering depends on the following factors:

- Wavelength
- Characteristics of the incident wave
- Albedo and scattering phase function of the single particle
- Particle concentration
- Path length
- Receiving antenna gain, that is, in optics, size, and field of view of the receiving aperture

Single-scattering approximation is valid if roughly $\tau < 1$ (Ishimaru, 1978), which is reasonable at microwaves where maximum specific attenuation are in the order of a few dB/km. In the optical region, the laser beam propagates only through thin clouds with small water content. In such cases, link geometry and the very narrow field of view of the receiver (of the order of 1 mrad or less) limit the impact of multiple scattering.

The full Mie's formulation for σ_{ext} is not reported here but full details can be found, for example, in Deirmedjian (1969). Here, we just point out that the calculation of σ_{ext} requires the evaluation of an infinite series. In practice, the number of terms to add in order to yield convergence of the series is roughly proportional to the size parameter, kr. Numerical computations are cumbersome, especially if the procedure must be repeated over a wide number of particle sizes, and are not straightforward though the literature on the subject has solved the critical issues (Wiscombe, 1979, 1980; Du, 2004). Free-ware code for scattering calculations is available on the Internet as well (ScattPort, 2014). If the particle size is much larger or smaller than the wavelength, some simple approximations, mentioned in the following, can be used.

8.4.2.1 Rayleigh Approximation

The Rayleigh approximation applies to particles with arbitrary refraction index n and much smaller dimension than the wavelength, that is, for a sphere $kr \ll 1$. By virtue of this, the particle is excited by a uniform incident electric field. Furthermore, Rayleigh scatter requires that the polarization field is established in a short time compared to the period of the wave, which imposes the additional condition $|nkr| \ll 1$. If such is the case, $\sigma_{ext}(r,\lambda)$ takes the following expression derived from Rayleigh theory (McCartney, 1976):

$$\sigma_{ext} = \sigma_{abs} + \sigma_{sca} = \frac{8\pi^2 r^3}{\lambda} \mathrm{Im}\left(\frac{n^2-1}{n^2+2}\right) + \frac{128}{3}\frac{\pi^5 r^6}{\lambda^4}\left|\frac{n^2-1}{n^2+2}\right|^2 \tag{8.23}$$

Moreover, if absorption prevails over scattering, as it is the case for suspended liquid water in the microwave region, we get

$$\sigma_{ext} \simeq \sigma_{abs} = \frac{8\pi^2 r^3}{\lambda} \mathrm{Im}\left(\frac{n^2-1}{n^2+2}\right) \tag{8.24}$$

As σ_{ext} is proportional to the particle volume, by substituting Equation 8.24 into 8.22, we obtain that the extinction coefficient is proportional to the liquid water content w:

$$\beta_{ext} = \frac{0.0819}{\rho_w \lambda} \mathrm{Im}\left(\frac{n^2-1}{n^2+2}\right) w \quad (\text{dB/km}) \tag{8.25}$$

where ρ_w is the liquid water density in g/cm^3, λ is in m, and w is in g/m^3.

According to Ulaby et al. (1981), the above equation holds for droplet radii smaller than 100 µm and for frequencies below 100 GHz. Comparing the specific attenuation (in dB/km) obtained from the Mie's rigorous theory against Equation 8.25 for the cloud types in Table 8.2, relative errors are well within 1% at 100 GHz in all cases except C6 (i.e., a superior mode in a precipitating cloud), while remaining below 4% at 300 GHz. Therefore, Equation 8.25 can be safely used for cloud attenuation calculations at microwaves.

8.4.2.2 Optical Approximation

The geometrical optics representation of light can be used in solving the problem of light scattering by a particle only if the particle size is much larger than λ. If such is the case, rays can be distinguished in two types, according to the way they are produced:

- Reflection and refraction of rays hitting the surface of the particle. Part of the energy carried by these rays is absorbed by the body, and part of it emerges from the particle by direct refraction at its surface or after one or more internal reflections and forms a pattern of scattered light with a certain angular distribution.
- Diffraction of the incomplete wave front formed by the rays passing along the particle. The diffracted light results in a certain distribution of intensity, which, at sufficiently large distance, follows the Fraunhofer diffraction pattern.

An important difference between the above two processes is that the diffracted light depends on the shape and size of the body but it is independent of its nature. Furthermore, diffraction does not change the polarization state and the diffracted pattern is the same whatever the polarization of the incident wave.

If the particle size is much larger than the incident wavelength, the extinction cross section approaches twice the geometric cross section of the particles (Van de Hulst, 1957) hence for spheres:

$$\sigma_{ext} = 2\pi r^2 \tag{8.26}$$

This result can be justified by energetic arguments, provided the sphere is absorbing. In fact, by virtue of Babinet principle (Born and Wolf, 1999), the diffraction pattern is the same (in terms of intensity) as the one produced by replacing the particle with an indefinite black screen normal to the incident radiation with a hole of the same geometrical cross section as the particle. Hence, the energy flux removed by diffraction equals the one intercepted by the particle and lost due to scattering or absorption.

As a consequence, the extinction coefficient can be written as

$$\beta_{ext} = 6.51 \times 10^3 \frac{w}{\rho r_e} \quad (dB/km) \tag{8.27}$$

where w is in g/m^3, ρ is in g/cm^3, and r_e is in μm. Clouds with a moderate liquid water content of 0.1 g/m^3 and a 5 μm effective radius produce a specific attenuation around 130 dB/km in the visible range. Note that (a) for a given value of w, β_{ext} is inversely proportional to r_e, and (b) only two independent parameters are required to calculate β_{ext} in the optical limit (whereas it is not necessary to know the full microphysical details, such as the PSD).

The one in Equation 8.27 is an acceptable approximation for σ_{ext} of cloud droplets in the visible range up to the first and third optical windows. At longer IR wavelengths, the general expression in Equation 8.22 and Mie theory should be used.

8.4.3 Cloud Attenuation

In the microwave band, specific attenuation due to clouds can be estimated by the simple Equation 8.25. Figure 8.3 shows β_{ext} against the frequency in the 3–100 GHz range with the liquid water content as the parameter at two different temperatures (0°C and 25°C).

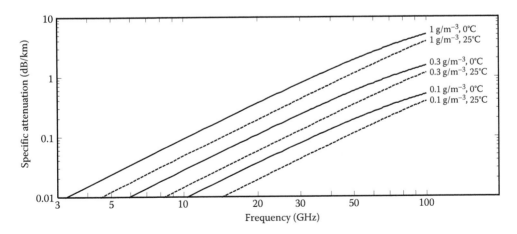

FIGURE 8.3 Specific attenuation as a function of frequency for several liquid water contents and temperatures.

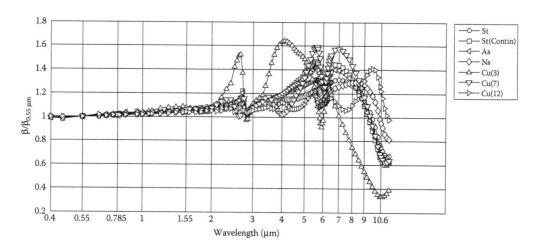

FIGURE 8.4 Normalized extinction coefficient against the wavelength for several cloud types.

On the other side, in the optical region, the microphysics must be taken into account. To get an insight into the dependence of β_{ext} on both λ and the PSD, Figure 8.4 shows the ratio between β_{ext} (λ) and β_{ext} at 0.550 μm (i.e., the middle of the visible window taken here as a reference), over a wide range of wavelengths, including the ones suitable for FSO for a subset of cloud types in Table 8.2. β_{ext} is independent of either λ or the PSD shape up to $\lambda \sim 1$ μm (which descends from Equation 8.27). In the first optical window (0.75–0.85 μm), the PSD dependence is negligible. β_{ext} at 1.550 μm is usually slightly larger than β_{ext} at 0.850 and 0.550 μm, the percent differences being within 10% for all PSDs. On the other hand, large differences are evident when comparing longer wavelengths with the visible range: for instance, the reduction in β_{ext} at 10.6 μm exceeds 20% in a majority of cases. The PSDs with small particles exhibit the higher transmission gain at 10.6 μm.

By rescaling each PSD with the correct value of liquid water content and particle concentration, we get some order of magnitude estimates of optical attenuation for the various cloud types. With this objective, every cloud type has been given a lower and upper bound for the liquid water content as drawn from measurements (Squires, 1958; Durbin, 1959; Zuev, 1970; Miles et al., 2000; Deng et al., 2009). Table 8.3 lists the denormalized PSDs that are a subset of the ones in Table 8.2 as PSDs with similar characteristic parameters, hence with similar attenuation values, have been decimated for the sake of visual clarity. The maximum particle concentration has been limited to 1000 particle/cm^3 (i.e., the maximum reported in the literature), when the liquid water content for a certain PSD would have unacceptably large values (this is the case of C1 and Cu(3) in Table 8.3).

As for ice particles, it is reasonable to state that the extinction coefficient is negligible at microwaves due to the small number concentration of ice particles. In the optical range, specific attenuation is not negligible even at small particle concentrations. The calculation of the extinction coefficient from Equation 8.22 is complicated by the several different crystal shapes (also named "habits") observed in ice clouds. In spite of the presence of diverse crystals, Platt (1997) has shown that in the optical range, the specific attenuation due to ice clouds can be derived from the ice water content w_i through the following power-law relationship:

$$\beta_{ext} = 40.26 w_i^{0.68} \quad (dB/km) \tag{8.28}$$

where w_i is in g/m^3.

TABLE 8.3

PSD Parameters, Lower and Upper Bounds for the Number Concentration and the Liquid Water Content, and Corresponding Extinction Coefficient at W-Band and at Two Optical Wavelengths for Several Cloud Types

Cloud Type	r_c (μm)	α	Γ	N (cm^{-3})	w (g/m^3)	93 (GHz)	0.550 (μm)	10.6 (μm)
Cu	6	3	1	10–210	0.03–0.86	0.1–4.0	17–487	17–479
St	3.33	2	1	10–220	0.01–0.26	0–1.2	8–215	6–160
St/StCu	2.67	2	1	20–750	0.01–0.45	0–2.1	10–468	6–283
As	4.5	5	1	10–100	0.01–0.1	0–0.5	10–96	6–64
Ns	4.7	2	1	3–30	0.01–0.1	0–0.5	6–58	6–55
Cl (Cu)	4.0	6	1	50–1000	0.03–0.62	0.1–2.9	35–723	19–398
Cu(3)	3.53	8	2.15	130–1000	0.03–0.23	0.1–1.1	53–407	18–135
Cu(5)	5.28	3	1.07	10–350	0.03–0.86	0.1–4.0	20–585	18–517
Cu(7)	6.92	5	6.6	20–650	0.03–0.86	0.1–4.0	29–844	20–567
Cu(10)	10.26	8	7.41	10–200	0.03–0.86	0.1–4.0	20–570	19–554
Cu(11)	10.40	4	2.34	5–130	0.03–0.86	0.1–4.0	16–459	17–496
Cu(12)	15.6	9	11.75	2–60	0.03–0.86	0.1–4.0	13–377	16–460
St (continental)	5.0	7.7	1	10–250	0.01–0.26	0–1.2	10–258	6–168

Note: Cu = cumulus, St = stratus, St/Sc = stratus/stratocumulus, As = altostratus, Ns = nimbostratus.

8.5 CLOUD ATTENUATION PREDICTION MODELS

Assessing the impact of clouds on wireless communication systems (microwave range or optical wavelengths) typically consists of predicting the long-term statistics of cloud attenuation along the whole Earth–space link, whereby the power margin to cope with such attenuation can be determined for a given system availability requirement. As a result, with this objective, cloud attenuation prediction models are required to combine the expressions derived so far for the calculation of the specific attenuation based on the microphysical properties of cloud bodies with additional inputs and methodologies specifically addressing the occurrence of clouds over the site of interest (as well as their water content amount), which is obviously tightly bound to meteorology. Indeed, as is clear in Figure 8.1, cloud cover varies significantly from site to site, and so does consequently the overall impact of clouds on the communication system.

Below is the description of some models proposed so far in the literature to estimate cloud attenuation. The overview is far from being comprehensive, as we have selected only those models that do not simply define relationships typically between the specific attenuation and liquid water content, but also take advantage of input meteorological data to predict yearly cloud attenuation statistics (typically, the complementary cumulative distribution function [CCDF]). Models are sorted based on their complexity, starting from semiempirical methodologies receiving as input simple local measurements such as the ground temperature and water vapor concentration to much more complex models exploiting global atlases of cloud cover and water content.

8.5.1 DINTELMANN AND ORTGIES MODEL

One of the early models aimed at predicting cloud attenuation statistics was proposed by Dintelmann and Ortgies (1989). The model relies on sound physical concepts on the formation of clouds associated with water vapor condensation but also includes an empirical component regarding the vertical extension of clouds.

The basic idea of the model is that clouds result from the condensation of water vapor, which takes place when the water vapor concentration at a given height H, $\rho(H)$, exceeds the saturation density ρ_S. Furthermore, the model assumes that $\rho(H)$ can be derived from meteorological ground measurements and that the clouds base lies around the 0°C isotherm height H, which, according to Ito (1989), is related to the ground temperature T_0 as follows (H in km and T_0 in K):

$$H = 0.89 + 0.165(T_0 - 273) \tag{8.29}$$

The water vapor density can be estimated using the equation state assuming an adiabatic process:

$$\rho(H) = \rho_0 \frac{T_0}{T} \left[1 - \frac{k-1}{k} \frac{gH}{RT_0} \right]^{k/k-1} \tag{8.30}$$

where ρ_0 (g/m³) and T_0 (K) are the ground water vapor concentration and temperature, respectively, g is the gravitational acceleration (approximately equal to 9.8 m/s²), R is the fundamental gas constant (for water vapor $R \approx 0.4615$ J/(K g)), k is 4/3 (specific heat ratio for the water vapor molecule), and T (K) is the absolute temperature close to H. In the model, the latter is assumed to be approximately equal to 270 K.

Once $\rho(H)$ is available, the cloud liquid water content w can be calculated as the difference between the water vapor density ρ and the water vapor saturation density ρ_S, whose value is about 3.82 g/m³ for $T \approx 270$ K $= -3.15$°C (Bolton, 1980). In mathematical terms

$$w(H) = \begin{cases} 0 & \rho(H) \leq 3.82 \text{ g/m}^3 \\ \rho(H) - 3.82 & \rho(H) > 3.82 \text{ g/m}^3 \end{cases} \tag{8.31}$$

From w, the specific cloud attenuation can be calculated using Slobin's approximated expression, that is (Slobin, 1982):

$$\alpha = \frac{5.038 \, w \, 10^{-0.744}}{\lambda^2} \tag{8.32}$$

where α is expressed in dB/km and λ is the wavelength in cm. The final step to derive the cloud attenuation A along a slant path with elevation angle θ is

$$A = \alpha \frac{D}{\sin(\theta)} \tag{8.33}$$

where D is the vertical extent of the cloud. While most of the concepts underpinning the Dintelmann and Ortgies model are physically sound, due to the scarce knowledge on D, the authors resorted to the following empirical expression for this parameter (D in km and w in g/m³):

$$D = 0.15 - 0.023 \, w + 0.0055 \, w^2 \tag{8.34}$$

Equation 8.34 was derived from radiometric measurements at 20 and 30 GHz and concurrent ground meteorological data, which were combined to estimate the slant path attenuation ($\theta = 30°$) due to clouds by removing the contribution of oxygen and water vapor.

TABLE 8.4

Inputs, Equations, and the Constants for the Practical Application of the Dintelmann and Ortgies Model

Inputs	Outputs
Surface temperature T_0	Slant path attenuation A
Surface water vapor concentration ρ_0	
Link elevation angle θ	
Wavelength λ	

Equations	Values of Constants	Units
$H = 0.89 + 0.165(T_0 - 273)$	–	T_0 in K H in km
$\rho = \rho_0 \dfrac{T_0}{T}\left[1 - \dfrac{k-1}{k}\dfrac{gH}{RT_0}\right]^{\frac{k}{k-1}}$	$k = 4/3$ $T = 270$ K $R = 0.4615$ J/(K g) $g = 9.8$ m/s^2	T_0 in K ρ and ρ_0 in g/m^3 H in km
$w = \begin{cases} 0 & \rho \le 3.82 \text{ g/m}^3 \\ \rho - 3.82 & \rho > 3.82 \text{ g/m}^3 \end{cases}$	–	w in g/m^3 ρ in g/m^3
$\alpha = \dfrac{5.038\, w\, 10^{-0.744}}{\lambda^2}$	–	α in dB/km w in g/m^3 λ in cm
$D = 0.15 - 0.023\, w + 0.0055\, w^2$	–	w in g/m^3 D in km
$A = \alpha \dfrac{D}{\sin(\theta)}$	–	A in dB α in dB/km D in km θ in rad

The formulation of the Dintelmann and Ortgies model is simple and effective, but, on the other hand, its main drawback comes from the empirical nature of Equation 8.34 whose validity for other sites is an open point. Moreover, an aspect to be further elucidated is when clouds are associated with large weather fronts rather than being formed locally.

Table 8.4 summarizes the inputs, equations, and constants for the practical application of the Dintelmann and Ortgies model.

8.5.2 ALTSHULER AND MARR MODEL

Altshuler and Marr (1989) developed a methodology to estimate cloud attenuation along slant paths from the sole knowledge of the surface absolute humidity ρ_{AH}. The model originates from an extensive measurement campaign performed in Massachusetts, using a dual-frequency radiometer.

The method starts from the investigation of specific attenuation due to fog as a function of the liquid water content presented in Altshuler (1989), where theoretical calculations of the extinction due to fog were given for different values of temperature ($-8°C < T < 25°C$), which the dielectric constant of water depends on according to the values tabulated in Rozenberg (1974), and wavelength, that is, 3 mm $< \lambda < 3$ cm (frequency between 10 and 100 GHz for vacuum). The results derived in Altshuler (1989) for fog are valid for clouds as well because both of them consist of small suspended water droplets, whose typical diameter (in the order of microns) is much smaller than the wavelength range considered. As discussed in Section 8.4.2.1, in this case, the Rayleigh approximation holds for the calculation of the droplets' extinction properties and the specific attenuation only depends on their concentration (number of particles per unit volume), not on their shape. The derived values (attenuation in [dB/km]/[g/m³]) were fitted using the following analytical formulation assuming $T = 10°C$ (same temperature assumed for all the droplets in the cloud):

$$\alpha_w = 1.2 + 0.0371\lambda + \frac{19.96}{\lambda^{1.5}} \tag{8.35}$$

where λ is in mm.

The analysis of the large set of data collected by the radiometers showed that the cloud attenuation and the surface absolute humidity ρ_{AH} measured by collocated ancillary sensors are linearly related. Combining this information with Equation 8.35, the authors derived the following expression for the zenith attenuation due to clouds A_z (dB):

$$A_z = \left(-0.0242 + 0.00075\lambda + \frac{0.403}{\lambda^{1.15}} \right)(11.3 + \rho_{AH}) \tag{8.36}$$

where the absolute humidity ρ_{AH} is expressed in g/m³.

Equation 8.36 is valid for zenithal paths; the vertical extent of clouds is empirically included in the expression because the equation was fitted to the radiometric derived data. The dependence of cloud attenuation A (full cloud cover case) on the elevation angle is further taken into account by complementing Equation 8.36 as follows:

$$A_{full} = A_z D(\theta) \tag{8.37}$$

where

$$D(\theta) = \begin{cases} \dfrac{1}{\sin(\theta)} & \theta > 8° \\ \sqrt{(a_e + h_e)^2 - a_e^2 \cos^2(\theta)} - a_e \sin(\theta) & \theta \le 8° \end{cases} \tag{8.38}$$

$D(\theta)$ is simply calculated as the cosecant of the elevation angle (assumptions of flat Earth and of horizontal homogeneity for clouds) for $\theta > 8°$, while for lower angles, the curved Earth model is needed. In the latter case, additional values are necessary to calculate $D(\theta)$, that is, the effective Earth radius $a_e = 8497$ km and the effective height of attenuating atmosphere h_e (km). The latter, derived in a previous study (Altshuler and Marr, 1988), is a function of the absolute humidity as well (ρ_{AH} in g/m³):

$$h_e = 6.35 - 0.302\rho_{AH} \tag{8.39}$$

The statistical comparison of the cloud attenuation data collected under partial (A_{part}) and full (A_{full}) cloud cover (visual classification) revealed that, while for very low humidities, values are almost coincident, for higher humidities, the partial cloud attenuations are approximately 85% of the attenuations experienced under full coverage. Hence, it is reasonable to assume that

$$A_{part} = 0.85\, A_{full} \tag{8.40}$$

The model proposed by Altshuler and Marr is of easy application as it requires only local surface absolute humidity values as input, but being its formulation partially empirical, its validity in other sites is yet to be evaluated.

TABLE 8.5

Inputs, Equations, and the Constants for the Practical Application of the Altshuler and Marr Model

Inputs	Outputs
Absolute humidity ρ_{AH}	Slant path attenuation A
Link elevation angle θ	
Wavelength λ	

Equations	Values of Constants	Units
$A_z = \left(-0.0242 + 0.00075\lambda + \dfrac{0.403}{\lambda^{1.15}} \right)(11.3 + \rho_{AH})$	–	A_z in dB λ in mm ρ_{AH} in g/m^3
$h_e = 6.35 - 0.302\rho_{AH}$	–	h_e in km ρ_{AH} in g/m^3
$D(\theta) = \begin{cases} \dfrac{1}{\sin(\theta)} & \theta > 8° \\ \sqrt{(a_e + h_e)^2 - a_e^2 \cos^2(\theta)} - a_e \sin(\theta) & \theta \le 8° \end{cases}$	$a_e = 8497$ km	θ in rad h_e in km
$A = A_z D(\theta)$	–	A in dB A_z in dB

Table 8.5 summarizes the inputs, equations, and constants for the practical application of the Altshuler and Marr model.

8.5.3 SALONEN AND UPPALA MODEL (ITU-R P.840-6)

The model proposed by Salonen and Uppala (1991), currently adopted in Recommendation ITU-R P.840-6 (ITU-R P.840-6), relies on the method developed by Liebe et al. (1993), which estimates the path attenuation induced by clouds using microphysical concepts and vertical profiles of temperature (T) and liquid water content (w). According to Liebe et al. (1993), the specific attenuation γ_C (dB/km) due to clouds is expressed as

$$\gamma_C = K_l w \tag{8.41}$$

where w is the liquid water content (g/m^3), f is the frequency in GHz and

$$K_l = \frac{0.819 f}{\varepsilon''(1 + \eta^2)} \tag{8.42}$$

In Equation 8.42, $\eta^2 = (2 + \varepsilon')/\varepsilon''$, while ε' and ε'' are the real and imaginary parts of the electric permittivity of water, respectively. The latter are deduced from the double-Debye model (Manabe et al., 1987), whose Equations 8.7 and 8.8 have been already introduced in Section 8.3.

The significant contribution provided by Salonen and Uppala (1991) lies in the definition of a model to estimate, in a simple way, the distribution of the liquid water content along the path using vertical profiles of temperature (T), pressure (P), and relative humidity (RH). In turn, these quantities can be measured by radiosondes (typically launched at airports up to four times per day) or extracted from NWP products, for instance, made available by the ECMWF (Uppala et al., 2005). The cloud model developed by Salonen and Uppala, subsequently slightly modified in Martellucci et al. (2002) to introduce also the prediction of ice water content, detects clouds in all the layers for which the relative humidity, $RH(h)$, exceeds a critical humidity threshold defined as

$$RH_C(h) = 1 - \alpha\,\sigma(h)(1 - \sigma(h))[1 + \beta(\sigma(h) - 0.5)] \tag{8.43}$$

where $\alpha = 1.0$, $\beta = \sqrt{3}$, and $\sigma(h)$ is ratio of the pressure at the considered height h and at the surface level. If the measured relative humidity is higher than the critical one at the same pressure level ($RH_C(h)$), the level is assumed to be in the cloud. Once a level is evaluated to be in the cloud, the next phase is to predict the liquid water content w (g/m^3) at that level, which can be achieved using the following expression:

$$w = \begin{cases} w_0(1 + cT)\left(\dfrac{h_c}{h_r}\right)p_w(T) & T \ge 0°C \\[3mm] w_0\,e^{cT}\left(\dfrac{h_c}{h_r}\right)p_w(T) & T < 0°C \end{cases} \quad (g/m^3) \tag{8.44}$$

In Equation 8.44, T is the temperature (°C), h_c is the height from the cloud base (m), $w_0 = 0.17$ (g/m^3), $c = 0.04$ (1/°C), and $h_r = 1500$ (m). The cloud liquid water/ice fraction, $p_w(T)$, is given by

$$p_w(T) = \begin{cases} 1 & 0°C < T \\[2mm] 1 + \dfrac{T}{20} & -20°C < T \le 0°C \\[2mm] 0 & T \le -20°C \end{cases} \tag{8.45}$$

According to Equation 8.45, the model predicts that only liquid water and ice are present at high ($T > 0°C$) and at low ($T \le -20°C$) temperature, respectively, while both physical phases coexist for $0°C < T \le -20°C$.

As a result, using the equations above, it is possible to calculate the vertical profile of the cloud-specific attenuation γ_C, as well as the path attenuation A by simple integration. Although interesting and accurate, the methodology presented so far is of quite complex application as it requires the knowledge of the full vertical structure of the troposphere. As a further step, Salonen and Uppala proposed a viable way to simplify the calculation of A using vertically integrated quantities, much easier to handle than whole profiles, by defining the liquid water content reduced to a fixed temperature T_R, $w_R(T_R)$, such that

$$\gamma_C \approx \gamma'_C = K_l(T_R)\,w_R(T_R) \tag{8.46}$$

γ'_C in Equation 8.46 indicates that the information on the temperature (specific cloud attenuation) variation with height can be embedded in w_R. In other words, the specific attenuation due to clouds is calculated as if the temperature within the cloud were always T_R, regardless of the layers' height, and this deviation from the actual values of T is taken into account by modifying w into w_R. This in turn allows the use of the same $K_l(T_R)$ for all cloud layers, and, therefore, to consider the integrated liquid water content reduced to T_R, $W_R(T_R)$ (mm), for the calculation of the whole path attenuation due to clouds (L and l indicate the zenithal cloudy path length and cloud layer height, respectively):

$$A \approx A' = \frac{\displaystyle\int_L K_l(T_R)w_R(T_R,l)\,dl}{\sin(\theta)}$$

$$= K_l(T_R)\frac{\displaystyle\int_L w_R(T_R,l)\,dl}{\sin(\theta)} = \frac{K_l(T_R)\,W_R(T_R)}{\sin(\theta)} \tag{8.47}$$

In Equation 8.47, the dependence on the link elevation angle has been made explicit by the introduction of the cosecant of the elevation angle θ, which indicates that the Salonen and Uppala model considers a simple scaling of the attenuation assuming horizontal homogeneity of clouds. This might represent a limitation for very low values of θ (e.g., <10°), because the probability for the link to intersect different liquid water content values clearly increases.

The expression for w_R has been obtained in Salonen and Uppala (1991) by minimizing the overall discrepancy between γ_C and γ'_C over the frequency range $f_{int} = f_{min} - f_{max}$, which Equation 8.47 is expected to cover. This leads to

$$w_R(T_R) = \frac{w \int_{f\,min}^{f\,max} \gamma_C(w,f,T)\,\gamma_C(w,f,T_R)df}{\int_{f\,min}^{f\,max} [\gamma_C(w,f,T_R)]^2\,df} \tag{8.48}$$

The ITU-R has adopted in Recommendation P.840-6 (ITU-R P.840-6) the approach proposed by Salonen and Uppala to calculate cloud attenuation using W_R, whose statistics are attached to the recommendation as calculated according to Equation 8.48 from *P-RH-T* profiles extracted from the ERA40 database with latitude/longitude grid resolution equal to $1.125° \times 1.125°$ (Uppala et al., 2005); in this case, $\Delta f = 10–60$ GHz and $T_R = 0°C$ were chosen. Indeed, the application of Equation 8.47 is tightly linked to such database, as this simplified approach cannot readily take advantage of alternative (and possibly more accurate) inputs, such as integrated liquid water content data provided by Earth observation sensors (e.g., moderate-resolution imaging spectroradiometer [MODIS, 2012]) and cloud profiling radar (CPR) (Stephens et al., 2002) on board the Aqua and CloudSat satellites, respectively).

Table 8.6 summarizes the inputs, equations, and constants for the practical application of the Salonen and Uppala model adopted in Recommendation ITU-R P.840-6.

TABLE 8.6

Inputs, Equations, and the Constants for the Practical Application of the Salonen and Uppala Model Adopted in Recommendation ITU-R P.840-6

Inputs	Outputs	
Local statistics of reduced integrated liquid water content W_R (ITU-R P.840-6) Link elevation angle θ Frequency f	Slant path attenuation A	

Equations	Values of Constants	Units
$\varepsilon'(f) = \varepsilon_2 + \dfrac{(\varepsilon_0 - \varepsilon_1)}{1 + (f/f_D)^2} + \dfrac{(\varepsilon_1 - \varepsilon_2)}{1 + (f/f_S)^2}$	$\varepsilon_0 = 87.81$	f in GHz
$\varepsilon''(f) = \dfrac{f(\varepsilon_0 - \varepsilon_1)}{f_D[1 + (f/f_D)^2]} + \dfrac{f(\varepsilon_1 - \varepsilon_2)}{f_S[1 + (f/f_S)^2]}$	$\varepsilon_1 = 5.48$ $\varepsilon_2 = 3.51$	
$\eta^2 = \dfrac{2 + \varepsilon'}{\varepsilon''}$	$f_D = 8.97$ GHz $f_S = 442.55$ GHz	
$K_l = \dfrac{0.819 f}{\varepsilon''(1 + \eta^2)}$	–	f in GHz K_l in (dB/km)/(g/m³)
$A = \dfrac{K_l W_R}{\sin(\theta)}$	W_R extracted from the global maps of reduced integrated liquid water content attached to Recommendation ITU-R P.840-6	A in dB K_l in (dB/km)/(g/m³) W_R in mm θ in rad

8.5.4 MASS ABSORPTION COEFFICIENTS TO CALCULATE CLOUD ATTENUATION

The Salonon and Uppala model relies on physically sound concepts and, thus, it is expected to show good prediction accuracy, as indeed verified, for example, in Davies et al. (1998) and Salonen and Uppala (1991). On the one hand, the high-accuracy results can be obtained only using full vertical profiles of temperature and liquid water content (following the approach proposed in Liebe et al. [1993]), while, on the other hand, the simplified version of the methodology (through Equation 8.47) requires as input global maps of integrated reduced liquid water content W_R, which are calculated from NWP products. This poses two main problems: first, a change in T_R (as suggested in Luini et al. [2013] to reduce the approximation error inherent in the use of w_R instead of w) and/or Δf (e.g., to address frequencies higher than 60 GHz with minimum loss of accuracy) would require a full (and computationally heavy) reprocessing of the entire set of ERA40 *P-RH-T* vertical profiles (or of any equivalent global database); second, the accuracy of the resulting W_R (and, hence, of the predicted attenuation using Equation 8.47) might be negatively affected by the coarse spatial and temporal resolution of NWP products.

An alternative approach to efficiently estimate cloud attenuation has been recently proposed in Luini and Capsoni (2014a). The method does not requires the knowledge of the vertical structure of the troposphere, but it involves the use of the mass absorption coefficient for liquid water, $a_W(f)$ in Equation 8.49, widely employed in remote sensing applications to linearly relate the integrated liquid water content W (mm) to the associated attenuation at a given frequency f, A (θ in the elevation angle) (Luini et al., 2007):

$$A(f) = \frac{a_W(f)W}{\sin(\theta)} \tag{8.49}$$

As a matter of fact, a_W can be calculated as the slope of line fitting the values on the W/A plane, which, as mentioned in the previous sections, can be obtained from radiosonde observations (RAOBS) by first applying the Salonen and Uppala model to the vertical profiles of *P-RH-T* for cloud detection and integrated liquid content quantification, and, afterward, by using the model in Liebe et al. (1993) to calculate γ_C (Equation 8.41) and, hence, A. In this way, the variation of the temperature within the cloud profile (and of the associated specific attenuation) is no more embedded in W_R but is taken into account by the mass absorption coefficient a_W.

Luini and Capsoni (2014a) derived a_W from an extensive set of RAOBS data collected routinely twice a day for 10 years (1980–1989) in 14 sites ranging from northern (e.g., Sodankyla, Finland) to southern (e.g., Trapani, Italy) Europe, that is, subject to very different climates. The differences in a_W from site to site were found to be negligible as regards the calculation of cloud attenuation. This suggests that, although both the type and probability of occurrence of clouds differ from site to site, their vertical structure, in terms of relationship among pressure, temperature, and relative humidity, is alike. This allows deriving a simple expression for a_W (dB/mm) as a sole function of frequency (Luini and Capsoni, 2014a):

$$a_W^*(f) = \frac{0.819\,(0.0155 f^{1.668} + 14.8523 f^{0.3885} - 27.4863)}{\varepsilon''(1 + \eta^2)} \tag{8.50}$$

where 20 GHz $\leq f \leq$ 200 GHz, while the real and imaginary parts of the electric permittivity of water (see Equations 8.7 and 8.8) are calculated for $T = 0°C$. The accuracy of Equation 8.50 was assessed against the RAOBS dataset mentioned above, showing a maximum relative approximation error of 6% (Luini and Capsoni, 2014a).

As a result, Equation 8.50, which is a deviation from K_l in Equation 8.42, can be used in temperate/mid-latitude sites to calculate the cloud attenuation along the path A, for frequencies between 20 and 200 GHz, as a function of local W values derived from different sources/models.

TABLE 8.7

Inputs, Equations, and the Constants to Calculate Cloud Attenuation Using the Liquid Water Mass Absorption Coefficient

Inputs	Outputs
Local statistics of integrated liquid water content W	Slant path attenuation A
Link elevation angle θ	
Frequency f	

Equations	Values of Constants	Units
$\varepsilon'(f) = \varepsilon_2 + \dfrac{(\varepsilon_0 - \varepsilon_1)}{1 + (f/f_D)^2} + \dfrac{(\varepsilon_1 - \varepsilon_2)}{1 + (f/f_S)^2}$	$\varepsilon_0 = 87.81$	f in GHz
	$\varepsilon_1 = 5.48$	
$\varepsilon''(f) = \dfrac{f(\varepsilon_0 - \varepsilon_1)}{f_D[1 + (f/f_D)^2]} + \dfrac{f(\varepsilon_1 - \varepsilon_2)}{f_S[1 + (f/f_S)^2]}$	$\varepsilon_2 = 3.51$	
	$f_D = 8.97$ GHz	
$\eta^2 = \dfrac{(2 + \varepsilon')}{\varepsilon''}$	$f_S = 442.55$ GHz	
$a_W = \dfrac{0.819\,(0.0155\,f^{1.668} + 14.8523\,f^{0.3885} - 27.4863)}{\varepsilon''(1 + \eta^2)}$	–	f in GHz a_W in (dB)/(mm)
$A = \dfrac{a_W\,W}{\sin(\theta)}$	–	A in dB a_W in (dB)/(mm) W in mm θ in rad

Table 8.7 summarizes the inputs, equations, and constants to calculate cloud attenuation using the liquid water mass absorption coefficient.

8.5.5 DAH Model

The DAH model was preliminary presented in Dissanayake et al. (1997) and extended in Dissanayake et al. (2001) for a more accurate prediction of cloud attenuation along low-elevation paths.

The main assumption of the DAH model is that cloud attenuation A follows the lognormal distribution (as verified from measurements for instance in Al-Ansafi et al., 2003), according to which, the probability that A (dB) exceeds a given threshold A^* in a year is

$$P(A > A^*) = 1 - \frac{P_C^{tot}}{2}\,\mathrm{erfc}\left(\frac{\ln A^* - \mu}{\sqrt{2}\sigma}\right) \qquad (8.51)$$

In Equation 8.51, P_C^{tot} is the yearly probability to have cloud attenuation, μ and σ are the mean and standard deviation values of $\ln A$ while erfc is the complementary error function.

A key feature of the DAH model is that it takes into account the attenuation induced by four different types of cloud, whose average properties are listed in Table 8.8.

TABLE 8.8

Main Features of the Four Cloud Types Used in the DAH Model

Cloud Type	Vertical Dimension L_V (km)	Horizontal Dimension L_H (km)	Liquid Water Content (g/m³)
Cumulonimbus	3.0	4.0	1.0
Cumulus	2.0	3.0	0.6
Nimbostratus	0.8	10.0	1.0
Stratus	0.6	10.0	0.4

The spatial distribution of the liquid water content w within real clouds is rather inhomogeneous, but, for simplicity and for ease of application, the DAH model sketches each cloud as a cylinder (with horizontal and vertical extent equal to L_H and L_V, respectively) filled with the same w value. Depending on the site, each of the cloud types reported in Table 8.8 is associated with a different probability of occurrence, as it is reasonable to expect moving from one climatic region to another other one. In the DAH model, these occurrence probabilities are extracted from a cloud cover atlas developed on the basis of more than 4000 World Meteorological Organization (WMO) stations where visual observations of different cloud types were carried out for more than 10 years at least four times per day. Such data were assembled to create a global cloud cover atlas providing the occurrence probability for several cloud types (Warren et al., 1986) on a uniform $5° \times 5°$ latitude/longitude grid.

The specific cloud attenuation γ_C (dB/km) is calculated according to the same Rayleigh approximation already introduced while describing the Salonen and Uppala model (Section 8.5.3). As a result, for each cloud, the zenith path attenuation (dB) is obtained as

$$A_Z = 0.52\,\gamma_C\,w\,L_V = 0.52\left[0.4343\left(\frac{3\pi}{50\lambda\rho_d}\right)\mathrm{Im}\left(\frac{1-\varepsilon}{\varepsilon+2}\right)\right]w\,L_V \tag{8.52}$$

where the liquid water content w (g/m³) and the vertical extent L_V (km) are extracted from Table 8.8. The formulation of γ_C (dB/km) reported in Dissanayake et al. (2001) is equivalent to Equation 8.42 but is expressed as a function of the wavelength λ (m) and the density of water ρ_d (g/m³), which, in the temperature range $-20°C \leq T \leq 10°C$, can be reasonably taken as 1. For simplicity, the temperature within the whole cloud is assumed to be $T = 0°C$ (which the electric permittivity of water ε depends on) and the adjustment factor 0.52, empirically determined against real cloud attenuation measurements, is introduced to account for the fact that the cloud cover values in the cloud atlas applies to the complete sky while the cloud attenuation is calculated along a given direction. Thus, the four cloud types (attenuation and probability of occurrence) and the total cloud cover (extracted from the atlas as well) provide five points of the CCDF curve, and the best-fit curve (refer to Equation 8.51) is found from these points.

While Equation 8.52 is valid for zenithal links, for slant paths, the effective length through each cloud L_i must replace L_V. Making reference to Figure 8.5, if the elevation angle θ is larger than ϕ (case a), the path length is fully contained within the cylindrical cloud. For θ smaller than ϕ (case b), the probability to intersect more than one cloud along the path increases, which can be taken into account by using an effective horizontal path length L_{eff} similar to the one used for some rain attenuation models (ITU-R P.618-11).

To sum up (L^i in km, i being the i index for the cloud type reported in Table 8.8):

$$L^i = \begin{cases} \dfrac{L_V^i}{\sin(\theta)} & \theta \geq \phi^i \\[2ex] \dfrac{L_{eff}^i}{\cos(\theta)} & \theta < \phi^i \end{cases} \tag{8.53}$$

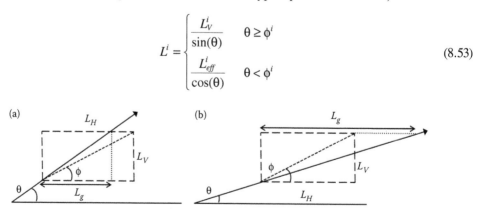

FIGURE 8.5 Reference scheme for the calculation of cloud attenuation statistics on slant paths using the DAH model.

where

$$L_{eff}^i = L_H^i + \frac{(L_g^i - L_H^i)}{1 + 0.78\sqrt{L_g^i} - 0.38[1 - \exp(-2L_g^i)]} \tag{8.54}$$

The DAH model is inherently global and in fact it has proved to predict cloud attenuation statistics collected in different sites (Dissanayake et al., 2001) with good accuracy. On the other hand, its applicability is tightly linked to cloud cover atlases, not always easily retrievable, containing information on the four types of clouds defined in the model.

Table 8.9 summarizes the inputs, equations, and constants for the practical application of the DAH model.

TABLE 8.9
Inputs, Equations, and the Constants for the Practical Application of the DAH Model

Inputs

Latitude and longitude of the site
Link elevation angle θ
Frequency f

Outputs

CCDF of cloud attenuation A

Equations	Values of Constants	Units
$\varepsilon'(f) = \varepsilon_2 + \dfrac{(\varepsilon_0 - \varepsilon_1)}{1 + (f/f_D)^2} + \dfrac{(\varepsilon_1 - \varepsilon_2)}{1 + (f/f_S)^2}$ $\varepsilon''(f) = \dfrac{f(\varepsilon_0 - \varepsilon_1)}{f_D[1 + (f/f_D)^2]} + \dfrac{f(\varepsilon_1 - \varepsilon_2)}{f_S[1 + (f/f_S)^2]}$ $\varepsilon = \varepsilon' + j\varepsilon''$ $\gamma_C = \left[0.4343\left(\dfrac{3\pi}{50\lambda\rho_d}\right)\mathrm{Im}\left(\dfrac{1-\varepsilon}{\varepsilon+2}\right)\right]$ $L_g^i = L_V^i \cot(\theta)$ $L_{eff}^i = L_H^i + \dfrac{(L_g^i - L_H^i)}{1 + 0.78\sqrt{L_g^i} - 0.38[1 - \exp(-2L_g^i)]}$	$\varepsilon_0 = 87.81$ f in GHz $\varepsilon_1 = 5.48$ $\varepsilon_2 = 3.51$ $f_D = 8.97$ GHz $f_S = 442.55$ GHz $\rho_d = 1$ g/m³ γ_C in dB/km λ in m i index of the cloud type (see Table 8.8)	L_{eff}, L_g, and L_H (extracted from the third column of Table 8.8) and L_V (extracted from the second column of Table 8.8) in km θ in rad
$A_Z^i = 0.52\,\gamma_C\,w^i\,L_V^i$	i index of the cloud type (see Table 8.8)	A_Z in dB γ_C in dB/km w (extracted from the fourth column of Table 8.8) in g/m³ L_H (extracted from the third column of Table 8.8) in km
$\phi^i = \tan^{-1}(L_V^i/L_H^i)$ $L^i = \begin{cases} L_V^i/\sin(\theta) & \theta \ge \phi^i \\ L_{eff}^i/\cos(\theta) & \theta < \phi^i \end{cases}$	i index of the cloud type (see Table 8.8)	L_s, L_{eff}, and L_H (extracted from the third column of Table 8.8) and L_V (extracted from the second column of Table 8.8) in km ϕ and θ in rad
$A^i = A_Z^i L^i$	i index of the cloud type (see Table 8.8)	A and A_Z in dB L in km
Extract P_C^i (percentage of cloud of type i, see Table 8.8) and P_C^{tot} (probability of cloud coverage) for the site of interest from the cloud cover atlas	i index of the cloud type (see Table 8.8)	–
Fit the lognormal CCDF to the 4 (P_C^i, A^i) points plus the $(P_C^{tot},0)$ point	i index of the cloud type (see Table 8.8)	–

8.5.6 STOCHASTIC MODEL OF CLOUDS

As is clear from the cloud attenuation methods described so far, accurate predictions of A can only be achieved by combining physical models for the interaction between electromagnetic waves and suspended liquid water droplets with proper knowledge of the spatial distribution of the liquid water content w or, at least, of the integrated liquid water content W. This is a key aspect gaining more and more importance as the complexity of communication systems progressively increases. In fact, for systems implementing site diversity schemes for the mitigation of high fades (Luini and Capsoni, 2013a) or for low Earth orbit (LEO) satellite applications where the ground antenna changes elevation (from very low to high) and azimuth angles in a few minutes, the spatial correlation of clouds plays a relevant role both because it is implicitly involved in the design of the diversity schemes, but also because at low elevations the assumption of horizontal homogeneity for clouds is no longer acceptable and thus the full vertical and horizontal distribution of w needs to be taken into account.

All these aspects are addressed by SMOC, a methodology first described in Luini and Capsoni (2012a) and further extended in Luini and Capsoni (2014b), to synthesize high-resolution three-dimensional (3D) cloud fields. The model takes advantage of the stochastic approach proposed by Bell (1987), which, as an intermediate step, generates random spatially correlated Gaussian fields. The synthesis of spatially correlated cloud fields relies on the knowledge of the fractional cloud cover f_W and the average integrated cloud liquid water E_W over the "target area," typically derivable from global meteorological products (e.g., NWP from the ECMWF [Uppala et al., 2005]), from which, in addition, the standard deviation of the cloud liquid water content over the target area (a necessary input), S_W, can be derived as explained in the following.

The main idea of SMOC is that, by introducing suitable statistical properties of the integrated liquid water content W (first-order statistics and spatial distribution) observed in real cloud fields, it is possible to deintegrate average cloud quantities regularly provided worldwide over a coarse latitude/longitude grid (at least for radio wave propagation applications) with long sampling time (NWP), and, in practice, to synthesize realistic fields of w with fine spatial resolution (1 km × 1 km × 100 m) over areas in the order of 200 km × 200 km (the "target area").

8.5.6.1 Model Development

The statistical properties of the integrated liquid water content were inferred from an extensive set of cloud fields observed by the MODIS on board the LEO Aqua satellite, which acquires radiance data in 36 optical channels (wavelength in the 0.4–14.4 μm range) with high spatial resolution (from 250 m to 1 km footprint, linear size). Among the atmospheric high-resolution products made freely available by the National Aeronautics and Space Administration (NASA), maps of W (dimension of 200 km × 200 km, 1 km × 1 km spatial resolution, 40,000 pixels) were extracted to investigate the main statistical features of W (MODIS, 2012).

First, the analysis of the W maps revealed a strict correlation between E_W and S_W: the probability density function (PDF) of S_W conditioned to E_W, $p(S_W|E_W)$, follows the lognormal law and, in addition, its main parameters, μ_p and σ_p in Equation 8.55, depend on E_W. In mathematical terms

$$p(S_W|E_W) = \frac{1}{S_W \, \sigma_p \sqrt{2\pi}} \exp\left[-\frac{(\ln S_W - \mu_p)^2}{2\sigma_p^2} \right] \tag{8.55}$$

where

$$\mu_p(E_W) = -5.61 E_W^{-0.076} + 4.69$$
$$\sigma_p(E_W) = 1.03 E_W^{-0.029} - 0.81 \tag{8.56}$$

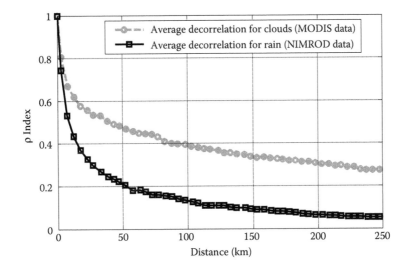

FIGURE 8.6 Average decorrelation with distance of the integrated cloud liquid water W calculated from MODIS data (gray line with circle markers) and of rain rate R calculated from NIMROD data (black solid line with square markers).

In turn, the distribution of W within each 200 km × 200 km map (all the samples in the map conditioned to having $W > 0$ mm) was found to be lognormal as well. Also, the spatial correlation of clouds was investigated from MODIS data using the customary spatial correlation index

$$\rho(\mathbf{x}, \mathbf{y}) = \frac{E[W(\mathbf{x}) \cdot W(\mathbf{y})] - E[W(\mathbf{x})] E[W(\mathbf{y})]}{\sigma[W(\mathbf{x})] \ \sigma[W(\mathbf{y})]} \tag{8.57}$$

$E[\cdot]$ and $\sigma[\cdot]$ are the mean and standard deviation operators, while $W(\mathbf{x})$ and $W(\mathbf{y})$ are the cloud liquid water content time series, respectively, relative to pixels $\mathbf{x}$ and $\mathbf{y}$ in each 200 km × 200 km cloud map. Figure 8.6 depicts the spatial correlation of W obtained by averaging ρ values relative to all the couples of pixels at the same distance (gray line with circle markers).

For comparison, the average spatial correlation of rainfall as obtained from a set of rain fields derived by the NIMROD weather radar network (Luini and Capsoni, 2012b) has been added to Figure 8.6 in order to show the much higher spatial variability of precipitation with respect to clouds. The results shown in Figure 8.6 indicate a decorrelation trend for cloud that is similar to the one reported in Garcia et al. (2008) (slow decay with distance), where the spatial correlation of clouds was investigated from 5 years of data collected every 6 h in 33 sites across Spain; however, the two sets of results are not directly comparable because in Garcia et al. (2008), the statistical dependence index was used instead of the correlation index and, moreover, the quantity investigated is not W but the cloud cover f_W.

As clarified in Bell (1987), the stochastic approach that SMOC relies on to synthesize realistic cloud fields starts from the generation of random Gaussian fields, whose spatial correlation $\rho_G(d)$ needs to be known. This was estimated as well by first turning each MODIS cloud field into a truncated Gaussian field, which, under the assumption of lognormal distribution for W, corresponds to inverting Equation 8.63 reported in Table 8.10. Afterward, the spatial correlation of the random Gaussian process has been calculated from converted maps using the same definition of ρ as in Equation 8.57. The resulting average $\rho_G(d)$ is well represented by the following analytical expression (the distance d is expressed in km):

$$\rho_G(d) = 0.35\, e^{-d/7.8} + 0.65\, e^{-d/225.3} \tag{8.58}$$

TABLE 8.10

Step-by-Step Procedure for the Practical Application of SMOC to Predict Cloud Attenuation Statistics (CCDF)

Inputs	Outputs
Latitude and longitude of the site	CCDF of cloud attenuation A
Link elevation angle θ	
Frequency f	

Step	Comments
1. Extract the fractional cloud cover (f_{ERA}) and the spatial average of the integrated cloud liquid water (W_{ERA}) from the ERA40 database. Thus, $E_W = W_{ERA}$ (mm) and $f_W = f_{ERA}$ (0–1 range)	For a given site with coordinates (lat, lon), f_{ERA} and W_{ERA} will result from the bilinear interpolation of the values relative to the four surrounding grid pixels, as suggested by the ECMWF
2. From E_W, derive μ_p and σ_p as from the expressions in Equation 8.56	μ_p and σ_p completely define $p(S_W \mid E_W)$ in Equation 8.55
3. Randomly extract S_W from the lognormal distribution $p(S_W \mid E_W)$ derived at step 2	–
4. Generate a random Gaussian field $g(x,y)$ with the spatial correlation ρ_G in Equation 8.58	Full details on the procedure to generate spatially correlated random Gaussian fields are provided in Bell (1987)
5. Calculate μ_{LN} and σ_{LN} of the lognormal distribution characterizing the cloud map to be generated $$(\text{for } W > 0 \text{ mm}): \begin{cases} \mu_{LN} = \ln\left[\dfrac{1}{f_W^{1.5}} \dfrac{E_W^2}{\sqrt{E_W^2 + S_W^2}}\right] \\ \sigma_{LN} = \sqrt{\ln\left[f_W\left(\dfrac{S_W^2}{E_W^2} + 1\right)\right]} \end{cases}$$	μ_{LN} and σ_{LN} derive from the inversion of the following equation system: $$\begin{cases} E_W = f_W \exp(\mu_{LN} + \sigma_{LN}^2/2) \\ S_W = \sqrt{f_W \exp(2\mu_{LN} + 2\sigma_{LN}^2) - f_W^2 \exp(2\mu_{LN} + \sigma_{LN}^2)} \end{cases}$$ While E_W and S_W in the equations above come from the NWP database, the right-hand sides express the mean and standard deviation values of a random variable whose value is 0 with probability $1 - f_W$ (cloud-free fraction of the map) and is extracted from a lognormal distribution (with parameters μ_{LN} and σ_{LN}) with probability f_W
6. Turn the Gaussian field $g(x,y)$ into a lognormal (cloud) field $C(x,y)$: set $C(x,y) = 0$ if $g < g_{th}$, otherwise: $$C(x,y) = \exp\left\{\mu_{LN} + \sqrt{2}\sigma_{LN}\text{erfc}^{-1}\left[\frac{1}{f_W}\text{erfc}\left(\frac{g(x,y)}{\sqrt{2}}\right)\right]\right\}$$	$g_{th} = \sqrt{2}\,\text{erfc}^{-1}(2f_W)$ and erfc is the complementary error function.
7. Randomly extract the cloud base height h_0 from the generalized extreme value PDF in Equation 8.62	–
8. For each pixel of the generated synthetic map (i.e., every W value), derive a and b from Equations 8.60 and 8.61 and, hence, generate the vertical profile of w using Equation 8.59	–
9. For every position of the ground station across the synthetic cloud field, integrate w along the radio link to obtain the integrated liquid water content W	–
10. Calculate the mass absorption coefficient a_w in Equation 8.50	As in Table 8.7, the liquid water temperature is assumed to be 0°C
11. Calculate the slant path attenuation in dB using a_w and the W values obtained from synthetic cloud fields: $A = a_w W$	–

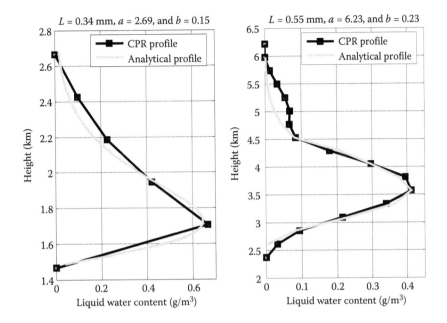

FIGURE 8.7 Examples of vertical profiles of the liquid water content $w(h)$ as measured by the CPR on board the CloudSat satellite and as fitted using the expression in Equation 8.59.

A key point for the correct prediction of cloud attenuation is the vertical development of clouds, which was investigated in Luini and Capsoni (2014b) by taking advantage of another Earth observation product, that is, vertical profiles of w (from ground up to 25 km) collected with high spatial resolution (footprint of 1.4 km × 1.7 km and profile sampled every 240 m) by the 94-GHz nadir-looking CPR on board the CloudSat satellite (Stephens et al., 2002). The analysis of several profiles (samples of which are reported in Figure 8.7) pointed out that, for most clouds, the trend of $w(h)$ with height h is asymmetric (the peak value of the liquid water content being typically closer to the cloud base) and that profiles slowly decay to zero with increasing height.

According to these features, the following analytical expression was proposed in Luini and Capsoni (2014b) to model $w(h)$ (h in km a.m.s.l.):

$$\tilde{w}(h) = \begin{cases} \dfrac{W}{b^a\,\Gamma(a)}(h-h_0)^{a-1}e^{-(h-h_0)/b} & \text{for } h \geq h_0 \\ 0 & \text{for } h < h_0 \end{cases} \tag{8.59}$$

In Equation 8.59, a and b are parameters regulating the shape of $\tilde{w}(h)$, h_0 is the cloud base height, Γ is the gamma function, and W is the integrated liquid water content of the cloud. The analytical expression in Equation 8.59, used to fit the profiles in Figure 8.7, is constrained to the liquid water content W and needs to be truncated to model real clouds by setting $\tilde{w}(h) = 0$ for $\tilde{w}(h) < 0.06\,W$. Further investigation of all the cloud profiles showed that the parameters a and b in Equation 8.59 can be calculated using the following expressions:

$$a = 4.27\,e^{-4.93(W+0.06)} + 54.12\,e^{-61.25\,(W+0.06)} + 1.71 \tag{8.60}$$

$$b = 3.17\,a^{-3.04} + 0.074 \tag{8.61}$$

As a result, exploiting Equations 8.60 and 8.61 to estimate a and b in Equation 8.59, a realistic cloud vertical profile can be derived from the simple knowledge of W.

The final information extracted from CloudSat data is the cloud base height h_0. SMOC assumes that the base height of all the clouds in a target area (200 km × 200 km) is fairly constant and the values of h_0 turned out to follow the generalized extreme value PDF:

$$p(h_0) = \frac{1}{\sigma}t(h_0)^{\xi+1}e^{-t(h_0)} \quad \text{with} \quad t(x) = \begin{cases} \left[1+\left(\frac{x-\mu}{\sigma}\right)\xi\right]^{-1/\xi} & \xi \neq 0 \\ e^{-(x-\mu)/\sigma} & \xi = 0 \end{cases} \tag{8.62}$$

where $\xi = 0.484$, $\sigma = 0.582$, and $\mu = 0.987$.

8.5.6.2 Procedure for Cloud Field Synthesis

According to SMOC, the horizontal cloud field synthesis in the target area can be achieved from the knowledge of E_W and f. In turn, this information can be extracted from NWP products. To illustrate the synthesis procedure, we will refer to the ECMWF ERA40 dataset (Uppala et al., 2005), in particular, to E_W and f extracted with temporal sampling of 6 h (i.e., nearly instantaneous values every 6 h) and spatial resolutions of 2° × 2° (latitude × longitude), respectively, the latter approximately corresponding to 200 km × 200 km at mid-latitude sites.

Table 8.10 summarizes the step-by-step procedure for the practical application of SMOC to predict cloud attenuation statistics (CCDF).

As a result, the steps in Table 8.10 allow generating a synthetic 3D distribution of the liquid water content w maintaining the basic field integral information (f_{ERA} and W_{ERA}) and reproducing the vertical and horizontal spatial correlation observed in real cloud fields.

Using multiple couples of E_W and f_W extracted from the NWP database (e.g., 5 years, corresponding to ~7300 samples), it is thus possible to generate a statistically meaningful set of synthetic cloud fields whose ensemble reflect the local statistics of integrated liquid water W and liquid water content w. This has been in fact verified with satisfactory performance in Luini and Capsoni (2014b) using the same set of RAOBS data mentioned in the previous section. Figure 8.8 (right side) shows an example of the cloud attenuation impairing an Earth–space link operating at 20 GHz with an

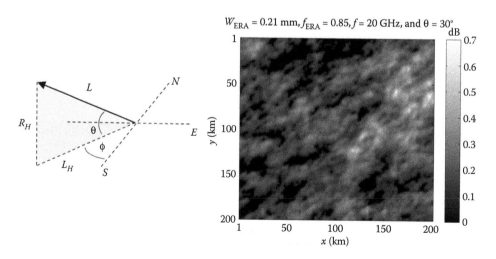

FIGURE 8.8 Left side: Reference scheme for the Earth–space link affected by clouds. Right side: Example of the cloud attenuation field obtained from the interaction between the link ($f = 20$ GHz and $\theta = 30°$) and the cloud field generated using SMOC (ERA40 inputs: $W_{ERA} = 0.21$ mm, $f_{ERA} = 0.85$).

elevation of $\theta = 30°$. The link is pointed toward the south ($\phi = 0°$ on the left side of Figure 8.8) and the position of the ground station was moved across the whole cloud field to calculate A for every pixel of the map.

8.5.7 COMPARISON OF CLOUD ATTENUATION MODELS

A comprehensive comparison of the cloud attenuation models described so far is not an easy task, not only because the inputs to such models are diversified and not always easily retrievable, but also because of the difficulty in finding reliable data against which the models' accuracy can be evaluated. The latter can be typically obtained from radiometers after proper calibration of the instrument and processing of brightness temperatures to extract the attenuation only due to clouds. As a simpler alternative, reference cloud attenuation statistics can be accurately estimated using RAOBS data as input to the Salonen and Uppala cloud detection model (to quantify the vertical profile of w) and, afterward, to the Liebe's MPM93 model (to calculate the specific cloud attenuation using the Rayleigh approximation, i.e., Equations 8.41 and 8.42). In this way, each set of P-RH-T profiles collected during a radiosonde ascent can be turned into path attenuation only due to clouds.

In order to provide at least a hint of the cloud models' performance, we resorted to the extensive set of RAOBS data made freely available by the College of Engineering of the University of Wyoming (RAOBS data). As an example, we selected the Nottingham/Watnall station (WMO code 03354) located in England (latitude 53°N and longitude −1.25°E, altitude 117 m a.m.s.l.) and downloaded RAOBS profiles of pressure, temperature, and relative humidity for 10 years (2004–2013) relative to 0 and 12 coordinated universal time (UTC). The whole RAOBS database includes 6049 sets of P-RH-T profiles, which were all used to calculate the CCDF of the cloud attenuation (Earth–space link with elevation $\theta = 40°$ and frequency $f = 30$ GHz) reported in Figures 8.9 and 8.10 (black solid line with asterisks) using the procedure just mentioned. Indeed, to ease the readability of the results, the curves predicted by the cloud attenuation models have been split into two figures.

Figure 8.9 includes the outputs from the Altshuler–Marr and Dintelmann–Ortigies models. The CCDFs were calculated using as input to the two models the meteorological data (absolute humidity ρ_{AH} for the former model, and water vapor content ρ_0 plus temperature T_0 for the latter one)

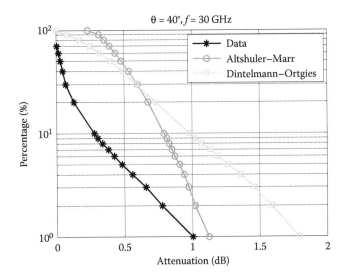

FIGURE 8.9 Comparison of cloud attenuation models: reference CCDF obtained using RAOBS data coupled with the Salonen and Uppala cloud detection method plus the Liebe's MPM93 mass absorption model (black solid line with asterisks); also, curves predicted by the Altshuler–Marr (solid line with circles) and the Dintelmann–Ortigies (solid line with squares) models.

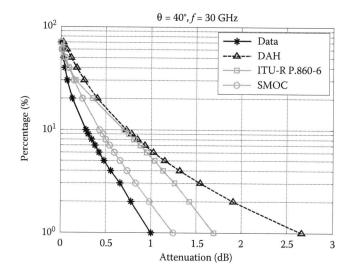

FIGURE 8.10 Comparison of cloud attenuation models: reference CCDF obtained using RAOBS data coupled with the Salonen and Uppala cloud detection method plus the Liebe's MPM93 mass absorption model (black solid line with asterisks); also, curves predicted by the DAH (dashed line with triangles), the ITU-R P.840-6 (solid line with squares), and SMOC (solid line with circles) models.

recorded by the radiosonde near the ground (altitude of first layer between 0 and 100 m). Both models show a marked overestimation of cloud attenuation and predict occurrence of clouds for 100% of the yearly time, while the reference curve indicates cloud-free conditions for 30% of the time.

Figure 8.10 complements the results on the models' accuracy by depicting the predictions provided by the DAH (dashed line with triangles), the ITU-R P.840-6 (solid line with squares), and SMOC (solid line with circles) models. In all these cases, the inputs to the models do not come from RAOBS data but originate from independent cloud datasets: the cloud cover atlas for the DAH model, the global map of integrated reduced liquid water content W_R for the method included in Recommendation ITU-R P.840-6 and the ERA40 database for SMOC. All models provide approximately the same probability to have cloud attenuation (quite close to 70% of the time) and tend to overestimation. The most accurate predictions are given by SMOC, followed by the ITU-R P.840-6 model and, finally, by the DAH model, whose overestimation increases significantly as the percentage P decreases.

Although the results provided in this section are just an example and do not allow drawing general conclusions, the comparisons shown in Figures 8.9 and 8.10 reflect the increasing complexity of the considered prediction methodologies. The Altshuler–Marr and the Dintelmann–Ortgies models receive as input quite limited ground meteorological information to infer the occurrence of clouds over the site and quantify their impact on the link; moreover, they include empirically derived expressions, whose validity in other sites remains an open point. On the other hand, the three models whose predictions are reported in Figure 8.10 are inherently applicable worldwide, as they all rely on global datasets providing simple, yet fundamental, information on clouds. The DAH model likely owes its limited accuracy to the very coarse spatial resolution of the input cloud atlas ($5° \times 5°$) and to the rather strong simplifying assumptions on the horizontal and vertical extent of clouds, as well as on their average liquid water content (see Table 8.8). The ITU-R model represents a valuable step forward in cloud attenuation modeling because it takes advantage of the full vertical structure of clouds provided by the ERA40 database. In this case, the limitations in the model's accuracy do not come from the liquid reduction approximation (yielding very small errors as shown in Luini et al. [2013]) but, most probably, from the averaging effects associated with the coarse spatial resolution of the ERA40 database ($1.125° \times 1.125°$).

This issue is overcome by SMOC, which, at the cost of an increased complexity, takes advantage of the same NWP dataset, but operates the deintegration of such data (e.g., the area-averaged integrated liquid water content E_W) to recover the embedded original high-resolution information on clouds. Additional contributions to the enhanced prediction accuracy of SMOC come from the representation of the full 3D distribution of the liquid water content w, which allows more accurate simulations of the interaction between clouds and electromagnetic waves in a variety of scenarios, from those involving classical geostationary satellites to those including LEO satellites and deep-space probes.

REFERENCES

Abreu, L.W. and G.P. Anderson (eds.), *The MODTRAN 2/3 Report and LOWTRAN 7 MODEL*, Ontar Corporation for Phillips Laboratory, North Andover, MA, Geophysics Directorate, 1996.

Ahrens, C.D., *Meteorology Today: An Introduction to Weather, Climate, and the Environment*, 10th Ed., Pacific Grove, California: Brooks/Cole, 640pp., 2013.

Al-Ansafi, K., P. Garcia, J.M. Riera, and A. Benarroch, One-year cloud attenuation results at 50 GHz, *Electronics Letters*, 39(1), 136–137, 2003.

Altshuler, E. A simple expression for estimating attenuation by fog at millimeter wavelengths, *IEEE Transactions on Antennas and Propagation*, AP-32(7), 757–758, 1989.

Altshuler, E. and R. Marr, Cloud attenuation at millimetre wavelengths, *IEEE Transactions on Antennas and Propagation*, 37, 1473–1479, 1989.

Altshuler, E. and R.A. Marr, A comparison of experimental and theoretical values of atmospheric absorption at the longer millimeter wavelengths, *IEEE Transactions on Antennas and Propagation*, 36(10), 1471–1480, 1988.

Andrews, L.C. and R.L. Phillips, *Laser Beam Propagation through Random Media*, Bellingham, Washington: SPIE Optical Engineering, 1998.

Bell, T.L., A space–time stochastic model of rainfall for satellite remote-sensing studies, *Journal of Geophysical Research*, 92(D8), 9631–9643, 1987.

Bolton, D., The computation of equivalent potential temperature, *Monthly Weather Review*, 108, 1046–1053, 1980.

Born, M. and E. Wolf, *Principles of Optics*, Cambridge, UK: Cambridge University Press, 1999.

Davies, O.T., R.G. Howell, and P.A. Watson, Measurement and modelling of cloud attenuation at millimetre wavelengths, *Electronics Letters*, 34(25), 2433–2434, 1998.

Deirmedjian, D., *Electromagnetic Scattering on Spherical Polydispersions*, New York: American Elsevier Publishing, 1969.

Deirmendjian, D., Far-infrared and submillimeter wave attenuation by clouds and rain, *Journal of Applied Meteorology*, 14, 1584–1593, 1975.

Deng, Z., C. Zhao, Q. Zhang, M. Huang, and X. Ma, Statistical analysis of microphysical properties and the parameterization of effective radius of warm clouds in Beijing area, *Atmospheric Research*, 93(4), 888–896, 2009. Online publication date: August 1, 2009.

Dintelmann, F. and G. Ortgies, Semi-empirical model for cloud attenuation prediction, *Electronics Letters*, 25, 1487–1479, 1989.

Dissanayake, A., J. Allnutt, and F. Haidara, A prediction model that combines rain attenuation and other propagation impairments along Earth-satellite paths, *IEEE Transactions on Antennas and Propagation*, 45(10), 1546–1558, 1997.

Dissanayake, A., J. Allnutt, and F. Haidara, Cloud attenuation modelling for SHF and EHF applications, *International Journal of Satellite Communications*, 19, 335–345, 2001.

Du, H., Mie-scattering calculation, *Applied Optics*, 43, 1951–1956, 2004.

Durbin, W.G., Droplet sampling in cumulus clouds, *Tellus*, 11, 202–215, 1959.

ECMWF, The European Centre for Medium-Range Weather Forecasts (ECMWF) website, http://www.ecmwf.int/, Accessed July 2014.

EUMETSAT, http://www.eumetsat.int/, website of the European Organisation for the Exploitation of Meteorological Satellites, Accessed July 2014.

Garcia, P., A. Benarroch, and J.M. Riera, Spatial distribution of cloud cover, *International Journal of Satellite Communications*, 26, 141–155, 2008.

Hale G.M. and M.R. Querry, Optical constants of water in the 200 nm to 200 μm wavelength region, *Applied Optics*, 12, 555–563, 1973.

Hemmati H., A. Biswas, and I.B. Djordjevic, Deep-space optical communications: Future perspectives and applications, in *Proceedings of the IEEE*, 99(11), 2020–2039, 2011.

Hess, M., P. Koepke, and I. Schult, Optical properties of aerosols and clouds: The software package OPAC, *Bulletin of the American Meteorological Society*, 79, 831–844, 1998.

Ippolito, L.J., *Satellite Communications Systems Engineering Handbook: Atmospheric Effects, Satellite Link Design and System Performance*, Hoboken, New Jersey: Wiley, 2008.

Ishimaru, A., *Wave Propagation and Scattering in Random Media*, London, UK: Academic Press, 1978.

Ito, S., Dependence of 0°C isotherm height of temperature at ground level in rain, *Transactions of IEICE*, E72(2), 98–100, 1989.

ITU-R Recommendation P.618-11, *Propagation Data and Prediction Methods Required for the Design of Earth–Space Telecommunication Systems*, Geneva, Switzerland, 2013.

ITU-R Recommendation P.840-6, *Attenuation Due to Clouds and Fog*, Geneva, Switzerland, 2013.

Kim, I.I., B. McArthur, and E. Korevaar, Comparison of laser beam propagation at 785 nm and 1550 nm in fog and haze for optical wireless communications, in E. Korevaar (ed.), *Optical Wireless Communications III*, Proceedings of SPIE, Vol. 4214, pp. 26–37, 2001.

King, M.D., W.P. Menzel, Y.J. Kaufman, D. Tanre, B.-C. Gao, S. Platnick, S.A. Ackerman, L.A. Remer, R. Pincus, and P.A. Hubanks, Cloud and aerosol properties, precipitable water, and profiles of temperature and water vapor from MODIS, *IEEE Transactions on Geoscience and Remote Sensing*, 41(2), 442–458, 2003.

Liebe, H.J., G.A. Hufford, and M.G. Cotton, Propagation modeling of moist air and suspended water/ice particles at frequencies below 1000 GHz, in *Proceedings of the AGARD 52nd Specialists Meeting EM Wave Propagation Panel*, Palma De Maiorca, Spain, pp. 3.1–3.10, 1993.

Luini, L. and C. Capsoni, A methodology to generate cloud attenuation fields from NWP products, in *European Conference on Antennas and Propagation 2012*, Prague, Czech Republic, pp. 1–5, March 26–30, 2012a.

Luini, L. and C. Capsoni, The impact of space and time averaging on the spatial correlation of rainfall, *Radio Science*, 47, RS3013, doi:10.1029/2011RS004915, 2012b.

Luini, L. and C. Capsoni, A rain cell model for the simulation and performance evaluation of site diversity schemes, *IEEE Antennas and Wireless Propagation Letters*, 12(1), 1327–1330, 2013a.

Luini, L. and C. Capsoni, On the relationship between the spatial correlation of point rain rate and of rain attenuation on earth–space radio links, *IEEE Transactions on Antennas and Propagation*, 61(10), 5255–5263, 2013b.

Luini, L. and C. Capsoni, Efficient calculation of cloud attenuation for earth–space applications, *IEEE Antennas and Wireless Propagation Letters*, 13(1), 1136–1139, December 2014a.

Luini, L. and C. Capsoni, Modeling high resolution 3-D cloud fields for earth-space communication systems, *IEEE Transactions on Antennas and Propagation*, 62(10), 5190–5199, 2014b.

Luini, L., C. Riva, and C. Capsoni, Reduced liquid water content for cloud attenuation prediction: The impact of temperature, *Electronics Letters*, 49(20), 1259–1261, 2013.

Luini, L., C. Riva, C. Capsoni, and A. Martellucci, Attenuation in non rainy conditions at millimeter wavelengths: Assessment of a procedure, *IEEE Transactions on Geoscience and Remote Sensing*, 45(7), 2150–2157, 2007.

Manabe, T., H.J. Liebe, and G.A. Hufford, Complex permittivity of water between 0 and 30 THz, in *12th International Conference on Infrared and Millimeter Waves*, Orlando, FL, pp. 229–230, 1987.

Marshall, J.S. and W. Mc K. Palmer, The distribution of raindrops with size, *Journal of Meteorology*, 5, 165–166, 1948.

Martellucci, A., J.P.V. Poiares-Baptista, and G. Blarzino, New climatological databases for ice depolarisation on satellite radio links, Paper presented at *COST Action 280, 1st International Workshop, PM3037*, Malvern, July 2002.

McCartney, E.J., *Optics of the Atmosphere: Scattering by Molecules and Particles*, New York: Wiley, 1976.

Members of the MODIS Characterization Support Team, *MODIS Level 1B Product User's Guide*, NASA/Goddard Space Flight Center Greenbelt, July 20, 2012. Available at: http://mcst.gsfc.nasa.gov/sites/mcst.gsfc/files/file_attachments/M1054.pdf

Miles, N.L., J. Verlinde, and E.E. Clothiaux, Cloud droplet size distributions in low-level stratiform clouds, *Journal of the Atmospheric Sciences*, 57, 295–311, 2000.

NASA, Lunar Atmosphere and Dust Environment Explorer (LADEE). January 14, 2014, http://www.nasa.gov/mission_pages/ladee/main/#.U9nhb7E2STp (July 31, 2014).

Platt, C.M.R., A parameterization of the visible extinction coefficient of ice clouds in terms of the ice water content, *Journal of the Atmospheric Sciences*, 54, 2083–2098, 1997.

Petty, G.W. and W. Huang, The modified gamma size distribution applied to inhomogeneous and nonspherical particles: Key relationships and conversions, *Journal of the Atmospheric Sciences*, 68, 1460–1473, 2011.

RAOBS data, http://weather.uwyo.edu/upperair/sounding.html, Accessed July 2014.

Riva, C., C. Capsoni, L. Luini, M. Luccini, R. Nebuloni, and A. Martellucci, The challenge of using the W band in satellite communication, *International Journal of Satellite Communications and Networking*, 32, 187–200, 2014.

Rozenberg, V.I., Scattering and attenuation of electromagnetic radiation by atmospheric particles, NASA TT F-771, February 1974.

Salonen, E. and S. Uppala, New prediction method of cloud attenuation, *Electronics Letters*, 27(12), 1106–1108, 1991.

ScattPort, The Light Scattering Information Portal website. http://www.scattport.org/, Accessed July 2014.

Shettle, E.P., Models of aerosols, clouds and precipitation for atmospheric propagation studies, *AGARD Conference*, 454(15), 1–13, 1989.

Slingo, A. and H.M. Schrecker, On the shortwave radiative properties of stratiform water clouds, *Quarterly Journal of the Royal Meteorological Society*, 108, 407–426, 1982.

Slobin, S.D., Microwave noise temperature and attenuation of clouds: Statistics of these effects at various sites in the United States, Alaska, and Hawaii, *Radio Science*, 17, 1443–1454, 1982.

Squires, P., The microstructure and colloidal stability of warm clouds. I. The relation between structure and stability, *Tellus*, 10, 256–261, 1958.

Stephens, G.L. et al., The CloudSat mission and the A-TRAIN: A new dimension to space-based observations of clouds and precipitation, *Bulletin of the American Meteorological Society*, 83, 1771–1790, 2002.

Strickland, B.R., M.J., Lavan, E. Woodbridge, and V. Chan, Effects of fog on the bit-error rate of a free-space laser communication system, *Applied Optics*, 38, 424–431, 1999.

Tampieri, F. and C. Tomasi, Size distribution models of fog and cloud droplets in terms of the modified gamma function, *Tellus*, XXVIII, 333–347, 1976.

Ulaby, F.T., R.K. Moore, and A.K. Fung, *Microwave Remote Sensing: Active and Passive, Vol. I—Microwave Remote Sensing Fundamentals and Radiometry*, Reading, Massachusetts: Addison-Wesley, Advanced Book Program, 456pp., 1981.

Uppala, S.M. et al., The ERA-40 re-analysis, *Quarterly Journal of the Royal Meteorological Society*, 131(612), 2961–3012, Part B, 2005.

Van de Hulst, H.C., *Light Scattering by Small Particles*, New York: John Wiley & Sons, 1957.

Warren, S.G. and R.E. Brandt, Optical constants of ice from the ultraviolet to the microwave: A revised compilation, *Journal of Geophysical Research*, 113, D14220, 2008.

Warren, S.G. et al., Global distribution of total cloud cover and cloud type amounts over land, National Center for Atmospheric Research (NCAR) Technical Notes, NCAR/N-273, October 1986.

Wiscombe, W.J., Improved Mie scattering algorithms, *Applied Optics*, 19, 1505–1509, 1980.

Wiscombe, W.J., Mie scattering calculations: Advances in technique and fast, vector-speed computer codes. NCAR Technical Note, June 1979 (edited/revised August 1996).

Zuev, V.E., *Atmospheric Transparency in the Visible and the Infrared*, Jerusalem: Israel Program for Scientific Translations, 1970.

9 Aeronautical Communications Channel Characteristics and Modeling

From Legacy toward Future Satellite Systems

Ana Vazquez Alejos, Manuel Garcia Sanchez, and Edgar Lemos Cid

CONTENTS

9.1 INTRODUCTION

Unstoppable air traffic growth and the increasing demands of global and harmonized aeronautical services make the need to enhance the legacy of mobile communication services evident. In its conclusions, ANConf/11 agreed that the aeronautical mobile communication infrastructure had to evolve in order to accommodate new functions and to provide the adequate capacity and quality of services (QoS) required to support evolving air traffic management (ATM) requirements within the framework of the global ATM operational concept. In addition, the importance of harmonization and global interoperability, particularly for air/ground communications, was stressed (ICAO, 2005; Pouzet, 2007, 2008; Phillips et al., 2007).

In this chapter, we focus on the long communication systems used in civil aviation and analyze the radio channel characteristics and modeling. We consider the different frequency bands and services used at present, as well as the future communication systems identified as potential candidates to be integrated into the future communication infrastructure foreseen to be operative for beyond 2020.

9.2 FROM LEGACY TO FUTURE

Today, most options for on-board and external data communications offer limited capacity and versatility, thus explaining that the exchange of information is restricted to a limited volume of short messages with predetermined formats (Durand and Longpre, 2014). The very first aeronautical data communications system was ACARS (aircraft communications addressing and reporting). It was developed in 1978 by ARINC Inc. ACARS operated in high frequency (HF), very high frequency (VHF), and SATCOM bands and used analog radio with amplitude modulation for data link services (Jain et al., 2011).

In the 1990s, efforts were made to transition to digital radio and the resulting technologies were called VHF digital link (VDL). Four versions (or modes) of VDL were developed sequentially, called VDL1, VDL2, through VDL4. VDL2 allows only aircraft-to-ground communication, while VDL4 added support for aircraft-to-aircraft communication. Both systems have seen very limited deployment. Since the VHF band was getting congested, L band versions of VDL2 and VDL4 have also been proposed and are known as LDL2 and LDL4, respectively. EUROCONTROL—the European Organization for the Safety of Air Navigation—has funded two parallel projects and developed two proposals called L-DACS1 and L-DACS2 (Jain et al., 2011).

It is therefore evident that new air-to-ground wireless data links are needed to supplement existing civil aviation technologies and to find new spectrum allocation for them. The subject matter of most interest is probably the VHF communication band, HF band, and satellite bands, but the future communication bands should also be stressed, which could likely be VHF (108–137 MHz), L band (960–1215 MHz), S band (2.7–3.1 GHz), and C band (5.000–5.250 GHz) or a hybrid of these.

All of the previously existent technologies used single-carrier modulation and time division multiple access (TDMA). The first aeronautical data link to use multicarrier modulation was B-VHF (broadband very high frequency) proposal funded by the European 6th Framework (FP6) program. It was designed for 118–137 MHz VHF band using MC-CDMA (multicarrier code division multiple access) and time division duplexing (TDD). In MC-CDMA, each bit is encoded as a sequence of chips (code bits) and then code bits are used to modulate subcarriers of OFDM (orthogonal frequency division multiplexing). The subcarrier spacing was 2 kHz.

Since VHF band was congested, B-VHF was updated to operate in L band and the resulting design was called B-AMC (broadband aeronautical multicarrier system). The CDMA was dropped, leaving only OFDM. The subcarrier spacing was increased to 10 kHz (to account for increased Doppler spread at higher frequency). To get a reasonable capacity, the required band was increased to two channels of 500 kHz (50 subcarriers 10 kHz apart). The two channels are used for frequency division duplexing (FDD).

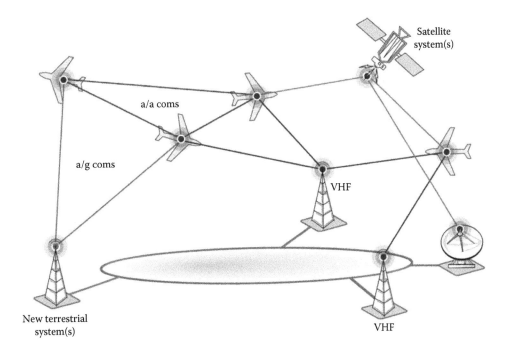

FIGURE 9.1 Future communication infrastructure for beyond 2020.

Another relevant wireless standard is P34 (Project 34) developed by EIA (Electronic Industry Association) and TIA (Telecommunications Industry Association) for public safety radio. It covers 187.5 km sectors and uses 50, 100, 150 kHz channels in the L band and uses OFDM.

In summary, a reliable and efficient communication infrastructure will have to serve all airspace users in all types of airspace and phases of flight, providing the appropriate QoS needed by the most demanding applications. The mobile part of this infrastructure will be based on a multilink approach composed of three different subnetworks, as shown in Figure 9.1 (Durand and Longpre, 2014):

- *LDACS:* A ground-based line-of-sight (LOS) data link as the main system in continental airspace and supporting air/ground services and possibly air/air services, offering a high QoS, which will be necessary in high-density areas; two systems are under consideration (LDACS 1 and 2) with the objective to select one for implementation. Both operate in the L band and are based on modern and efficient protocols.
- *Satellite:* A satellite-based system providing the required capacity and QoS to serve oceanic airspace while complementing ground-based continental data link as a way of improving the total availability. The system is being defined in close cooperation with the European Space Agency. The type of satellite constellation to be used (dedicated or commercial) is still under consideration.
- *AeroMACS:* A system dedicated to airport operations, based on mobile WiMAX 802.16e, providing a broadband capacity to support the exchanges of a significant amount of information such as the uploading of databases or maps in the aircraft.

Several research programs have been launched to define, develop, and validate these new solutions, and prepare the aeronautical community to transition to these new access networks. These activities are handled within SESAR program. The SANDRA project also takes into account the integration aspects of these new solutions, and the networking environment (IPv6 will be introduced in place of IPv4) (Durand and Longpre, 2014).

9.3 CHARACTERISTICS OF THE AERONAUTICAL RADIO PROPAGATION CHANNEL

From a radio propagation point of view, the aeronautical channel shows complicated features for modeling, given that this channel combines mobility and a set of different factors of influence (Haque, 2011), with some of them, for instance, as follows:

1. The flight phase: The radio channel for standing on the runway is different from the takeoff and cruise phases, or cruising at a constant altitude, and descent phase for landing.
2. Doppler spread due to the aircraft speed that might be different for each flight phase.
3. Shadowing due to horizon, terrain obstacles as mountains or buildings, even due to other aircrafts and the own aircraft body or its maneuvers.

The special history or legacy of aeronautical communications is responsible for the lack of rigorous, harmonized, and standardized radio characterization of the different frequency bands allocated for it. The majority of the models found in the early 1960s are narrowband for HF and VHF bands. No comprehensive, validated, wideband models existed for time-varying air–ground aeronautical channels till almost 2000. In NASA (2006) and Matolak (2014), NASA, as the leader of the Future Communication Study, reported the lack of uniformity and rigor in aeronautical radio channel modeling and characterization. Since then, a relevant effort has been deployed to address this deficiency.

In general terms, for planning radio link budgets, some of International Telecommunication Union (ITU) regulations apply and we can list some of them:

1. Recommendation ITU-R P.370-7 provides guidance on the prediction of field strength for the broadcasting service for the frequency range 30–1000 MHz and for the distance range up to 1000 km.
2. Recommendation ITU-R P.452 provides guidance on the detailed evaluation of microwave interference between stations on the surface of Earth at frequencies above about 0.7 GHz.
3. Recommendation ITU-R P.528 provides guidance on the prediction of point-to-area path loss for the aeronautical mobile service for the frequency range 125 MHz to 30 GHz and the distance range up to 1800 km.
4. Recommendation ITU-R P.529 provides guidance on the prediction of point-to-area field strength for the land mobile service in the VHF and UHF bands.
5. Recommendation ITU-R P.617 provides guidance on the prediction of point-to-point path loss for trans-horizon radio-relay systems for the frequency range above 30 MHz and for the distance range 100–1000 km.

However, specific modeling for each service and frequency band has been considered of most interest, especially to decide on potential candidate technologies to be integrated in the future communication infrastructure. However, general radio characteristics applicable to any service, frequency, and flight phase can be inferred from the literature.

An interesting compilation of the general radio channel characterization can be found in Haque et al. (2010) and Haque (2011). The aeronautical channel can be broken into three segments: takeoff/landing, en route, and taxiing/parked. Since an aircraft spends most of its time in the en route segment, this dissertation will focus on the en route channel. For the en route environment, most of the current research assumes a two-ray model (Bello, 1973; Haque et al., 2010). The remaining two segments fall within the scope of the non-line-of-sight (NLOS) dispersive channel. A two-ray channel in an aeronautical environment will experience a narrow Doppler spread bandwidth and shifts, that is, less than 360° (Haas, 2002). Each ray in the two-ray channel will experience significantly different Doppler shifts. Also, the en route channel experiences a different condition between air to ground and ground to air.

Surface communication is broken down into large and small scale, differentiating between path loss and multipath (Rappaport, 2003).

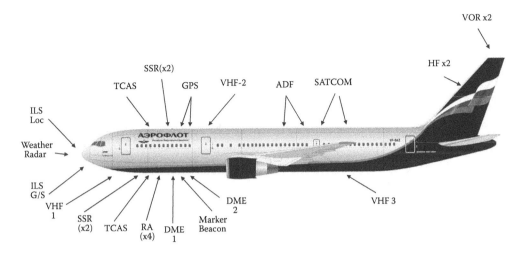

FIGURE 9.2 Aeronautical services available on a civil aircraft. ADF: automatic direction finder; DME: distance measuring equipment; GPS: global positioning system; G/S: glide slope; HF: high frequency; ILS: instrument landing system; Loc: localizer; RA: radio altimeter; SATCOM: satellite communication; SSR: secondary surveillance radar; TCAS: traffic collision avoidance system; VOR: VHF omnidirectional range. (Adapted from Hoeher, P. and E. Haas, *IEEE Vehicular Technology Conference*, 4, 1961–1966, 1999.)

The modeling can be divided into two types of fading effects that characterize mobile communications: large-scale and small-scale fading effects. Propagation models that determine the mean signal strength for an arbitrary transmitter–receiver separation over larger distances are useful in estimating the radio coverage area of a transmitter and are called large-scale propagation models. Aircraft experience both large and small channel conditions due to their changing altitude and speed; therefore, Doppler shifts play a dominant role in aircraft channel modeling (Neskovic et al., 2000).

We should note that the Doppler frequency spread depicts different characteristics compared to terrestrial networks. The arrival/takeoff, taxi, and parking scenarios depict different multipaths and received angle spreads (Hoeher and Haas, 1999). These different scenarios have different channel parameters. For instance, dual Doppler shifts occur in the channel for air-to-ground and ground-to-air aeronautical communications in an en route scenario.

In Figure 9.2, we show the aeronautical systems commonly found on board an aircraft. The intercompatibility and interference of those systems should be another topic of study. In Figure 9.3, we show the aeronautical services grouped by category and frequency band (Mettrop, 2009), and in Figure 9.4, the spectrum spaces allocated for the different aeronautical systems.

In the following sections, we will describe the radio characterization of the main frequency bands and systems used at present and to be incorporated in future for long-distance communications: HF, VHF, L band, and satellite communications. In Table 9.1, we summarize most of the present aeronautical services and frequency bands (ICAO, 2012).

9.4   AERONAUTICAL COMMUNICATION SERVICES AT HF BAND

As described in Table 9.1 (ERC, 2001), in the HF band, the use of nondirectional radio beacon (NDB)/locator beacons is allocated, that in general is stabilized in number and may be reduced over time as a result of the incorporation of ongoing global navigation satellite system (GNSS) and area navigation (Random NAVigation, RNAV) system implementation.

Even at present, aircraft utilize HF communications (3–30 MHz) when VHF and satellite communications require far beyond LOS with land-based ground stations (GSs). The primary usage

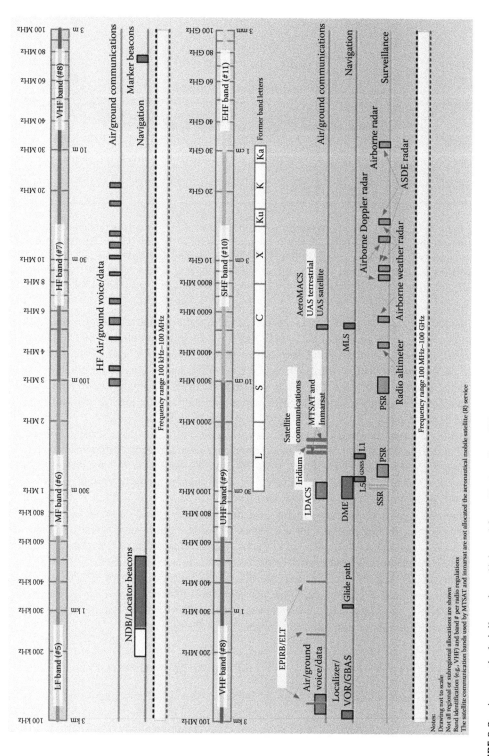

FIGURE 9.3   Aeronautical civil services. (Adapted from Hoeher, P. and E. Haas, *IEEE Vehicular Technology Conference*, 4, 1961–1966, 1999.)

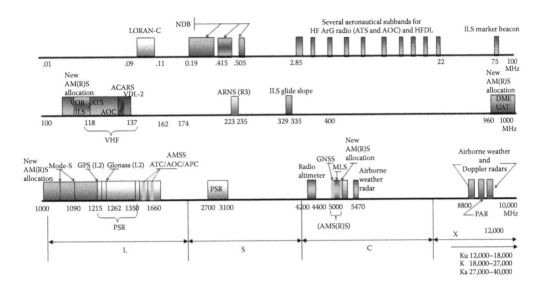

FIGURE 9.4   Spectrum allocation to civil aeronautical services.

TABLE 9.1
Aeronautical Systems on Aircraft

Frequency Band	Aviation Use
HF	Communication (voice and data), NDB/locator
VHF	ILS, VOR, GBAS, communication (voice and data)
960–1215 MHz	DME, future communication system (FCS), SSR, GNSS, ACAS, ADS-B
1215–1350 MHz	L band radar
1545–1555/1646.5–1656.5 MHz	Satellite communications
1559–1610 MHz	GNSS
2700–3100 MHz	S band radar
4200–4400 MHz	Radar altimeter
5000–5250 MHz	MLS, UAS, satellite communications, aeronautical telemetry
5350–5470 MHz	Airborne weather and ground mapping radar
9000–9500 MHz	Precision approach radar, airborne weather and ground mapping radar

Source: Adapted from European Radiocommunications Committee (ERC). Current and future use of frequencies in the LF, MF and HF bands. ERC Report 107, 2001.

Note: ILS: instrument landing system; VOR: VHF omnidirectional range; GBAS: ground-based augmentation system; ACAS: airborne collision avoidance system; MLS: microwave landing system; UAS: unmanned aerial system.

of HF is for voice and data mobile communications over oceans—transoceanic flights that communicate with GSs via HF for position reports and other purposes. Another utilization of HF communications is for the mobile service called high-frequency datalink (HFDL). Military aircraft (MILCRAFT) also utilizes HF communications for operational and training purposes (ERC, 2001; Tooley and Wyatt, 2007).

HF communications are handled on a single-channel simplex basis, mainly voice based, with the aircraft and the GS using the same frequency for transmission and reception (Tooley and Wyatt, 2007). The frequencies in use will depend upon the time of day or night and conditions that affect radio wave propagation especially on HF frequencies (ICAO-APO, 2010). The stations will remain on continuous watch for aircraft within their communication areas, and when practicable, will transfer this guard to another station when the aircraft reaches the limit of the communications area. In the early years, HF communications supported double-sideband amplitude modulation (DSB-AM); however, they are now single sideband amplitude modulation (SSB)-AM (Tooley and Wyatt, 2007).

The HF frequencies currently used by the aeronautical mobile services for distress, safety, and other communications, including allotted operational frequencies, suffer from harmful interference and are often subject to difficult propagation conditions (ERC, 2001; WRC, 2003). Unauthorized operations using aeronautical frequencies in the HF bands are continuing to increase and are already a serious risk to HF distress, safety, and other communications. Enforcing compliance with these regulatory provisions is becoming increasingly difficult with the availability of low-cost HF SSB transceivers (ERC, 2001; WRC, 2003).

In certain situations, HF radio is the only means of communication for the aeronautical mobile (R) service and given that it is a safety service, WRC-2000 has reviewed the use of HF bands by aeronautical mobile (R) services with a view to protecting operational, distress, and safety communications, thereby indicating the employment of as many interference mitigation techniques are appropriate for aeronautical mobile (R) services (WRC, 2003).

Even though HF communications provide the main means for long distance and beyond the radio horizon air/ground voice and data communications, the introduction of satellite communication systems appears as an alternative to the use of HF bands in aviation to provide those long-distance communications. However, the use of HF bands for long-distance aeronautical voice and data communications are expected to continue to be required in the short term, although it is not expected to significantly increase in use, and future requirements are projected to be able to be met in the currently available frequency bands. The need for a continuous availability for safe use of air/ground communications is, on a global basis, in the foreseeable future to support HF voice and data (ICAO, 2012). In the long term, it is foreseen that the HF communication services will be closed down by 2020 in the civil aeronautical sector, but not in the military sector where its use is becoming more extensive.

9.4.1 HF Medium Characteristics

Various media impairments, including rainfall attenuation and ionospheric scintillation, can affect the availability and reliability of VHF and satellite communications, thus turning HF communications into the sole option to continue connecting a flying aircraft in oceanic or transpolar areas. However, the HF radio channel has also been considered unreliable due to its intrinsic properties that considerably affect HF communications availability (Goodman et al., 1997).

For instance, owing to a refraction mechanism, the path of an HF signal may vary its propagation direction. Refraction mechanisms occur in coastal, atmospheric, and ionospheric propagation, and the percentage of signal propagation variation produced due to the refraction fluctuates considerably, depending on different environmental conditions. Those conditions could propitiate a direction change when a signal crosses a coastline (coastal refraction), a direction change due to a variation in temperature, pressure, and humidity, particularly at low altitude (atmospheric refraction), or a direction change if the radio wave passes through an ionized layer (ionospheric refraction) (ICAO-APO, 2010).

HF communications use the ionosphere as a propagation medium, which in some way results in a chaotic channel (Goodman et al., 1997) due to the natural disturbances associated directly or indirectly with solar flares, geomagnetic storms, atmospheric tidal forces, and atmospheric gravity waves, all of them characterized by abnormal values of temporal and spatial variability.

It is known that several ionized layers exist within the ionosphere. During daytime hours, there are four main ionization layers designated D, E, F1, and F2 in ascending order of height. At night, when the sun's radiation is absent, ionization still persists but it is less intense, and fewer layers are found (D and F layers). The factors that affect the ionosphere layers are strength of the sun's radiation, since it varies with latitude, causing the structure of the ionosphere to vary widely over Earth's surface, and the state of the sun, since sunspots affect the amount of ultraviolet radiation (ICAO-APO, 2010).

For good long-range HF reception, a frequency must be chosen that will not undergo too much attenuation. If a relatively high frequency is used, most of the energy will pass through the E layer and be reflected from the more intensely ionized F layer. The higher the frequency, the greater the degree of ionization required to obtain reflection. As the frequency is reduced and the attenuation of the E layer reflections increases, a limit is reached called the lowest usable frequency (LUF) and below this frequency, the attenuation is too great for the signal to be usable (ICAO-APO, 2010).

Thus, for least attenuation, and so the highest received signal strength for a given transmitter power, a frequency is chosen, which is as high as possible without exceeding the maximum usable frequency (MUF) for the path between the transmitter and distant receiver. The MUF is the frequency, for the prevailing conditions, that produces a skip zone extending just short of the distant receiver. Any higher frequency would give a higher critical angle and a greater skip distance exceeding beyond the receiver, which would then lose the sky wave contact with the transmitter (ICAO-APO, 2010).

The MUF value at night is much less than that by day because the intensity of ionization in the layer is less, so that lower frequencies have to be used to produce the same amount of refractive bending and give the same critical angle and skip distance as by day. However, the signal attenuation in the ionosphere is also much less at night, so the lower frequency needed is still usable. Hence, the night frequency for a given path is about half of the day frequency, and shorter distances can be worked at night than by day while still using a single reflection from the F layer (ICAO-APO, 2010).

The MUF varies not only with path length and between day and night, but also with season, meteor trails, sunspot state, and sudden ionospheric disturbances produced by eruptions on the sun. Because of the variations of MUF, HF transmitting stations have to use frequencies varying widely between about 2 and 20 MHz (ICAO-APO, 2010).

The theoretical range for HF frequencies varies, depending on the propagation path used, ground, or sky waves. Ground waves can usually reach up to 100 nm and sky waves longer distances; however, sky waves will not be received within the skip distance (probably several hundred miles from the transmitter). The theoretical maximum range obtained by means of a single reflection from the E layer is about 1300 nm, and from the F layer about 2500 nm. This theoretical maximum range is achieved with the transmitted signal leaving Earth's surface tangentially. Ranges of 8000 nm or more may be achieved by means of multiple reflections, mainly from the F layer, being the signal alternately refracted down from the layer and reflected up again from Earth's surface until it becomes too weak to use. Different websites provide real-time predictions that summarize the propagation conditions to allocate the required HF frequencies (ICAO-APO, 2010).

Basically, an HF radio link fits a large-scale path loss model that may match the free-space path loss model or Fresnel approach, or a modification to it. According to Haque (2011), elemental geometric relations are observed between an aircraft station (AS) at an aircraft's altitude (h_1) with a GS. The LOS communication distance, without considering path loss by the Fresnel model and other parameters, from AS to GS can be calculated using Pythagoras' theorem if a flat Earth model is assumed, as follows:

$$d_1 = \sqrt{(h_1 \cdot (2R + h_1))} \tag{9.1}$$

where R is the radius of Earth (ranging from 6336 to 6399 km, but widely assumed to be 6370 km). For distances between the two nodes above sea level, the above formula needs additional steps for calculating the communication distance, such as Earth's curvature. Then the formula must be calibrated by the parameter K statistically measured by the ITU, which indicates the effect of Earth's curvature on the LOS distance, as indicated in Equation 9.2:

$$d_1 \cong \sqrt{(2R \cdot h_1)} \qquad (9.2)$$

For a 2-km altitude, as for a very-low-orbit AS, the maximum communication distances that can be achieved between AS and AS/GS for LOS communication can reach a value of $d_1 = 120$ km. A commercial flying altitude of 9 km can potentially reach communication zones as far as $d_1 = 250$ km assuming a typical value of $K = 0.5$ factor. The communication distance between two ASs could reach up to $d_1 = 480$ km, if $K = 1/2$ is assumed (Haque, 2011).

ASs could be used as a back haul or relay for wireless infrastructures, since they have the capability of communicating long distances as compared to wireless ground back hauls. Aeronautical network (AN) will have a substantial lower round trip delay, which would allow for a low delay telephone and voice over IP services (Haque, 2011).

9.4.2 HF Aeronautical Communication Services

A study by WRC (2003) indicates that the aeronautical HF radiocommunication services continue to be the only means of communications for some aircrafts on the intercontinental routes even if the mobile satellite communication has overtaken some of the HF traffic. The radionavigation and radiolocation services both within the aeronautical radionavigation service and the maritime radionavigation service will continue to be in operational use in Europe. It is foreseen that these services will closed down by 2020. In the following, we describe the main of these HF mobile services: aeronautical mobile (R) service, aeronautical mobile (Or) service, aeronautical radionavigation service, and HF data link. Finally, military use of the HF band is briefly described.

9.4.2.1 Aeronautical Mobile Radiocommunication (R) Service

This service covers a total of 1301 kHz within the band 2850–22,000 kHz. The frequency allotment plan is used for the civil on-route HF radiocommunication (R) with airplanes and for flight communications where VHF communication systems are not practical (WRC, 2003).

The base stations are typically situated at major civil airports, which may be in suburban or rural areas. The power levels are in the order of 1 kW with mobile stations using 400 W output power (WRC, 2003).

The HF traffic within the aeronautical mobile (R) service has decreased in Europe and the mobile satellite service (MSS) provides an alternative communication. The HF service is, however, still the only means of long-distance communication for some airplanes. The HF service is extensively used for air traffic control (ATC) and weather and airline communications over North Atlantic routes. The service is also used over remote areas or northern land when the VHF service is not available. EUROCONTROL indicates that the voice frequencies over North Atlantic are already congested with a yearly 9% traffic increase due to the implementation of reduced vertical separation. As recognized at the ITU WRC-2000 Atlantic, there are increasing cases of harmful interference due to unauthorized sources all over the world, including the North Atlantic (ERC, 2001; WRC, 2003).

9.4.2.2 Aeronautical Mobile Off-Route Service

This service covers a total of 1125 kHz within the frequency range 3025–23,350 kHz. The frequency allotment plan in the radio regulations (3025–18,030 kHz) is used for off-route service (OR), which typically covers noncivil aviation communication requirements but also civil use in, for instance, helicopter operations (WRC, 2003).

The base stations are situated close to military installations and airports, typically in rural areas. The power levels are expected to be comparable to the civil (R) service (WRC, 2003).

In addition, the frequency band 23,200–23,350 kHz is also allocated to the aeronautical mobile (OR) service. This band is also allocated to the aeronautical fixed service but is not used for this purpose (WRC, 2003).

The HF frequencies within the OR are still in use for both civil and military air–ground–air communications. The service includes elements of life safety service and should be protected (WRC, 2003).

9.4.2.3   Aeronautical Radionavigation Service

This service, still in use in Europe, covers a total of 191 kHz within the band 255–526.5 kHz (WRC, 2003).

The nondirectional aeronautical radiobeacons (NDBs) are situated in both rural and urban areas with a power level of around 100 W. The NDB's have a typical range of 25–100 nautical miles (NM) and locators a typical range of 15 NM. The service is used over sea and land routes and is used by general aviation at airfields. At present, a large number of general aviation aircrafts are equipped with automatic direction finders (ADF) and adequate accuracy can be achieved with low-cost equipment (ERC, 2001; WRC, 2003).

The aviation strategy is to require NDBs for international operations until at least 2010 to 2015. ICAO has retained 2015, while globally European Civil Aviation Conference (ECAC) expects to continue the use of NDBs in international operation until 2010. Some countries for cost reasons may continue national operation of NDBs until 2015 (WRC, 2003).

9.4.2.4   HF Data Link

The evolution of aircraft communications from voice to data to transport ACARS communications has motivated the installation of HFDL system, with a global coverage provided by ARINC (Tooley and Wyatt, 2007). HFDL is a highly cost-effective data link capability for aircrafts on remote oceanic routes, becoming the only data link technology that works over the North Pole, providing continuous, uninterrupted data link coverage on the popular polar routes between North America and Eastern Europe and Asia, where SATCOM coverage is unavailable. It has been found to provide better availability than HF voice on the routes over the poles beyond the 80° north/south limit of Inmarsat satellite coverage. HFDL also represents a further means of data linking with an aircraft, supplementing VDL, global positioning system (GPS), and SATCOM systems (Tooley and Wyatt, 2007).

The main disadvantage of HFDL is very low data rates, thus being unsuitable for high-speed wideband communications. However, the advantages of HFDL are relevant (Tooley and Wyatt, 2007):

- Wide coverage due to the extremely long range of HF signals
- Rapid network acquisition favored by simultaneous coverage on several bands and frequencies (currently 60)
- Exceptional network availability due to multiple ground GSs (currently 14) at strategic locations around the globe

HFDL uses phase shift keying (PSK) at data rates of 300, 600, 1200, and 1800 bps. The rate used is dependent on the prevailing propagation conditions. HFDL is based on frequency division multiplexing (FDM) for access to GS frequencies and time division multiplexing (TDM) within individual communication channels (Tooley and Wyatt, 2007).

The new HF avionics radios can switch between voice and data mode using the same aerial, but they are required to give voice communications precedence over data link, which limits the HFDL availability. The HFDL system capacity is limited by the spectrum availability in the HF band whose scarcity will collapse the capacity of this system (Durand and Longpre, 2014).

9.4.2.5 Military Use of HF Spectrum

The military frequency requirements within the HF band are increasing. The technologies used and the propagation conditions for those bands provide a lot of military communication opportunities for land, air, and maritime forces. NATO and its partners have a particular interest in the use of the range below 12 MHz. In this range, fixed and mobile applications belong to the main usage, including aeronautical mobile military services. An adequate spectrum support is required for networks on a continuous 24-h-a-day basis. In this respect, several essential military functions critically depend on the use of this part of the HF spectrum (WRC, 2003).

Until recently, the HF band was questioned as a medium for strategic and tactical communications because of its unreliability and limitations. However, modern technology has made HF communications considerably more robust than in the past. Consequently, this band is now being utilized to provide reliable communications for many requirements, at both short-range and beyond LOS distances, and features strongly in the overall planning for communications (WRC, 2003).

HF communications are used between air command and control ground elements and aircraft for exchanging mission control and surveillance/sensor data at extended ranges and when other communications are not available due to equipment or interference. HF is also used for ATM purposes when facilities are needed beyond the range of VHF and SATCOM (WRC, 2003).

The use of aeronautical OR channels constitutes another essential spectrum resource for strategic and tactical military employment of air forces and maritime air components (WRC, 2003). In view of an increase in use, any reduction of existing aeronautical HF OR allocations would have detrimental consequences for the execution of military missions (WRC, 2003).

The high-frequency global communications system (HF-GCS) is a military long-range AN of single-sideband shortwave transmitters of the United States Air Force, which is used to communicate with aircraft in flight, GSs, and some United States Navy surface assets (Ham Universe, 2015). HF-GCS complements the use of satellite communications, and digital modes between aircraft and GSs. HFGCS stations operate on specific frequencies providing global coverage. This service operates under encrypted or coded voice transmissions, often single-sideband, although the use of the automatic link establishment (ALE), a type of digital transmission mode, is commonly extended.

HF-GCS stations may operate in the aviation bands clustered around 5, 8, and 11/12 MHz, although other frequencies are also in use. The primary HF-GCS voice frequencies are 4724.0, 6739.0, 8992.0, 11,175.0, 13,200.0, and 15,016.0 kHz.

While no new technologies are planned in the HF band, there is a continued need for the services offered. This is in the case of aeronautical mobile (R) service frequencies, and also for the aeronautical radionavigation service, that being a life safety service must be protected until a sufficiently reliable means of radionavigation can be developed (ERC, 2001; WRC, 2003).

For the purpose of information, we can mention that ICAO adopted the Standards and Recommendations Practices for HF Data Link in 1998, operating in the bands (WRC, 2003). Data links are, however, not expected to replace HF voice due to the lack of available frequency spectrum in the HF band. Growing data traffic and further development of the HF Data Link system may result in further frequency requirements above 5 MHz.

9.5 AERONAUTICAL SERVICES AT VHF AND L BAND

The portion of VHF band ranging from 108 to 137 MHz constitutes jointly with HF frequencies 3–30 MHz the core legacy of aeronautical communications. The VHF airband uses the frequencies between 108 and 137 MHz, phonetically called *victor*. The upper band is divided for AM voice transmissions using a channel spacing originally of 200 kHz until 1947, providing 70 channels from 118 to 132 MHz. As of 2012, increasing air traffic congestion has led to further subdivision into narrowband 8.33 kHz channels in the ICAO European region (Pouzet and Rees, 2007; Stacey, 2008; Neji et al., 2012).

The VHF band ranging from 960 to 1215 MHz is extensively used for air/ground voice communications, as well as air/ground and air/air data links. Even with a channel spacing of 8.33 kHz in Europe, saturation of this band is foreseen around 2025. The VHF band is used by several aeronautical systems—distance measuring equipment (DME), secondary surveillance radar (SSR) (including automatic-dependent surveillance–broadcast [ADS-B]), universal access transceiver (UAT), and global navigation satellite system (GNSS)—and its availability therefore needs to be secured by a long-term strategy. This band is also planned to be used by the future communication system for air/ground and air/air data link systems (Pouzet and Rees, 2007; Stacey, 2008; Neji et al., 2012).

Technologies that currently provide or are planned for aeronautical communications in the VHF band, providing dedicated voice and data services, should be used to their fullest extent. Owing to congestion in the VHF band, the provision of future communication services outside the VHF band must be considered. For the VHF aeronautical spectrum, the band will continue to be used to provide DSB-AM voice communications and an initial data link capability (Pouzet and Rees, 2007; Stacey, 2008; Neji et al., 2012).

The aeronautical L band spectrum ranged from 960 to 1024/1164 MHz has been proposed as a candidate band for supporting a new data link communication capability, given that there is a potential large spectral region to support future aeronautical communication systems. However, it is a challenging environment for aeronautical communications due to the aeronautical channel characteristics and the current usage of the band (Pouzet and Rees, 2007; Stacey, 2008; Neji et al., 2012):

- Estimated root mean square delay spreads (RMS-DSs) for this channel, on the order of 1.4 μs, can lead to frequency selective fading performance for some technologies.
- Cochannel interference to and from existing aeronautical L band systems for a proposed communication technology requires measurements and testing.

The aeronautical L band (960–1024/1164 MHz) spectrum provides an opportunity to support the objectives for future global communication systems; however, no evaluated technology for supporting data communication in this band fully addresses all requirements and limitations of the operating environment. Desirable features for an aeronautical L band in the range 960–1024/1164 MHz technology include (Pouzet and Rees, 2007; Stacey, 2008; Neji et al., 2012)

- Multicarrier modulation to provide power-efficient modulation against signal fading
- Low duty cycle waveform with narrow-to-broadband channels to achieve successful compatibility with legacy L band systems without clearing spectrum
- Adaptable/scalable features, improving flexibility in deployment and implementation, and adaptability to accommodate future demands

Two options were identified for an L band digital aeronautical communication system (L-DACS) at 960–1024/1164 MHz. The first option for L-DACS includes a frequency division duplex (FDD) configuration utilizing OFDM modulation technique, reservation-based access control, and advanced network protocols. This solution is a derivative of the B-AMC and TIA-902 (P34) technologies. The second L-DACS option, L-DACS2, includes a time division duplex (TDD) configuration utilizing a binary modulation derivative of the implemented UAT system (continuous-phase frequency shift keying [CPFSK] family) and of existing commercial (e.g., GSM) systems and custom protocols for lower layers providing high QoS management capability (Pouzet and Rees, 2007; Stacey, 2008; Neji et al., 2012).

9.5.1 VHF AND L BAND RADIO CHANNEL CHARACTERISTICS

Here, we will consider the communication systems (voice and data) operating at VHF (108–137 MHz) and L band (960–1215 MHz). Even though some early research work can be found around 1960, the majority of models assumed to characterize VHF band were narrowband due to the use of communication bandwidths of few tens of kilohertz. Owing to the low latency of the voice and telemetry systems used in this band before the adoption of future radio system (FRS), few wideband models were experimentally derived. We should indicate that the L band was used for satellite mobile communications from aircraft in the past; however, here we only consider the new functionalities of this band.

Generally speaking, propagation models estimated for this band are typically large-scale models that predict the mean signal strength for an arbitrary transmitter–receiver separation distance to facilitate estimation of radio coverage area and link budget. These models are characterized by a slow change in average received power with increasing distance from the transmitter, averaging the received power in a local area over tens of wavelengths.

On the other side, we find small-scale fading models valid to characterize the rapid fluctuations of the received signal strength over very short distances or short time durations, mainly due to motion over short distances. An appropriate small-scale model should help us to determine proper waveform design and optimize receiver implementation (Matolak, 2014).

A channel model must indicate the type of fading, DS, and Doppler power spectrum (DPS), and their statistics. DS is due to multipath propagation conditions and it causes dispersion or distortion on the received signal and it will require the use of compensation techniques, such as equalization, multicarrier, or diversity. The Doppler time variation is important for frequency carrier tracking, intercarrier interference mitigation, and so on (Matolak, 2014).

Existing measurements are sparse, for very different environments, and the derived models are not parameterized as the function of parameters, such as radiolink elevation angle, involved antenna heights and beamwidths, or local environment (Matolak, 2014).

The variant nature of the aeronautical channel generates different scenarios for channel estimation, showing different types of fading, DS and DPS. The aeronautical channel can be first divided into ground-to-ground, air-to-ground, and air-to-air links. This division is combined with the dynamics of the aircraft: taxi or park, takeoff or landing, and en route. The geographic features of the underlying scenario is also a factor to be considered, and then we can find over ocean, over flat ground, mountainous terrain, or airport environments that influence the radio propagation differently (Haque, 2011).

For both VHF and L bands, the radio channel characterization scenarios can be classified into the following categories (Karasawa and Shiokawa, 1984; Braun and Dersch, 1991; Raleigh et al., 1994; Rice et al., 2000, 2004; Zhang and Liu, 2002; Jayaweera and Poor, 2005; Lei and Rice, 2009; Rice and Jensen, 2011; Wu et al., 2011; Matolak et al., 2014), all of them summarized in Table 9.2 (Haque, 2011).

9.5.1.1 Taxi/Park at the Airport

This scenario, also called surface area, offers ground-to-ground radio links with NLOS conditions for the propagation due to the multiple obstacles blocking the direct component, if an urban environment is assumed.

TABLE 9.2
L Band Path Loss Parameters

Setting	$A_{0,L,s}$ (dB)	$n_{L,s}$	$\sigma_{X,L,s}$ (dB)	$L_{L,s}$ (dB)	L_0 (dB)	$\sigma_{X,L,l}$ (dB)	$L_{L,l}$ (dB)	d_t (km)
Freshwater	57.7	1.4	2.8	1.8	1.1	3.2	1.8	6.6
Seawater	50.6	1.5	2.8	1.2	1.0	4.8	1.1	9.1

Source: Adapted from Matolak, D.W., AG channel sounding for UAS in the UAS, Graduate Thesis, University of South Carolina, 2014.

As in many terrestrial channel models, there are LOS and NLOS regions in airport surface areas.

A narrowband characterization was done by Wu et al. (2011) in the band of 118–137 MHz. These models estimate the path loss and are basically small scale. Narrowband measurements were made in the aeronautical VHF band at Detroit Metropolitan Airport (DTW). A continuous-wave (sinusoidal) signal was transmitted from existing VHF radio transmitter–receiver (RTR) sites, and a mobile receiver moved about the airport surface area in a prescribed path. Propagation path loss was computed using basic link budget analyses. Results for both LOS and NLOS conditions are provided using the log-distance path loss model. The mean propagation path loss exponents are found to be approximately 5.6 and 4 for NLOS and LOS regions, respectively. Results for LOS regions also mostly agree with the two-ray or plane-Earth models (Lei and Rice, 2009).

The mean path loss is well described using the log-distance model, where path loss L is given by the following equation, in dB: $L_d = A \cdot n \cdot 10 \log_{10}(d/d_0) + X$, where A is the measured path loss in dB at the reference distance d_0, $d > d_0$ is the link distance in meters, X is a zero mean Gaussian random variable in dB with a standard deviation σ_X, and n is the dimensionless propagation path loss exponent. The reference distance d_0 is generally chosen for convenience, within the far field of antennas, for example, a value of 5–50 m for the airport surface area. A larger value of d_0 might be employed depending upon antenna heights, accessibility of areas, etc.

For the NLOS situations, the fitted value of the path loss exponent $nN_{LOS} = 5.6$ dB and the standard deviation of X (zero mean Gaussian variable) resulted to be $\sigma_X = 8.3$ dB. For LOS, the obtained path loss exponent n_{LOS} was 4 dB and a standard deviation σ_X of 3.7 dB. This model applies for a range of distances in the range 1722–4853 m.

A wideband model was proposed by Hoeher and Haas (1999) in the band of 118–137 MHz. This study concludes that during this flight phase, a classical Gaussian type of Doppler spectrum as defined by COST-207 (1989) applies given that the echoes arrive equally distributed from all directions, that is, a scattering beamwidth of 360°, achieving a maximum Doppler frequency of $f_{Dmax} = 2.5$ Hz for the parking scenario for a typical speed case of 15 m/s, and $f_{Dmax} = 23$ Hz for the taxi scenario for the worst allowed speed case of 50 m/s, both at 137 MHz.

From the DS perspective, for the parking scenario, a typical urban environment was proposed; however, for the taxi scenario, a rural area was assumed. The power delay profiles (PDP) for both cases were defined by COST-207 (COST 207, 1989). In the urban environment (parking), a Rayleigh fading can be assumed, with typical maximum delay of 7 µs that corresponds to a path excess of 2100 m. For this case, the Rayleigh PDP shows a delay slope decay of 1 µs that indicates when the power amplitude of the received diffuse components reaches −30 dB for a normalized PDP function.

Rician fading is assumed in the rural area environment (taxi) with a Rice factor of 6.9 dB, given by the square ratio between the power of the LOS component (0.83) and the power of the diffuse or multipath components (0.17). For this case, the maximum delay is 0.7 µs for a path excess of 210 m, and the Rice PDP shows a delay slope decay of 1/9.2 µs.

It is mentioned in Hoeher and Haas (1999) that for systems operating at this portion of the aeronautical channel, such as VDL Mode 3 and Mode 4 for digital data links, the main counteractions to compensate the taxi/parking scenario propagation impairments consist of receiver antenna diversity given that frequency hopping, channel coding, or interleaving does not apply here.

9.5.1.2 En Route

For this scenario, air–ground and air–air links are present. This portion of the channel shows a fast fading type due to the LOS component and other diffuse components (multipath, reflected, and delayed paths) that can be characterized by a two-ray model (Hoeher and Haas, 1999). The direct component is assumed to fit a Rice process and the diffuse components a Rayleigh process. Rice factors range from 2–20 dB, with 2 dB the worst case, and 15 dB the typical value (Rice et al., 2000).

In the PDPs, the delays approximate 200 µs, equivalently 60 km of path difference, for air–ground links; and 1 ms, or 300 km, for ground–air links (Hoeher and Haas, 1999).

For air–air links, fast fading occurs and the path difference can be assumed as $2h/c$, with h being the flight altitude, if there is one dominant reflector, and then the two-ray model fits. For a typical maximum altitude of 10 km, air–air links find maximum delay values of 66 μs if no ducting is considered, which would affect the wave speed (Hoeher and Haas, 1999).

The maximum Doppler frequencies are 200 Hz, for the air–ground link at 440 m/s, and 280 Hz for air–air link at a speed of 620 m/s. The scattered components are not isotropically distributed, assuming a uniform distribution of the scatterers with a beamwidth of 3.5° is typical. The LOS direction is along the aircraft heading, whereas the scattered components come from the rear (Hoeher and Haas, 1999).

We can conclude that the en route channel is frequency-selective, particularly for small Rice factor values.

Recent measurements (Walter et al., 2014) indicate that the wide-sense stationary uncorrelated scattering (WSSUS) assumption is not valid for air-to-air channels, and therefore, purely stochastic channel models cannot be used. In order to meet the nonstationarity condition, a geometric component must be included. As stated in Walter et al. (2014), the joint delay-Doppler probability density functions (PDFs) should exactly describe the influence of scattering on aeronautical air-to-air channels by calculating the time-variant delay-dependent Doppler PDFs and taking the exact flight trajectories into account. The developed new PDF generalizes the von Mises distribution for an ellipse. This directional distribution of the scatterers influences the amplitudes of the joint delay-Doppler PDF but not the shape.

The results indicated in Walter et al. (2014) show that the specific flight trajectories have a direct influence on the shape of the joint delay-Doppler PDF, demonstrating that the Jakes and Gaussian Doppler spectra are not exact for nonstationary air-to-air channels.

In European Radiocommunications Committee (ERC) (2001) and NASA (2006), NASA introduced a report on the air–ground (AG) channel characterization in the L band obtained from flight tests in 2013 for over-sea (salt water) conditions over the Pacific Ocean (Oxnard, California, USA), and flight tests for over-freshwater conditions over Lake Erie (Cleveland, Ohio, USA). Based upon data gathered from these over-water flight tests, path loss and dispersion models have been derived. These flight tests measured channel impulse responses for both straight and oval-shaped flight paths.

Path loss generally follows s curved-Earth two-ray (CE2R) model, with deviation due to a number of factors, including water surface roughness, nonideal antenna patterns, intermittent multipath components, and some slow variation of received power due to channel measurement equipment imperfections (NASA, 2006; Matolak, 2014). Path loss models for the over-water settings are as follows, with path loss in units of dB:

$$PL(d,\theta) = \begin{cases} A_{0,L,s} + 10n_{L,s}\log(d) + X_{L,s} + \zeta_{L,s} & \theta > \theta_t(d > d_t) \\ CE2R + L_0 + X_{L,l} + \zeta_{L,l} & \theta < \theta_t(d < d_t) \end{cases} \qquad (9.3)$$

where threshold elevation angle $\theta_t = 5°$, $\theta_t = -1$ for travel toward the GS and $+1$ for travel away from the GS, and distance ranges from $1 \leq d \leq 28$ km for freshwater and $2 \leq d \leq 24$ km for sea water. The variable X is a zero-mean Gaussian random variable, with standard deviation σ_X dB, and CE2R denotes the curved-Earth two-ray model. Table 9.3 summarizes the parameters of Equation 9.3 fitted for the performed measurements.

The over-sea (or over-freshwater) AG channel is predominantly an LOS channel with a surface reflection (Karasawa and Shiokawa, 1984). This is well approximated by the deterministic CE2R model. Other channel propagation mechanisms, including discontinuous multipath components and random magnitude and phase components of the sea-surface reflection, can be statistically quantified. A path loss model as in Equation 9.3 embodies the losses of signal variation in the linear fit of the loss versus distance. If it is a small-scale fading model, the loss variation must be specified by means of the statistical distribution of the amplitude of the channel gain, once large-scale effects

TABLE 9.3

Aeronautical Channel Characteristics

Parameter	Parking Scenario	Taxi Scenario	Arrival Scenario	En Route (Air–Air)	En Route (Air–Ground)
Aircraft velocity v (mph)	0–5.5	0–15	25–150 typ. 85	440 17–440 typ. 250	620
Maximum delay t_{max} (s) worst case	7.0×10^{-6}	7.0×10^{-6}	7.0×10^{-6}	$66 \times 10^{-6} - 200 \times 10^{-6}$	$33 \times 10^{-6} - 1 \times 10^{-3}$
Number of echo paths	20	20	20	20	20
f_{DLOS}/f_{DMax}	–	0.7	1	1	1
Rice factor K_{Rice} (dB)	–	6.9	15 (mean) 9 – 20	15 (mean) 2 – 20	15 (mean) 2 – 20
Start angle of beam (degree)	0.0	0.0	–90	178.25.0	178.25.0
End angle of beam	360	360	90	181.75	181.75
Exponential or two ray	Exponential	Exponential	Two ray	Two ray	Two ray

Source: Adapted from Haque, J., An OFDM based aeronautical communication system, Graduate Thesis, University of South Florida, 2011.

are removed. In the LOS case, the most logical choice for modeling the amplitude distribution is the Ricean distribution, a two-parameter distribution characterized by the strength of the LOS (non-fading) component and the strength of the remaining (assumed Rayleigh distributed) components.

Dispersion was found to be fairly small, as expected in these open over-water environments, with RMS-DSs typically less than 50 ns, with the occasional value up to nearly 250 ns. DS statistics have been reported in a prior NASA report (Rice et al., 2004; Lei and Rice, 2009; Matolak, 2014; Matolak et al., 2014).

Flight tests were also conducted over hilly terrain near Latrobe (Pennsylvania, USA) on April 15, 2013. Path loss versus link distance in the L band fits the CE2R model and free-space path loss for comparison. Linear fits to the hilly terrain measured path loss on the log–log scale show standard deviations ranging from 3.2 to 3.6 dB for the L band. When comparing to the over-sea results, it is also clear that the two-ray effect is more pronounced in the over-sea setting than in the hilly terrain setting. This is as expected since water surfaces will generally be much more reflective than Earth surfaces. Nonetheless, the surface reflection in the hilly terrain does cause path loss peaks as seen in "two-ray" channel models; however, these peaks are just not as predictable as in the over-water cases. Specific path loss model for the hilly terrain setting, analogous to the model in Equation 9.3, is not yet available (Matolak et al., 2014).

9.5.1.3 Takeoff or Landing

Air-to-ground links are established during this phase of the flight. For this situation, the radio propagation conditions are a combination of taxi/parking and en route scenarios (Hoeher and Haas, 1999). For this portion of the aeronautical radio channel, we find sparse models. First, we observe that the air–ground portion of the aeronautical is largely dependent on the environment.

Rice (Lei and Rice, 2009; Rice and Jensen, 2011) performed measurements for telemetry down-link transmitting at frequencies of 2.225, 2.2345, and 2.3455 GHz, for rural/suburban flat terrains, that concluded that this portion of the radio channel is composed of an LOS signal, a specular reflection whose strength is 20%–80% that of the LOS signal, and a diffuse multipath component whose power is 10–20 dB less than that of the LOS signal. Thus the channel is close to additive white Gaussian noise (AWGN) (but not quite) with strong specular interference. When the channel is good (i.e., a low value of Γ), the channel is essentially an AWGN channel, and when the channel is bad (i.e., a high value of Γ), both the specular multipath interference and diffuse multipath interference degrade system performance, although the effects of the specular reflection tend to dominate. We observe that the fading events in this data set can be well approximated by an AWGN channel with strong specular reflection whose amplitude is 50%–90% that of the LOS signal.

In Lei and Rice (2009), for over-sea air-to-ground link, the effect of multipath fading was analyzed by applying the Kirchhoff approximation theory, especially paying attention to the coherent and incoherent components, geometrical shadowing, and sea surface conditions. As for the incoherent component, the analysis was made using scattering cross sections based only on the wave height and steepness among the parameters indicating these conditions.

With regard to the sea conditions, the rough sea condition where the incoherent component is dominant is the most important for estimating the fading depth in the aeronautical satellite communication because this condition has a high probability of occurrence and produces a large fading magnitude (Neji et al., 2012).

Fading depth under the rough sea condition has little dependence upon the wave height and the fading depth for 99% of the time is estimated to be 4.5–6.5 dB (gain: 20 dBi), 7–9 dB (15 dBi), and 8–10.5 dB (10 dBi) at an elevation angle of 5°. These theoretical values agree well with different experimental results (Lei and Rice, 2009).

The fading under the sea condition of wave height greater than 4 m was not analyzed.

9.6 SATELLITE COMMUNICATION SERVICES

With the forecast traffic growth and the limitations of the existing communication systems, significant effort has been allocated to investigating the future communication infrastructure for aviation. It is expected that this future infrastructure will include both terrestrial and satellite components to meet aviation requirements (Panagopoulos et al., 2004; Phillips et al., 2007; Pouzet, 2007, 2008; Sheriff and Hu, 2010; Van Wambeke and Gineste, 2010; Durand and Longpre, 2014; Richharia, 2014; ICAO, 2015).

Currently, satellite-based communications technology is limited to oceanic or remote areas of the world. In the longer term, there appears to be potential for satellite communications to be used in higher-density airspace to complement terrestrial systems provided that the QoS required for safety-related services can be achieved. For the new satellite systems, commercial or custom applications technology satellites (ATS), satellite-based communications will continue to offer great benefits to aviation requirements (Panagopoulos et al., 2004; Phillips et al., 2007; Pouzet, 2007, 2008; Sheriff and Hu, 2010; Van Wambeke and Gineste, 2010; Durand and Longpre, 2014; Richharia, 2014; ICAO, 2015).

Aeronautical satellite systems offer unique services that can be applied to large and/or remote geographic areas and provide supplemental coverage to terrestrial communication infrastructure requirements as follows (Panagopoulos et al., 2004; Phillips et al., 2007; Pouzet, 2007, 2008; Sheriff and Hu, 2010; Van Wambeke and Gineste, 2010; Durand and Longpre, 2014; Richharia, 2014; ICAO, 2015):

- Communication capability in oceanic, remote, and polar regions, where typically, there is no other alternative that provides the needed capacity and performance.
- Communication coverage to en route domains with historically sparse aircraft densities where it may be more cost effective.

Aeronautical radio frequency spectrum has been targeted for use by nonaeronautical services, in particular to satisfy requirements for mobile and satellite communications. This has led to the loss of some spectrum that was allocated exclusively for aeronautical mobile satellite (AMS) communications but authorized for nonaeronautical use. This has introduced potential interferences and the loss of spectrum capacity needed to satisfy current and future aeronautical requirements for the communications, navigation, and surveillance (CNS) system, in line with the CNS roadmap included in the Global Air Navigation Plan (GANP) requirements (Panagopoulos et al., 2004; Phillips et al., 2007; Pouzet, 2007, 2008; Sheriff and Hu, 2010; Van Wambeke and Gineste, 2010; Durand and Longpre, 2014; Richharia, 2014; ICAO, 2015).

Different frequency bands have been allocated for satellite communication. One of them is in the range 5000–5250 MHz, also dedicated to several aeronautical systems: the microwave landing system (MLS), unmanned aircraft systems (UAS), and aeronautical telemetry. The strategic objectives for the long-term use of this band are as follows (Panagopoulos et al., 2004; Phillips et al., 2007; Pouzet, 2007, 2008; Sheriff and Hu, 2010; Van Wambeke and Gineste, 2010; Durand and Longpre, 2014; Richharia, 2014; ICAO, 2015):

1. Secure the continuing availability of the frequency band 5030–5091 MHz, which is allocated to the aeronautical radio navigation service for use by the MLS and to support global air/ground communications for UAS.
2. Secure the continuing availability of the frequency band 5091–5150 MHz, which is allocated to the aeronautical mobile (R) service for use by airport communications on a global basis.
3. Secure future implementation of the aeronautical telemetry in the band 5091–5250 MHz.

There is current interest in using frequencies near 1.5 GHz for AMS systems. The band range 1545–1555/1646.5–1656.5 MHz is also allocated for satellite communications, mainly radionavigation systems. During the WRC/12, access to these bands for aeronautical satellite communications has been improved, something that will support the requirements for aeronautical satellite communications for long-term requirements (Panagopoulos et al., 2004; Phillips et al., 2007; Pouzet, 2007, 2008; Sheriff and Hu, 2010; Durand and Longpre, 2014; Van Wambeke and Gineste, 2010; Richharia, 2014; ICAO, 2015).

In the time frame of 2020+, new satellite-based communication technologies are expected to emerge, which can be used for aeronautical operational control (AOC) and air traffic services (ATS) communications. A range of options for satellite communication using low-, medium-, and geostationary orbit satellites are expected to be available, offering mobile communication services to aircraft. These could range from commercially operated systems offering a generic service to all mobile users (land, maritime, and aviation) to systems targeted to meet specific aviation requirements (Panagopoulos et al., 2004; Phillips et al., 2007; Pouzet, 2007, 2008; Sheriff and Hu, 2010; Van Wambeke and Gineste, 2010; Durand and Longpre, 2014; Richharia, 2014; ICAO, 2015).

The genesis of the future communication system can be traced to the Eleventh Air Navigation Conference (ANC), held in Montreal from 22 September through 3 October 2003. One of the highlights of that ICAO conference was the official report of the Technical and Operational Matters in Air Traffic Control Committee (Committee B). That report noted the current state of aviation communications and made several recommendations to advance this state. The observations included these requirements (Panagopoulos et al., 2004; Phillips et al., 2007; Pouzet, 2007, 2008; Sheriff and Hu, 2010; Van Wambeke and Gineste, 2010; Durand and Longpre, 2014; Richharia, 2014; ICAO, 2015):

1. The aeronautical mobile communication infrastructure has to evolve to accommodate new functions.
2. This evolution would likely require the definition and implementation of new terrestrial and/or satellite systems that operate outside the VHF band.
3. A variety of views had been presented with regard to the future evolution of aeronautical mobile communications.

4. The universally recognized benefits of harmonization and global interoperability of air–ground communications should not be lost sight of when pursuing solutions.
5. The successful gradual introduction of data communications should be continued to complement and replace voice for routine communications.

Based on these observations, several recommendations were provided:

1. Develop an evolutionary approach for global interoperability of air–ground communications.
2. Conduct an investigation of future technology alternatives for air–ground communications.
3. Prove compliance with certain minimum criteria before undertaking future standardization of aeronautical communication systems.

Subsequent to the 11th ANC, the Federal Aviation Administration (FAA) and European Organisation for the Safety of Air Navigation (EUROCONTROL) embarked on a cooperative research and development program in part to address these ICAO recommendations and in part to deal with frequency congestion and consequent spectrum depletion in both core Europe and dense United States airspace. By agreement, joint FAA and EUROCONTROL research and development activities required terms of reference, which are referred to as "action plans" and are numbered sequentially. The terms of reference for the Future Communication Study are detailed in Action Plan 17; and NASA, European Space Agency (ESA), the FAA, and EUROCONTROL all have defined roles in the research and development activity requirements (Panagopoulos et al., 2004; Phillips et al., 2007; Pouzet, 2007, 2008; Sheriff and Hu, 2010; Van Wambeke and Gineste, 2010; Durand and Longpre, 2014; Richharia, 2014; ICAO, 2015).

EUROCONTROL, through its NexSAT Steering Group and a broader international cooperation with the International Civil Aviation Organization (ICAO), FAA, ESA, NASA, etc., is interested in further investigating the technical, institutional, and financial aspects of using satellite communications for ATM purposes. The EUROCONTROL investigations include dedicated systems and commercial satellite services requirements (Panagopoulos et al., 2004; Pouzet, 2007, 2008; Sheriff and Hu, 2010; Van Wambeke and Gineste, 2010; Durand and Longpre, 2014; Richharia, 2014; ICAO, 2015).

Example systems that have potential as future satellite systems have been identified in the Future Communication Study. These examples are as follows (Panagopoulos et al., 2004; Phillips et al., 2007; Pouzet, 2007, 2008; Sheriff and Hu, 2010; Van Wambeke and Gineste, 2010; Durand and Longpre, 2014; Richharia, 2014; ICAO, 2015):

- *ATM SATCOM:* The ESA ATM SATCOM system can be described as a modernized version of the classic ICAO Aero SATCOM System (or AMSS). ATM SATCOM reuses some concepts of the AMSS, such as use of geostationary satellites, while overcoming the legacy system limitations with the aim to support future ATM mobile communication services with the required performance level.
- *Iridium NEXT:* Iridium LLC is embarking on the design of the next-generation of the Iridium satellite constellation. This new system—currently known as Iridium NEXT is proposed to seamlessly replace satellites in the current constellation and will be backward compatible with present applications and equipment. It will provide new and enhanced capabilities with greater speed and bandwidth, and which are expected to be available to aviation.

9.6.1 SATCOM Applicable Technology Analysis

For the Satellite and Over Horizon technology family, two technology inventory candidates emerged from the preliminary technology screening: Inmarsat SBB and Custom Satellite Solution. The Inmarsat SBB candidate differs from many of the other candidates considered in that it is an operational system with a defined service architecture and a defined set of service offerings. For this candidate, the ability to meet Concept of Operations and Communication Requirements (COCR) for the FRS performance requirements was the focus of detailed analysis requirements (Panagopoulos et al., 2004; Phillips et al., 2007; Pouzet, 2007, 2008; Sheriff and Hu, 2010; Van Wambeke and Gineste, 2010; Durand and Longpre, 2014; Richharia, 2014; ICAO, 2015).

COCR performance requirements are specified for data capacity, latency, QoS, and maximum number of users, but there are also availability and integrity requirements for aeronautical services. The performance of Inmarsat SBB with regard to capacity, latency, QoS, and number of users was evaluated as part of an initial technical evaluation iteration along with other screened technologies. This being the case, the selected focus for the detailed analysis was other COCR performance requirements, specifically availability performance. Availability was selected as it was considered as a potential shortfall of the satellite candidate solutions requirements (Panagopoulos et al., 2004; Phillips et al., 2007; Pouzet, 2007, 2008; Sheriff and Hu, 2010; Van Wambeke and Gineste, 2010; Durand and Longpre, 2014; Richharia, 2014; ICAO, 2015).

The Custom Satellite Solution is a technology concept that includes the fielding of a custom satellite or custom satellite payload specifically designed for aeronautical communications. This concept is being explored by several civil aviation authorities and related organizations. Japan has launched an aeronautical communication satellite and is exploring performance of next-generation satellite systems. The FAA's Global Communication, Navigation, Surveillance System (CGNSS) contract Phase I study explored the definition of a satellite architecture for providing aeronautical safety services. And finally, there are consortiums that are working to define aeronautical satellite specifications, such as the satellite data link system (SDLS). Within this body of work, the need to accommodate all communication safety services with associated performance requirements has been considered. It has been found that to meet these requirements, a highly reliable, highly available architecture is required, such as the five satellite architecture proposed by Boeing in the global communications navigation and surveillance system (GCNSS) study requirements (Panagopoulos et al., 2004; Phillips et al., 2007; Pouzet, 2007, 2008; Sheriff and Hu, 2010; Van Wambeke and Gineste, 2010; Durand and Longpre, 2014; Richharia, 2014; ICAO, 2015).

Availability arises again as an important issue. In order to provide required availability, a highly redundant custom satellite system architecture is needed. As this issue is similar to that noted above for Inmarsat, a separate study of availability for Custom Satellite Solution was not performed. Rather, it was considered to be more instructive to estimate the availability of two existing, operational satellite systems, Inmarsat SBB and Iridium, that provide services in protected aeronautical spectrum (AMS(R)S) requirements (Panagopoulos et al., 2004; Phillips et al., 2007; Pouzet, 2007, 2008; Sheriff and Hu, 2010; Van Wambeke and Gineste, 2010; Durand and Longpre, 2014; Richharia, 2014; ICAO, 2015).

In summary, the focus of the detailed analysis for satellite technologies was availability performance. The performance of existing AMS(R)S systems (namely, Inmarsat SBB and Iridium) was examined. Based on the study results, recommendations to be considered for evaluation of the Inmarsat SBB and Custom Satellite Solution technology candidates in support of the technology evaluation process were made requirements (Panagopoulos et al., 2004; Phillips et al., 2007; Pouzet, 2007, 2008; Sheriff and Hu, 2010; Van Wambeke and Gineste, 2010; Durand and Longpre, 2014; Richharia, 2014; ICAO, 2015).

9.6.2 SATELLITE COMMUNICATION AVAILABILITY ANALYSIS

The objective is to examine the availability performance of Inmarsat SBB and Iridium, two current satellite service offerings in AMS(R)S spectrum, and to provide a high-level comparative analysis of the calculated performance to a representative VHF terrestrial data communication architecture. The comparison can be observed in Table 9.4.

TABLE 9.4

Availability Performance Comparison of SATCOM Technologies

	System Component Failures				Fault-Free Rare Events			
	Ground Station	Control Station	Aircraft Station	Satellite	RF Link	Capacity Overload	Interference	Scintillation
Inmarsat	~1	~1	~1	0.9999	0.95	~1	~1	~1
Iridium	0.99997	~1	~1	0.99	0.995	—[a]	0.996	~1
VHF terrestrial	0.99999	N/A	~1	N/A	0.999	—[b]	~1	N/A

Note: N/A—data not available.

[a] Iridium capacity overload availability of AES to SATCOM traffic is essentially one (1) (for both ATS only and ATS and AOC). No steady state can be achieved for SATCOM to AES traffic.

[b] Terrestrial capacity overload availability is for VHF band reference architecture business case; for L band, terrestrial capacity overload availability would be essentially one (1).

Inmarsat SBB is a fourth-generation service offering of Inmarsat. It includes two I-4 geostationary satellites (positioned over the Atlantic and Indian Oceans) with the potential to include a third satellite over the Pacific Ocean (although this third satellite may remain as a ground spare for the first two spacecraft). A depiction of the proposed coverage of the SBB satellites and associated spotbeams is provided in Figure 9.5.

The SBB services include both circuit and packet switch connections, including a guaranteed "streaming" data service. The ground infrastructure is European based, including a satellite access station (SAS) in Berum, Belgium and Fucino, Italy. The ground infrastructure includes internal routing between these SAS sites to accommodate rerouting of traffic in the event of a SAS gateway failure. A depiction of the Inmarsat SBB ground infrastructure is provided in Figure 9.2.

To support the availability analysis, overall availability performance, as well as performance specific to a representative region of the U.S. National Airspace System (NAS), were considered. The geographic region associated with three air route traffic control centers (ARTCCs) was selected

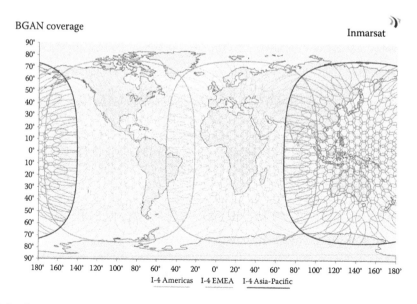

FIGURE 9.5 Proposed Inmarsat SBB coverage.

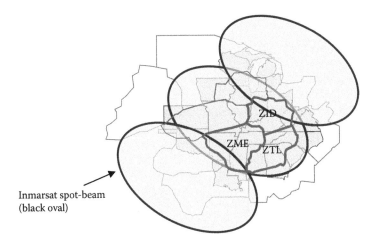

FIGURE 9.6 Inmarsat SBB spot-beam coverage of representative NAS ARTCCs.

as this representative region. These areas encompass the Memphis (ZME), Atlanta (ZTL), and Indianapolis (ZID) ARTCC regions. A view of Inmarsat SBB spot-beam coverage of the representative NAS region is shown in Figure 9.6.

Iridium is a second AMS(R)S system that offers two-way global voice and data aeronautical communication services. The Iridium architecture consists of 66 fully operational satellites and 11 in-orbit spares. Its full constellation life is designed to last through mid-2014 with plans in place to extend the constellation to beyond 2020. The satellites are organized into six planes of near-polar orbit, where each satellite circles Earth every 100 min. Iridium offers full duplex 2400 bps user channels for provision of voice and data services.

A representative view of Iridium coverage of the NAS reference region is provided in Figure 9.7. In this figure, a portion of the NAS reference area falls within view of two Iridium orbital planes (approximately 20% of the reference region).

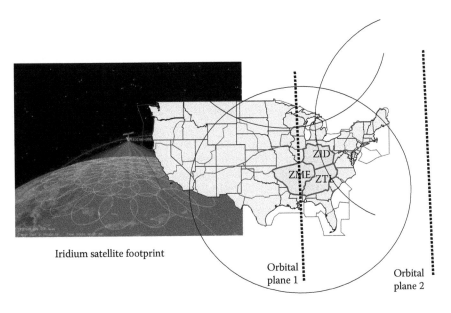

FIGURE 9.7 Iridium coverage of representative NAS ARTCCs.

9.6.3 Satellite Radio Channel Characterization

Generally speaking, propagation effects of aeronautical channels differ from maritime and land mobile propagation because of the high velocity of airplanes, their distance from ground, and influence of the aircraft body on antenna performance. Airplane maneuvers can affect signal under conditions when aircraft antenna is shadowed by the aircraft structure. When considering helicopters, the rotation of the rotor blades causes a cyclic interruption to the signal path (Richharia, 2014).

Propagation effects in the AMS service differ from those in the fixed satellite service (FSS) and other mobile satellite services because (ITU-R P.682-3, 2012)

- Small antennas are used on aircrafts, and the aircraft body may affect the performance of the antenna.
- High aircraft speeds cause large Doppler spreads.
- Aircraft terminals must accommodate a large dynamic range in transmission and reception.
- Aircraft safety considerations require a high integrity of communications, making even short-term propagation impairments very important, and communications reliability must be maintained in spite of banking maneuvers and three-dimensional operations.

Radio channel models are specifically required to characterize path impairments including (ITU-R P.682-3, 2012):

- Tropospheric effects, including gaseous attenuation, cloud and rain attenuation, fog attenuation, refraction, and scintillation
- Ionospheric effects such as scintillation
- Surface reflection (multipath) effects
- Environmental effects (aircraft motion, sea state, land surface type)

9.6.4 Aeronautical Radio Channel Characterization in L Band

The L band is defined as the MSS allocation in the frequency ranges 1525–1559 MHz and 1626.5–1660.5 MHz. Although the whole band is generically for MSS use, maritime safety-related services afforded a specific status in the ITU radio regulations in the subband 1646.5–1656.5 MHz and 1545–1555 MHz, with the communications in the AMS(R)S of larger priority over other types of communications.

Recommendation ITU-R P.682-3 regards the planning of Earth–space aeronautical mobile telecommunication systems for frequencies between 1 GHz and 3 GHz, covering the L band and partly the S band. The satellite azimuth can vary between 10° and 170°, or 190° and 350°. The elevation angle to the satellite can vary between 10° and 75° (Steingass et al., 2008; ITU-R P.682-3, 2012).

Characteristics of fading for aeronautical systems can be analyzed with procedures similar to those for maritime systems described in Recommendation ITU-R P.680, taking careful account of Earth's sphericity, which becomes significant with increasing antenna altitude above the reflecting surface (Steingass et al., 2008; ITU-R P.682-3, 2012).

According to the model in Steingass et al. (2008) and ITU-R P.682-3 (2012), the multipath propagation conditions of a receiving aircraft divide in two main parts:

- The aircraft structure
- The ground reflection

The aircraft structure shows significant reflections only on the fuselage (when the antenna is mounted at the top of the cockpit). This very short-delayed reflection shows little time variance and dominates the channel. A strong wing reflection was not observed (when the antenna is mounted at the top of the cockpit). The ground reflection shows high time variance and is Doppler-shifted according to the aircraft sink rate (ITU-R P.682-3, 2012).

Short-delayed multipath in aeronautical communication and navigation systems has to be considered, especially for broadband signals. The reflections on the aircraft structure produce significant disturbances. Especially during the final approach when communication availability and reliability, as well as navigation accuracy and integrity are most important, the ground reflection and the reflection on the fuselage generate significant propagation effects (ITU-R P.682-3, 2012).

Signal measurements on aircraft show that reflections and scattering from land is significantly lower than from the sea surface; furthermore, signals have elevation angle dependence over sea but not over land; as already mentioned, most studies have therefore concentrated on propagation behavior in flight paths over sea. Theoretical multipath estimation techniques applicable to the maritime environment can also be applied to the aeronautical channels taking into consideration Earth's curvature effects (Richharia, 2014). Measurements conducted to date demonstrate that (Richharia, 2014)

- Composite received signals follow a Ricean distribution.
- Multipath components exhibit a Rayleigh distribution.
- The magnitude of multipath components are elevation angle dependent.
- Multipath Doppler spectrum possesses a Gaussian distribution.
- Fading is frequency selective due to path delay associated with multipath components.
- The multipath signal is dominated by diffused components—specular components are likely to be negligible for the majority of the time.

In Richharia (2014), measured data illustrate that the power spectral density of multipath depends on aircraft speed, elevation angle, and ascent/descent angle of the flight. The aeronautical channel can be represented as the WSSUS channel due to the randomly changing nature of scatterers over Earth's surface (Haque et al., 2010). The model can be used to estimate a number of characteristics of the channel. The RMS Doppler spectrum is defined as twice the standard deviation of the DPS:

$$B_{rms} = 4 \cdot \alpha \cdot v \cdot \sin\left(\frac{\varepsilon}{\lambda}\right) \qquad (9.4)$$

where α is the RMS surface slope, v is the aircraft speed at constant elevation angle, ε is the elevation angle, and λ is the wavelength of the carrier.

9.6.5 Aeronautical Radio Channel Characterization at Frequencies above L Band

The conventional geostationary (GSO) satellite systems belonging to the FSS, gradually tend to employ higher-frequency bands to satisfy growing capacity requirements. Therefore, besides operation at the Ku band (12/14 GHz), the Ka band (20/30 GHz) and the V band (40/50 GHz) have been investigated or even adopted in satellite systems recently put into operation (Panagopoulos, 2004). However, crossing the 10 GHz frequency limit gives rise to signal fading due to physical phenomena related to the propagation of radiowaves through the atmosphere (Crane, 2003). The fade margin, that is, the system gain insuring the necessary QoS against various transmission and other impairments, must be significantly increased to compensate for the severe signal fading occurring at frequencies above 10 GHz. The larger fade margins required are not feasible either technically or economically. Under these conditions, it is more difficult for satellite systems to satisfy the availability and QoS specifications recommended by the radiocommunication sector of the ITU (ITU-R) (Crane, 2003).

The propagation phenomena concerning Earth–space links operating above 10 GHz mainly originate in the troposphere. The most important tropospheric phenomena affecting satellite communication systems at frequencies above 10 GHz are (Crane, 2003)

- *Attenuation due to precipitation:* It constitutes the main disadvantage of operating at the Ku, Ka, or V frequency bands. A variety of models exists for the prediction of the average rain attenuation on an annual basis: Rec. ITU-R P.618-7 (ITU-R.618-7, 2001), Leitao–Watson (Leitao and Watson, 1986), Lin (Lin, 1975), Morita and Higuti (Morita and Higuti, 1976), and the EXCELL (Capsoni et al., 1987) model.
- *Gaseous absorption:* Its contribution to the total attenuation is small compared to the attenuation due to rain. Water vapor is the main contributor to gaseous attenuation in the frequency range just below 30 GHz due to a maximum occurring at 22.5 GHz. The attenuation due to oxygen absorption exhibits an almost constant behavior for different climatic conditions, whereas the attenuation due to water vapor varies with temperature and absolute humidity. A complete method for calculating gaseous attenuation is given in Annex 1 of Rec. ITU-R P.676-4 (ITU-R P.676-4, 2001).
- *Cloud attenuation:* The liquid water content of clouds is the physical cause of cloud attenuation. Prediction models for this particular attenuation factor have been developed within the framework of ITU-R (ITU-R P.840-3, 2001) and elsewhere (Salonen and Uppala, 1991).
- *Sky noise increase:* As attenuation increases, so does emission noise (see Rec. ITU-R P.618-7). Scatter/emission from precipitation hydrometeors contribute to noise increase.
- *Signal depolarization:* Differential phase shift and differential attenuation caused by nonspherical scatterers (e.g., raindrops and ice crystals) cause signal depolarization. Depolarization results in cross-polar interference, that is, part of the transmitted power in one polarization interferes with the orthogonally polarized signal. In order to demonstrate the long-term statistics of hydrometeor-induced cross-polarization, the ITU-R method related to the attenuation estimation due to the hydrometeor must be considered (Crane, 2003).
- *Tropospheric scintillations:* Variations in the magnitude and the profile of the refractive index of the troposphere lead to amplitude and phase fluctuations called scintillations. These fluctuations increase with frequency and depend upon the length of the slant path decreasing with the antenna beamwidth. Models estimating the effects of scintillations on the received signal can be found in Mousley and Vilar (1982), ITU-R.618-7 (2001), and Van de Kamp et al. (1999).

Aeronautical communication systems used for the transport of ATC/AOC are considered as safety-critical in their frequency allocation by the ITU while systems used for aeronautical passenger communication (APC) communications are not. The principle of transmission in the safety satellite system is as follows (Durand and Longpre, 2014):

- The mobile link, between the satellite and the aircraft, is built on a safety satellite spectrum allocation, based on AMS(R)S standard.
- The satellite is in charge of signals frequency conversion, simultaneously from the C or Ku band to the L band for the forward link, and from the L band to the C or Ku band for the return link.
- The fixed link, between the ground and the satellite, is built on a fixed satellite spectrum allocation, based on the FSS standard.

Most of operational Ku band satellites operate in the Ku FSS unplanned band, from 10.7 to 12.95 GHz. Owing to a need for large amounts of spectrum for mobile communications, there has also been significant interest in utilizing the Ka band in the service link as the band offers large bandwidth and additionally offers the advantage of a lower-sized directive antenna. Radio signals in this band undergo considerable degradation in the troposphere and are affected more adversely by the local environment than in the L band, because the diffraction advantage is less and penetration loss is more severe (Richharia, 2014).

At present, the Ka band (20–30 GHz) propagation database for MSS is limited because the interest in this band for MSS is recent. Empirical models have been developed by scaling data from the L band to Ka band and measurement campaigns have been conducted for similar purposes (Richharia, 2014). Butt et al. (1992) reported multiband (L, S, and Ku bands) fade levels measured in southern England for 60, 70, and 80 elevation in three types of environment over a simulated satellite path illustrating trends in attenuation with an increase in frequency and variations in the environment and elevation angle. At an elevation of 60° and 5% probability of occurrence, the fade levels are 8, 9, and 19.5 dB for the L band (1297.8 MHz), S band (2450 MHz in summer and 2320 MHz in spring), and Ku band (10,368 GHz), respectively (Richharia, 2014).

The propagation conditions are relatively benign in open areas and high elevation angles where Ku band attenuation is 2.8 and 1.7 dB at 70° and 80°, respectively (Richharia, 2014).

Consider a representative Ka band measurement campaign in Europe, using 18.7 GHz propagation transmissions of Italsat F1 (Murr et al., 1995). Measurements were conducted in a number of European environments—open, rural, tree shadowed, suburban, urban, and mixed constituting four types of environment—and in the elevation angle range 30–35°, maintaining an orientation of 0°, 45°, and 90° with respect to the satellite (Richharia, 2014).

Radio wave propagation becomes more ray-like at Ku band and above in comparison to the L and S bands, and since the attenuation due to foliage and other obstructions is more severe, the transition between good and bad states is distinct (Butt et al., 1992). Therefore, the Markov chain model is suited for performance evaluation in these frequency bands as suggested by several authors (Scalise, 2008).

9.6.6 Satellite Radio Channel Characterization at Ku Band: Helicopter Case

Future aeronautical communications will provide new data services that are expected to be volume intensive (Phillips et al., 2007; Pouzet 2007, 2008; Durand and Longpre, 2014; ICAO, 2015). Current AMS systems operating in L band are not able to provide the required bandwidth, so the use of Ku band has been proposed for this purpose.

According to ITU-R radio regulations (ITU), several services have been allocated in the downlink portion of the available Ku band, from 10.70 to 12.75 GHz. The band from 10.70 to 11.70 GHz has been allocated for FSS, and the subbands from 10.70 to 10.95 GHz and from 11.20 to 10.95 GHz have a planned use according to Appendix 30B of the radio regulations (RR). However, other subbands are unplanned. Also, parts of the band from 11.7 to 12.95 GHz have been allocated for broadcast satellite service (BSS), and planned according to Appendix 30 of RR, while other subbands are unplanned.

These unplanned subbands may be used for AMS services as long as no higher requirements that those specified for FSS are needed.

The AMS radio channel differs from other satellite radio channels because

- The satellite LOS may be blocked due to the aircraft structure.
- The aircraft structure may produce multipath propagation.
- Doppler shift may be significant due to the higher speed of the aircraft.

In the following subsections, we will analyze these effects.

9.6.6.1 LoS Blocking by the Aircraft Structure

The received signal level at the aircraft receive antenna may suffer a significant fade if the direct propagation path between the satellite and the antenna is obstructed by part of the aircraft structure.

This may be the case of any airplane during flight maneuvers or a helicopter that may suffer the periodic obstruction of the LoS caused by the rotor blades.

Reductions of the mean signal level up to 8 dB for 5 ms and up to 14 dB for 2.1 ms in case of the H-500 due to the obstruction of the rotor blades have been found for different helicopters at Ku band (Lemos et al., 2014). Fade durations agree with the width of the helicopter blades. The effect of these obstructions could be mitigated if the dynamic range of the receiver is wide enough or if some kind of diversity technique is implemented.

9.6.6.2 Signal Doppler Shift

Another effect to be considered is the received signal Doppler shift due to the high-speed movement of the aircraft or parts of the aircraft. For instance, the rotor blades of a helicopter could produce a propagation component with a Doppler shift that would depend on the rotor speed and the blade length. Signals shifted up to 10 kHz have been reported to be produced by diffraction on helicopter rotor blades.

9.6.6.3 Multipath Propagation

Finally, multipath propagation due to scattering on the aircraft structure should also be considered. Multipath propagation gives rise to DS of the received signal, and this is a source of intersymbol interference (ISI). ISI degrades the system performance as it produces an irreducible bit error rate (BER) lower limit, that is, the BER cannot be reduced below this limit despite the fact that the signal-to-noise ratio is increased.

The DS would produce significant ISI if it is comparable to the symbol interval (Ts). If the DS is much lower than Ts, then ISI would be negligible. So, to avoid ISI, Ts could be increased by reducing the data rate. In other words, the DS limits the maximum data rate of the system.

To determine the DS of a radio channel experimentally, a wideband radio channel sounder is required. Results reported in Lemos et al. (2014) show that the DS produced by a helicopter structure is related to the maximum dimension of the structure. DSs up to 10 ns have been reported for helicopters at the Ku band. Coherence bandwidth values are around 64 MHz. This means that as far as satellite signal bandwidths much lower than 64 MHz are used, the DS could be neglected. However, if signals of comparable or larger bandwidths are used, the DS would be significant and its effects would have to be mitigated with the appropriate selection of the data modulation scheme, equalization, or some other technique.

9.6.6.4 Channel Modeling

As we have seen in the previous subsections, most of the undesired effects that we may find in the AMS radio channel depend on the specific airplane in terms of where reception of the signal takes place. If we want to determine the channel characteristics exactly to develop an empirical model, we would have to perform wideband measurements for each aircraft model we would like to study.

Fortunately, it has been demonstrated (Lemos et al., 2014) that the AMS radio channel at the Ku band can be modeled using high-frequency techniques. That is geometrical optics plus uniform theory of diffraction. Geometrical optics is an asymptotic approximation of Maxwell equations valid only when the wavelength of the radio signal is small compared to the obstacle size. This is the case for aircrafts at the Ku band where the wavelength is less than 3 cm while aircraft sizes are several meters.

REFERENCES

Bello, P., Aeronautical channel characterization, *IEEE Transactions on Communications*, 21(5), 548–563, 1973.

Braun, W. and U. Dersch, A physical mobile radio channel model, *IEEE Transactions on Vehicular Technology*, 40(2), 472–482, 1991.

Butt, G., B.G. Evans, and M. Richharia, Narrowband channel statistics from multiband measurements applicable to high elevation angle land-mobile satellite systems, multiband propagation experiment for narrowband characterisation of high elevation angle land mobile-satellite channels, *Electronics Letters*, 28(15), 1449–1450, 1992.

Capsoni, C., F. Fedi, C. Magistroni, A. Paraboni, and A. Pawlina, Data and theory for a new model of the horizontal structure of rain cells for propagation applications, *Radio Science*, 22(3), 395–404, 1987.

COST 207, Digital land mobile radio communications. Office for Official Publications of the European Communities, Final report, 1989.

Crane, R.K., *Propagation Handbook for Wireless Communication System Design*, Boca Raton: CRC Press, 2003.

Durand, F. and L. Longpre, Handling transition from legacy aircraft communication services to new ones— A communication service provider's view, in S. Plass (ed.), *Future Aeronautical Communications*, pp. 25–55. Rijeka: Intech Open, 2014.

European Radiocommunications Committee (ERC), Current and future use of frequencies in the LF, MF and HF bands. ERC Report 107, 2001.

Goodman, J., J. Ballard, and E. Sharp, A long-term investigation of the HF communication channel over middle- and high-latitude paths, *Radio Science*, 32(4), 705–1715, 1997.

Haas, E., Aeronautical channel modeling, *IEEE Transactions on Vehicular Technology*, 51(2), 254–264, 2002.

Haque, J., An OFDM based aeronautical communication system, Graduate Thesis, University of South Florida, 2011.

Haque, J., M. Erturk, and H. Arslan, Doppler estimation for OFDM based aeronautical data communication, in *Wireless Telecommunications Symposium*, Tampa, FL, 1–6, April 21–23, 2010.

Ham Universe, Aircraft frequencies. World Wide HF and VHF/UHF aircraft communications listening. www.hamuniverse.com/aerofreq.html (accessed December 2015).

Hoeher, P. and E. Haas, Aeronautical channel modeling at VHF-band, in *IEEE Vehicular Technology Conference*, 4, 1961–1966, 1999.

ICAO, Aeronautical radio frequency spectrum needs, in *Twelfth Air Navigation Conference*, Montréal, Canada, November 19–30, 2012.

ICAO, Networking the sky with new aircraft communication technology, 2015. http://cordis.europa.eu/news/rcn/36558_en.html (accessed December 5, 2015).

International Civil Aviation Organization (ICAO), Annex 10, Aeronautical Telecommunications, Volume I–V, 2005.

International Civil Aviation Organization Asia and Pacific Office (ICAO-APO), *High Frequency Management Guidance Material for the South Pacific Region*, 2010.

International Telecommunication Union (ITU), http://www.itu.int.

ITU-R.618-7, *Propagation Data and Prediction Methods Required for the Design of Earth–Space Telecommunication Systems*, 2001.

ITU-R P.676-4, *Attenuation by Atmospheric Gases in the Frequency Range 1–350 GHz*, 2001.

ITU-R P.682-3, *Propagation Data Required for the Design of Earth–Space Aeronautical Mobile Telecommunication Systems*, 2012.

ITU-R P.840-3, *Attenuation Due to Clouds and Fog*, 2001.

Jain, R., F. Templin, and K.-S. Yin, Analysis of L-band digital aeronautical communication systems: L-DACS1 and L-DACS2, in *IEEE Aerospace Conference*, 1–11, 2011.

Jayaweera, S. and H. Poor, On the capacity of multiple-antenna systems in rician fading, *IEEE Transactions on Wireless Communications*, 4(3), 1102–1111, 2005.

Karasawa, Y. and T. Shiokawa, Characteristics of L-band multipath fading due to sea surface reflection, *IEEE Transactions on Antennas and Propagation*, 32(6), 618–623, 1984.

Lei, Q. and M. Rice, Multipath channel model for over-water aeronautical telemetry, *IEEE Transactions on Aerospace and Electronic Systems*, 45(2), 735–742, 2009.

Leitao, M.J. and P.A. Watson, Method for prediction of attenuation on earth–space links based on radar measurements of the physical structure of rainfall, in *IEE Proceedings on Microwave, Antennas and Propagation*, 133(4), 429–440, 1986.

Lemos Cid, E., M. García Sánchez, A. Vazquez Alejos, and S. Garcia Fernandez, Measurement, characterization and modeling of the helicopter satellite communication radio channel, *IEEE Transactions on Antennas and Propagation*, 62(7), 3776–3785, 2014.

Lin, S.H., Method for calculating rain attenuation distribution on microwave paths, *Bell System Technical Journal*, 54(6), 1051–1086, 1975.

Matolak, D.W., AG channel sounding for UAS in the UAS, Graduate Thesis, University of South Carolina, 2014.

Matolak, D., K. Shalkhauser, and R. Kerczewski, *L-Band and C-Band Air-Ground Channel Measurement and Modeling*. ICAO, Aeronautical Communications Panel WG-F/31, 2014.

Mettrop, J., Radio spectrum. The invisible infrastructure. Civil aviation authority, aviation frequency spectrum, in *ITU World Recommendation Conferences*, 2009.

Morita, K. and I. Higuti, Prediction methods for rain attenuation distributions of micro and millimeter waves, *Review of the Electronics Communications Laboratory*, 24(7–8), 651–668, 1976.

Mousley, T.J. and E. Vilar, Experimental and theoretical statistics of microwave amplitude scintillations on satellite downlinks, *IEEE Transactions on Antennas and Propagation*, 30(6), 1099–1106, 1982.

Murr, F., B. Arbesser-Rastburg, and S. Buonomo, Land mobile satellite narrowband propagation campaign at Ka band, in *Fourth International Mobile Satellite Conference*, 134–138, 1995.

NASA. L-band channel modeling. *Future Communications Study Phase II*, 2006.

Neji, N., R. De Lacerda, A. Azoulay, T. Letertre, and O. Outtier, Survey on the future aeronautical communication system and its development for continental communications, *IEEE Transactions on Vehicular Technology*, 62(1), 182–191, 2012.

Neskovic, A., N. Neskovic, and G. Paunovic, Modern approaches in modeling of mobile radio systems propagation environment, *IEEE Communications Surveys Tutorials*, 3(3), 2–12, 2000.

Panagopoulos, A.D., P-D.M. Arapoglou, and P.G. Cottis, Satellite communications at Ku, Ka, and V bands: Propagation impairments and mitigation techniques, *IEEE Communications Surveys and Tutorials*, 6(3), 2–14, 2004.

Phillips, B., J. Pouzet, J. Budinger, and N. Fistas, Future communication study technology recommendations—Action Plan 17 final conclusions and recommendations report, *ICAO Aeronautical Communications Panel Working Group Technology*, Working Paper ACPWGT/1-WP/06_AP17, Montreal, Canada, October 2–5, 2007.

Pouzet, J., Review of the progress on the future communication study. ICAO, Aeronautical Communications Panel, 2007.

Pouzet, J., Air Traffic Management (ATM) communications and satellites: An overview of EUROCONTROL's activities, *Space Communication Magazine*, 21, 103–108, 2008.

Pouzet, J. and M. Rees, *Future Communications Infrastructure—Technology Investigations Step 1: Initial Technology Shortlist*. ICAO ACP/1-IP/4, 2007.

Raleigh, G., S. Diggavi, A. Naguib, and A. Paulraj, Characterization of fast fading vector channels for multi-antenna communication systems, in *Twenty-Eighth Asilomar Conference on Signals, Systems and Computers*, 2, 853–857, 1994.

Rappaport, T.S., *Wireless Communications: Principles and Practice*, New Jersey: Prentice-Hall, 2003.

Rice, M., A. Davis, and C. Bettweiser, Wideband channel model for aeronautical telemetry, *IEEE Transactions on Aerospace and Electronic Systems*, 40(1), 57–69, 2004.

Rice, M., R. Dye, and K. Welling, Narrowband channel model for aeronautical telemetry, *IEEE Transactions on Aerospace and Electronic Systems*, 36(4), 1371–1376, 2000.

Rice, M. and M. Jensen, A comparison of L-band and C-band multipath propagation at Edwards AFB. Report AFFTC-PA-11249, 2011.

Richharia, M., *Mobile Satellite Communications: Principles and Trends*, West Sussex: John Wiley & Sons, 2014.

Salonen, E. and S. Uppala, New prediction method of cloud attenuation, *Electronics Letters*, 27(12), 1106–1108, 1991.

Scalise, S., H. Ernst, and G. Harles, Measurements and modeling of the land mobile satellite channel at Ku-band, *IEEE Transactions on Vehicular Technology*, 57(2), 693–703, 2008.

Sheriff, R.E. and Y.F. Hu, *Mobile Satellite Communication Networks*, West Sussex: John Wiley & Sons, 2010.

Stacey, D., *Aeronautical Radio Communication Systems and Networks*, West Sussex: John Wiley & Sons, 2008.

Steingass, A., A. Lehner, F. Pérez-Fontán, E. Kubista, and B. Arbesser-Rastburg, Characterization of the aeronautical satellite navigation channel through high-resolution measurement and physical optics simulation, *International Journal of Satellite Communications and Networking*, 01(26), 1–30, 2008.

Tooley, M. and D. Wyatt, *Aircraft Communications and Navigation Systems*, Chapter 2, Oxford, United Kingdom: Butterworth-Heinemann, 2007.

Van de Kamp, M.M.J.L., J.K. Tervonen, E.T. Salonen, and J.P.V. Poirares Baptista, Improved models for long-term prediction of tropospheric scintillation on slant paths, *IEEE Transactions on Antennas and Propagation*, 47(2), 249–260, 1999.

Van Wambeke, N. and M. Gineste, The role of satellite systems in future aeronautical communications, in S. Plass (ed.), *Future Aeronautical Communications*, 187–200. Rijeka: Intech Open, 2010.

Walter, M., Shutin, D., and U.-C. Fiebig, Joint delay Doppler probability density functions for air-to-air channels, *International Journal of Antennas and Propagation*, 814218, 1–11, 2014.

World Radiocommunication (WRC). Measures to address unauthorized use of and interference to frequencies in the bands allocated to the maritime mobile service and to the aeronautical mobile (R) service. Resolution 207, 2003.

Wu, Q., D.W. Matolak, and R.D. Apaza, Airport surface area propagation path loss in the VHF band, in *Integrated Communications, Navigation and Surveilance Conference*, B4, 1–6, 2011.

Zhang, Q. and D. Liu, A simple capacity formula for correlated diversity Rician fading channels, *IEEE Communications Letters*, 6(11), 481–483, 2002.

10 Stratospheric Channel Models

Emmanouel T. Michailidis and Athanasios G. Kanatas

CONTENTS

10.1 INTRODUCTION

Terrestrial wireless and satellite systems represent two well-established infrastructures that have been dominant in the telecommunications arena for years. Terrestrial links are widely used to provide services in several areas, while satellite links are generally used to provide high-speed connections where terrestrial infrastructure is not available. In recent years, a new alternative wireless communications technology has emerged known as high-altitude platforms (HAPs) (Karapantazis and Pavlidou 2005a; Widiawan and Tafazolli 2007; Aragón-Zavala et al. 2008) and has attracted attention worldwide (Lee and Ye 1998; Colella et al. 2000; Thornton et al. 2001;

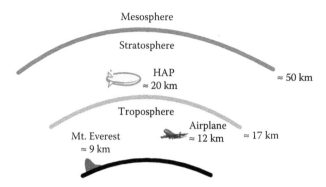

FIGURE 10.1 Troposphere, stratosphere, and mesosphere layers of the atmosphere.

ESA-ESTEC 2005). The term "HAPs" defines aerial platforms flying at an altitude between 17 and 22 km above Earth's surface, in the stratosphere. As shown in Figure 10.1, the stratosphere is the second major layer of Earth's atmosphere, just above the troposphere, and below the mesosphere. This operating altitude was chosen because it represents a layer of relatively mild wind and turbulence in most regions of the world. At this altitude, HAPs can succeed in maintaining station-keeping and flying against the wind without excessive power demands. Google has envisaged similar wireless communication payload systems based on advanced stratospheric balloon technology to create an aerial wireless network, which they call the "Loon project" (Google 2015). Each balloon can provide coverage of about 40 km in diameter at rates comparable to the third generation (3G) of mobile telecommunications technology. Both balloon-to-balloon and balloon-to-ground communications are considered and industrial, scientific, and medical radio frequency bands are utilized, specifically 2.4 and 5.8 GHz bands, which are available for anyone to use. This technology aims to bring Internet access to remote and rural areas poorly served by existing infrastructures and to restore communication services initially provided by terrestrial base stations in the affected area during natural disasters, for example, earthquakes and hurricanes.

Two major HAP structures are considered in the literature: the circling aircrafts (~30 m in length and a wingspan of about 35–70 m) and the lighter-than-air quasi-stationary airships (of about 150–200 m in length). The aircrafts fly in a roughly circular tight path of about 2 km radius or more above the targeted coverage area and can be either solar-powered and unmanned with continuous flight duration in the order of months, or manned with average flight duration of some hours due to fuel constraints and human factors. The airships use very large semirigid or nonrigid helium-filled containers and electric motors and propellers for station-keeping. Prime power required for propulsion and station-keeping as well as for the payload and applications is provided by lightweight solar cells in the form of large flexible sheets. The achievable mission duration for airships hopefully exceeds 5 years. The unmanned aerial vehicles (UAVs) are another type of small fuelled unmanned aircraft controlled and operated from distant locations using low frequencies (VHF, UHF). The UAVs are employed for military short-time surveillance (up to 40 h) at modest altitudes, surveillance of geographical boundaries, monitoring of agricultural areas, sensing, and image capturing. HAP stability problems and displacements, such as pitch, yaw, and roll, due to the winds or pressure variations of the stratosphere are a problem to be faced for both aircrafts and airships (Axiotis et al. 2004; Thornton and Grace 2005). Although, it is easier for airships than for aircrafts to remain quasi-stationary, it is rather difficult to remotely control the airship's position. Recently, advances in composite materials, computers, navigation systems, aerodynamics, and propulsions systems have made station-keeping and hence stratospheric systems feasible.

HAPs preserve some of the best characteristics of terrestrial and satellite communication systems, while avoiding many of their drawbacks. In comparison to terrestrial wireless technologies,

HAPs require considerably less communications infrastructure and they can serve potentially large coverage areas. When compared to satellite communication systems, HAPs provide lower propagation delays, a major issue for voice communications over satellite links, lower power loss, as well as easy maintenance and upgrading of the payload during the lifetime of the platform. In addition, the cost for the development of satellite systems is much greater, the eventual cost of HAPs is expected to be about 10% of that of a satellite, and it may be economically more efficient to cover a large area with many HAPs rather than with many terrestrial base stations or with a satellite system. Therefore, HAPs represent an economically attractive technology. HAPs are also well suited for temporary provision of basic or additional communications services due to their rapid deployment and displacement on demand, providing network flexibility. Finally, HAPs can use most of conventional mobile communications base station technology and terminal equipment, while solar-powered HAPs are nonpollutant and environment friendly. Although HAPs provide substantial advantages over terrestrial and satellite systems, their successful deployment requires integration of available and emerging technologies to make long-term operation feasible and profitable. Specifically, in the area of communication systems, there are issues of fundamental importance to be addressed, such as the design and implementation of onboard antennas, channel characterization and modeling, resource management, and coordination and interoperability between different systems.

10.2 DEVELOPMENT OF NEXT-GENERATION BROADBAND STRATOSPHERIC COMMUNICATION SYSTEMS

As new requirements for access to wireless networks emerge within the communications society, HAPs can play an important role in the evolution of current and future communication systems (Grace et al. 2005a; Karapantazis and Pavlidou 2005b; Falletti et al. 2006b). In particular, HAPs are expected to alternatively or complementarily fulfill the vision of optimal connectivity in the service area providing high data throughputs in virtually every possible scenario at low cost. The flexibility of stratospheric communication systems allows not only for carrying wireless communications payloads but also for serving other applications, such as navigation and positioning (Toshiaki and Masatoshi 2004), monitoring, remote sensing, and surveillance (Akalestos et al. 2005), disaster management and relief (Deaton 2008; Holis and Pechac 2008a), telemedicine (Lum et al. 2006), and military applications (Jamison et al. 2005). The major wireless communications applications to be offered by HAPs can be divided technologically into two types: the broadband fixed wireless access (BFWA) (Grace et al. 2001; Mohammed and Yang 2009) and the provision of mobile services (Mondin et al. 2001; Taha-Ahmed et al. 2005; Ahmed et al. 2006). As shown in Table 10.1, several frequency bands (ITU 2007a,b,c), have been licensed for communications through HAPs on

TABLE 10.1
Frequency Spectrum Available for Stratospheric Applications

Frequency Band (GHz)	Microwave Frequency Band	Area/Country	Type of Service
47.9–48.2	V	Global	Fixed (uplink)
47.2–47.5			Fixed (downlink)
31.0–31.3	Ka	40 countries (including all countries in North	Fixed (uplink)
27.5–28.35		and South America but excluding all of Europe)	Fixed (downlink)
2.160–2.170	S	Regions 1 and 3	Mobile
2.110–2.160		Global	
2.010–2.025		Regions 1 and 3	
1.885–1.985	L	Global	

a global, regional, or national basis, and for each frequency band, appropriate regulatory provisions have been established based on technical and operational studies (Park et al. 2008). The choice of the frequency bands at which HAPs can operate is mainly determined by the frequency sharing and compatibility with other existing services provided by terrestrial/satellite systems, which would cause interference problems if spectrum allocation is not carefully controlled (Milas and Constantinou 2005; Ku et al. 2008). Note that spectrum identification in the range between 5850 and 7075 MHz to facilitate gateway links to and from HAPs has been also envisioned, since WRC-07 agenda, Resolution 734 (ITU 2007d).

BFWA communications provide in general a high rate of data transmission. In 2006, the Organization for Economic Cooperation and Development has defined broadband as 256 kilobits per second (Kbps) in at least one direction and this bit rate is the most common baseline that is marketed as "broadband" around the world. The International Telecommunication Union (ITU) has defined broadband as a transmission capacity that is faster than 1.5–2 megabits per second (Mbps) (ITU-T 1992), while the U.S. Federal Communications Commission definition of broadband is at least 4 Mbps downstream and 1 Mbps upstream. However, the threshold of the broadband definition will rise, as higher data rate services become available. High-speed broadband services include but are not limited to conversational services (voice/video telephony, high-resolution video conferencing), streaming and broadcast services (real-time radio and television, video on demand), local multipoint distribution services, interactive data access (broadband Internet access, web browsing), and large file transfers. In addition, HAPs could potentially be used as an alternative solution for the provision of digital video broadcasting (DVB) and digital audio broadcasting (DAB). A feasibility study of HAPs-DVB/DAB has been conducted by the European Space Agency (ESA) under the STRATOS project (ESA-ESTEC 2005). These services are offered mainly to fixed terminals. However, mobile applications are also feasible (Morlet et al. 2007).

In addition to BFWA, ITU has also endorsed the use of HAPs in the IMT-2000 spectrum for the provision of 3G/UMTS mobile services, as well as enhanced HSDPA services offering data transfer speeds up to 14 Mbps on the downlink. These services are expected to have the same functionality, meeting the same service, and operational requirements as traditional terrestrial tower-based systems. HAPs can be designed to serve as the sole station in a stand-alone infrastructure, essentially replacing the tower base station network with a "base station network in the sky." In particular, a single HAP can replace a large number of terrestrial base stations and their backhaul infrastructure (microwave or optical links) and can serve a large city, a suburban area, a rural region (Holis and Pechac 2008d), or even a whole state. Moreover, HAPs could deliver Worldwide Interoperability for Microwave Access (WiMAX) (Palma-Lázgare et al. 2008; Thornton et al. 2008) mobile services based on the IEEE 802.16e standard at frequencies between 2 and 6 GHz and at speeds up to 40 Mbps (IEEE 2005; Andrews et al. 2007).

The growing demand for greater bandwidth, higher data rates, improved quality of service (QoS), and ubiquitous access have prompted the development of fourth-generation (4G) communication systems that employ a comprehensive and secure all-Internet-protocol (IP) network and support fixed and mobile ultra-broadband (gigabit speed) access and multicarrier transmission (Evans and Baughan 2000; Bria et al. 2001; Chen and Guizani 2006). New broadband applications have thrived with the advent of 4G and can be provided through stratospheric communications, such as multimedia broadcast and multicast services, IP telephony, high-quality voice, high-definition television, mobile television, ultra-broadband Internet access, gaming services, and high-quality streamed multimedia. Third Generation Partnership Project (3GPP) Long-Term Evolution (LTE) advanced technology fully supports 4G requirements offering data rates up to 1 gigabit per second (Gbps) (LTE-Advanced 2015), while next-generation Mobile WiMAX based on the IEEE 802.16m (WiMAX 2015) standard supports at least 100 Mbps at mobile and 1 Gbps at fixed-nomadic services, and hence are compliant with 4G next-generation mobile networks. The development of next-generation systems envisages the synergetic integration of heterogeneous terrestrial, satellite, and stratospheric networks with different capabilities, which gives rise

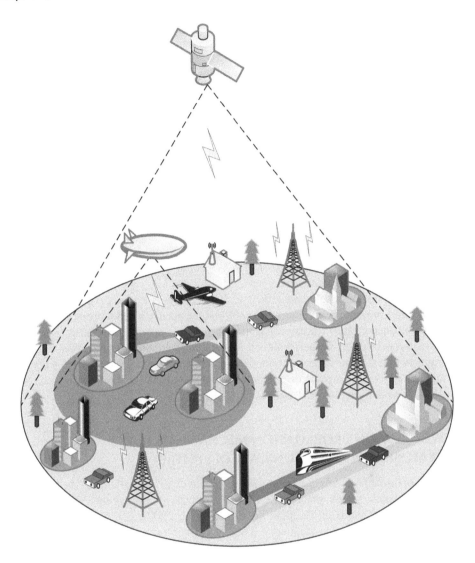

FIGURE 10.2 Visualization of an integrated satellite/HAP/terrestrial wireless communication system providing fixed and mobile broadband services.

to new services, architectures, and challenges (Evans et al. 2005; Giuliano et al. 2008; Paillassa et al. 2011). Indeed, the term "4G" is widely used to include several types of broadband fixed and mobile wireless communication systems. According to Figure 10.2, 4G and beyond 4G systems could seamlessly integrate the entire communications infrastructure (terrestrial base stations, satellites, and HAPs), modern communications terminals (mobile phones, personal digital assistants, tablet computers, laptops, etc.), transportation vehicles (cars, trains, airplanes, boats, etc.), available networks (body area networks, personal area networks, local area networks, metropolitan area networks, and wide area networks), and next-generation applications to satisfy increasing user demands. Hence, complex network topologies of heterogeneous nature can be deployed, such as stand-alone HAP networks, multiple-HAP networks, integrated terrestrial/HAP networks, integrated satellite/HAP networks, and integrated terrestrial/HAP/satellite networks. Each network has different capabilities, in terms of capacity, coverage, mobility support, cost, and is best suited for handling certain configurations. However, these different networks will have to coexist, and be complementarily and efficiently exploited. This can be achieved by introducing the reconfigurability concept (Demestichas et al. 2004; Stavroulaki et al. 2006), which offers service differentiation, customization, and personalization according to the environment requirements.

Nevertheless, wireless systems designers are facing a number of challenges, that is, the limited availability of the radio frequency spectrum and a complex space–time–frequency varying wireless propagation environment (Paulraj et al. 2003). In addition, next-generation wireless communications services will require increased network coverage, higher data rates, enhanced capacity, enriched QoS, and improved efficiency. Thus, new technologies have to be continuously under development in order to meet these growing demands. In recent years, the multiple-input multiple-output (MIMO) technology has emerged as the most promising technology in these measures (Biglieri et al. 2007). MIMO communication systems consider multiple antennas at both the transmitting and the receiving end. Different aspects of MIMO technology are being planned or have already been incorporated in wireless standards, such as the IEEE 802.11n, 802.16e, 802.16m, 802.20, 802.22, 3GPP Releases 7, 8, and 9, 3GPP2 ultra mobile broadband, and DVB-T2 among others. In an effort to remain competitive with terrestrial systems, stratospheric systems should follow the progress in MIMO technology (Arapoglou et al. 2011b,c) and profit from its significant enhancements. The successful application of MIMO technology requires accurate channel models incorporating all propagation phenomena.

To design and implement reliable and efficient channel models for next-generation stratospheric systems when single or multiple antennas are employed, an accurate and thorough characterization of the effects influencing radio waves propagation in the stratospheric channel is essential. These models are indispensable for the performance evaluation, parameter optimization, and testing of next-generation wireless stratospheric communication systems. The following section describes the mechanisms of radio wave propagation for the links between HAPs and terrestrial stations, which directly affect the underlying wireless stratospheric channel.

10.3 OVERVIEW ON PROPAGATION MECHANISMS FOR STRATOSPHERIC COMMUNICATION SYSTEMS

Communicating through a wireless stratospheric channel is a challenging task because the medium may introduce severe impairments (Pawlowski 2000; Rappaport 2002; Saunders and Aragón-Zavala 2007; Kandus et al. 2008; Smolnikar et al. 2009). The propagation effects may vary depending on the operating frequency. Thus, it is important to emphasize the considerable differences between the communication systems operating at low-frequency bands and the ones operating in upper bands. In particular, line-of-sight (LoS) conditions are required at the Ka and V frequency bands (Aragón-Zavala et al. 2008) due to the severe attenuation of possible obstructed signal components. Besides, both LoS and non-line-of-sight (NLoS) connections can be used and should also be evaluated for stratospheric communication systems operating at L and S frequency bands. In general, the wireless channel includes both additive and multiplicative effects. The additive effects arise from the noise generated within the receiver itself, although external noise contributions may also be significant, and their impact on the performance of a communication system is important, since they determine the received signal-to-noise ratio (SNR). On the contrary, the multiplicative effects arise from the various processes encountered by the transmitted waves on their way from transmit to receive end. It is conventional to further subdivide the multiplicative processes in the channel into two types of fading, the large- and the small-scale fading. The large-scale fading determines the cell coverage area, outage, and handoffs, and includes the path loss and the shadowing of the transmitted signal by objects and scatterers in the channel, that is, trees, buildings, hills, or mountains, when the receiver moves over distances greater than several tens of the carrier wavelength. Besides, the small-scale fading determines the link-level performance in terms of bit error rates (BERs) and average fade durations (AFDs) is caused mainly by the multipath propagation and is referred to as multipath fading.

Although radio signals coming from satellites are subject to propagation impairments in the ionosphere and troposphere, the signals transmitted from HAPs are distorted only by tropospheric effects. These effects include (a) clear air effects, that is, free-space loss (FSL), absorption due to atmospheric gases, beam spreading loss due to ray bending, and tropospheric scintillation and (b) hydrometeor effects, that is, absorption due to rain, liquid water clouds, and fog raindrops. Note that rain attenuation is the dominant mechanism that significantly affects the quality of the link at frequencies above 10 GHz (Aragón-Zavala et al. 2008). At these frequencies, the wavelength and raindrop size (about 1.5 mm) are comparable and the attenuation is quite large. Thus, stratospheric communication systems operating at Ka and V frequency bands are susceptible to rain, while the ones operating at L and S frequency bands are not significantly affected by rain.

With these in mind, this section aims to describe in detail the fundamental propagation mechanisms that influence the links between HAPs and terrestrial stations and present the large- and small-scale fading characteristics, as well as the rain effects.

10.3.1 PATH LOSS AND SHADOWING

For a link between HAPs and terrestrial stations, the minimum (reference) path loss is given by the FSL, which assumes an LoS link between the transmitter and the receiver and propagation in free space and is given by (Grace et al. 2001; Aragón-Zavala et al. 2008)

$$L_F(dB) = 32.4 + 20\log d_{km} + 20\log f_{MHz}, \tag{10.1}$$

where f_{MHz} is the carrier frequency in MHz and d_{km} is the elevation angle-dependent distance between the transmitter and the receiver in km. Note that the FSL increases by 6 dB for each doubling in either frequency or distance.

In practice, the signals do not experience free-space propagation due to other sources of loss, such as obstacles in the first Fresnel zone, rain attenuation (Spillard et al. 2004), gaseous absorption (oxygen absorption and water vapor) (Zvanovec et al. 2008), cloud attenuation, and attenuation due to vegetation (Agrawal and Garg 2007, 2009). Hence, the value of the path loss is highly dependent on many factors, such as the terrain contours, the propagation environment, the propagation medium (dry or moist air), the distance between the transmitter and the receiver, and the height and location of antennas. The summation of any losses caused by the aforementioned propagation effects is called the excess loss L_{ex}. Hence, the total loss is given by

$$L(dB) = L_F(dB) + L_{ex}(dB). \tag{10.2}$$

Iskandar and Shimamoto (2006b) described a path loss model regarding stratospheric systems for predicting path loss in urban environments. The impact of elevation and azimuth angles variation to the propagation mechanism was evaluated through a building block model and ray-tracing tools. Feng et al. (2006) presented statistical models for air-to-ground channels, which can be used for satellites and UAVs. The results show that air-to-ground channels have a much higher LoS probability, and less shadowing than terrestrial channels. Thus, airborne platforms may serve as relaying nodes to extend the range and improve the connectivity between terrestrial ad hoc terminals. These results also show that the diffraction loss is mainly caused by the buildings located above ground level. Simunek et al. (2011) presented results from a measurement campaign simulating the propagation conditions in UAV high-data rate communication links in urban areas and at frequencies of about 2 GHz. These results indicated strong signal variations due to multipath and shadowing and that the excess loss is mainly dependent on the elevation angle and fairly independent of the distance. A physical model for the excess loss was also developed that involves high elevation angles and combines diffracted and reflected components.

Shadowing controls the reliability of coverage based on the link budget and indicates random variations of the received power with respect to the mean predicted value given by the path loss models. The density of the obstacles causing shadowing significantly depends on the physical environment. The randomness in this environment is captured by modeling the density of obstacles and their absorption behavior as random numbers. The amplitude variations caused by shadowing and expressed in logarithmic scale is often modeled using a lognormal distribution with a standard deviation from the mean value predicted by log-distance path loss model. A detailed discussion of the shadow fading characteristics for stratospheric systems is included in Holis and Pechac (2008b) and Kong et al. (2008). Specifically, an empirical narrowband model for different types of built-up areas, that is, from suburban to high-rise urban, and frequencies, that is, from 2 to 6 GHz, was proposed in Holis and Pechac (2008b). This model can be used as a radio network planning tool for system-level simulations and availability estimations and is a function of the elevation angle considering both LoS and NLoS paths. A statistical method was exploited in order to simulate the propagation environment, while measurement data, obtained using a remote-controlled airship, were used to validate this model. Kong et al. (2008) characterized the shadowing characteristics caused by urban areas and developed an analytical model based on experimental data. By applying the statistical data of building distribution, which includes height, length, and density, and road characteristics of the corresponding propagation areas to the model, a reliable evaluation of the shadowing effects can be obtained.

10.3.2 Multipath Propagation

A signal propagating through a wireless channel usually arrives at its destination along a number of different paths, referred to as multipath components, which arise from single-bounce and/or multiple-bounce scattering, reflection and/or diffraction of the radiated energy by fixed and/or mobile intervening objects in the environment (Dovis et al. 2002). These local environment propagation effects influence the amplitude, direction, and phase of the propagating radio waves and this effect is known as multipath fading or fast (short-term) fading, a name given due to the random rapid fluctuations of the received wave over small displacements. The received wave is a superposition of the impinging multipath components. Depending on the phase of each partial wave, this superposition can be constructive or destructive. Figure 10.3 presents a typical scenario affected by multipath fading. One observes that the LoS component reaches the receiver directly. However, single-, double-, and/or multiple-bounce NLoS rays also arrive at the receiver causing fading. Note that several field measurements have been conducted in a large variety of propagation environments such as urban, suburban, tree-lined roads, rural, and open, where the effect of the LoS and NLoS components differ substantially. Hence, attention should be separately paid to each specific environment, when the stratospheric radio channel is studied. In addition, when a terrestrial mobile station (TMS) travels in a large area, shadowing and multipath may change abruptly. Note that strong multipath propagation is typical for indoor communication scenarios.

10.3.2.1 Time, Frequency, and Space Selectivity

The wireless channel may be described in the base domains, that is, time, frequency, and space. Then, the channel is either selective in these domains, that is, time and/or frequency and/or space selective/variant or coherent over a time–frequency-spatial range. Alternatively, the channel may be described in the spectral domains, that is, Doppler frequency, delay, and direction. These are related to the base domains through Fourier transform (FT). A selective wireless channel is dispersive in the spectral domains, that is, preserves frequency and/or delay and/or direction dispersion (Fleury 2000). Figure 10.4 illustrates the relationship through FT between the variances of time Δt, frequency Δf, and space (represented by the receive $\Delta \mathbf{x}_R$ and transmit $\Delta \mathbf{x}_T$ antenna position vectors) and the spreads of Doppler frequency shift v, propagation delay τ, and direction of arrival or departure (represented by the receive Ω_R and transmit Ω_T direction vectors, respectively).

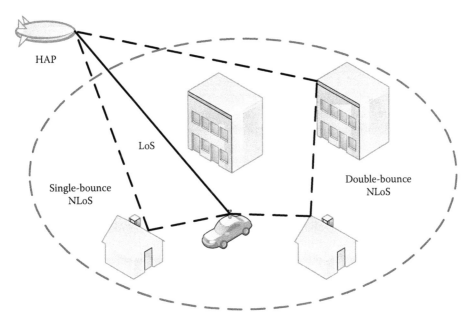

FIGURE 10.3 Visualization of a multipath propagation environment.

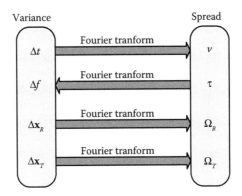

FIGURE 10.4 Relationships of the time, frequency, and space domains.

In single-antenna stratospheric systems, the transmitter and the receiver are each equipped with a single antenna and the underlying multipath channel is modeled as time and/or frequency variant. When viewed in the frequency domain by applying FT to the time correlation function, time selectivity appears as different Doppler (frequency) shifts of the individual multipath components over a finite spectral bandwidth that corresponds to a frequency dispersion of the Doppler power spectrum of the transmitted signal. Based on the rate with which the channel impulse response changes relative to the signal transmission rate, channels may be classified as fast fading, that is, the channel changes within the transmitted symbol duration, or slow fading, that is, the channel is approximately constant within symbol duration. A good measure of channel time selectivity is given by the channel coherence time, that is, the time duration for which the channel can be considered as approximately time invariant. The larger the coherence time, the slower the channel fluctuation, or equivalently the Doppler spread (in the frequency domain).

When viewed in the delay domain by applying inverse FT to the frequency correlation function, frequency selectivity appears as different time delays of the individual multipath components, which corresponds to a time dispersion of the transmitted signal. The span of the delays is called the delay spread and the channel acts like a tapped delay line (TDL) filter. Based on their degree of frequency selectivity, channels may be classified as frequency-flat or frequency-selective channels. In particular, if all the transmitted frequencies undergo approximately identical amplitude and phase changes, the channel is called frequency flat. Otherwise, the channel is called frequency selective and the transmission is wideband. Frequency selectivity is measured in terms of the coherence bandwidth, that is, the bandwidth over which the channel's frequency response remains constant. The channel can be considered frequency flat only if the transmission is narrowband compared to the channel's coherence bandwidth. Otherwise, the channel is frequency selective.

In multiple-antenna stratospheric systems, the transmitter and the receiver are equipped with multiple antennas. Then, apart from being time and/or frequency variant, the underlying multipath channel can be also space variant. When viewed in the angle domain by applying FT to the spatial correlation function, direction selectivity appears as different azimuth and/or elevation angles of arrival (or departure) of the multipath components at the receive (or transmit) antenna array over a finite spectral bandwidth, which corresponds to an angular dispersion of the power azimuth spectrum and/or the power elevation spectrum of the transmitted signal. Hence, the signal amplitude depends on the spatial location of the antennas. Space selectivity is measured in terms of the coherence distance, that is, the spatial separation for which the channel's spatial response remains constant, which is inversely proportional to the angle spread.

10.3.3 Rain Effects

The rain is confined to the first 2.5–5 km of the atmosphere depending on the latitude (Aragón-Zavala et al. 2008). Hence, an electromagnetic (EM) wave propagating in the troposphere is directly affected by rain effects. Rain effects primarily refer to the attenuation of a signal. The troposphere consists of a mixture of particles having a wide range of sizes and characteristics. The attenuation is the result of the conversion of EM energy to thermal energy within an attenuating particle. This attenuation increases with the number of raindrops along the path, the size of the drops, the length of the path through the rain, and the carrier frequency. The main particles of interest are hydrometeors, including raindrops, fog, and clouds. Note that attenuation is negligible for snow or ice crystals, in which the particles are tightly bound and do not interact with the waves.

The rain attenuation L_r can be empirically obtained using the specific rain attenuation γ_r (dB/km) (ITU-R 1997), which is defined as

$$\gamma_r = a_r R_r^{b_r}, \tag{10.3}$$

where R_r is the rain rate measured on the ground in millimeters per hour (mm/h) and is strongly dependent on the geographical location. Worldwide rain rate contour maps can be found in ITU-R (2003). Typical rain rate values for Europe are around 30 mm/h, while for some Mediterranean regions the rain rate exceeds 50 mm/h and for equatorial regions the rain rate may reach 150 mm/h. The rain rate corresponds to the measure of the average size of the raindrops. However, the period of time for which the rain rate exceeds a certain value is more important than the total amount of rain falling during a year. The values of the empirical regression coefficients a_r and b_r can be obtained from ITU-R (1997) and depend on the climatic zone, the transmission frequency, and the polarization. Table 10.2 shows typical values for a_r and b_r at various frequencies for horizontal and vertical polarization extracted from ITU-R (1997), while Figure 10.5 utilizes Equation 10.3 and demonstrates the specific rain attenuation γ_r as a function of the rain rate for different carrier frequency. One observes that the rain attenuation is only significant to communications systems operating above 10 GHz. At these frequencies, the wavelength and raindrop size (about 1.5 mm) are comparable and the attenuation is quite large. Thus, stratospheric communications systems operating at

TABLE 10.2

Regression Coefficients for Estimating Specific Rain Attenuation

Frequency (GHz)	Horizontal Polarization		Vertical Polarization	
	a_r	b_r	a_r	b_r
2	0.000154	0.963	0.000138	0.923
10	0.0101	1.276	0.00887	1.264
20	0.0751	1.099	0.0691	1.065
30	0.187	1.021	0.167	1.000
40	0.350	0.939	0.310	0.929
50	0.536	0.873	0.479	0.868

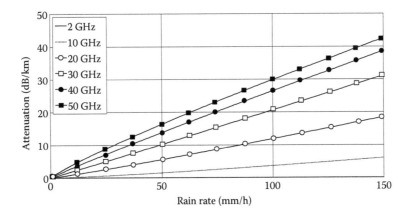

FIGURE 10.5 Specific rain attenuation as a function of the rain rate for different carrier frequencies.

Ka and V frequency bands are susceptible to rain, while the ones operating at L and S frequency bands are not significantly affected by rain. The total rain attenuation can be obtained as follows:

$$L_r = \gamma_r d_r, \tag{10.4}$$

where d_r is the total rainy path length and can be geometrically obtained as (see Figure 10.6)

$$d_r = (H_r - H_R)/\sin\beta_T, \tag{10.5}$$

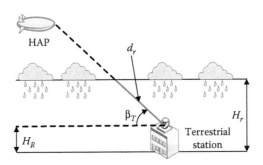

FIGURE 10.6 Rain attenuation path geometry.

where H_r is the effective rain height, H_R is the height of the terrestrial station, and β_T is the elevation angle of the platform. Representative values for H_r vary according to the latitude of the terrestrial station. In practice, high rain attenuations are sometimes avoided by using site diversity (Konefal et al. 2002; Panagopoulos et al. 2007), in which two widely separated terrestrial stations are used. Then, the probability that both terrestrial stations are within the same area of rain concentration is small. Alternatively, a portion of spectrum in a lower frequency band may be used.

The aforementioned empirical modeling of the long-term behavior of the rain attenuation effects does not convey any information on the time dynamics of these effects. An alternative and more accurate way of modeling the rain attenuation is through the use of time series (Cuevas-Ruíz and Delgado-Penín 2005). Since rain is characterized by a significant spatial inhomogeneity within the distances of interest, rain attenuation time series can be obtained using actual meteorological data accumulated over many years of carefully performed propagation measurements. The possibility of making a computer synthesis of these time series allows for the implementation of statistical channel models, which are characterized by short-term statistics.

Apart from introducing signal attenuation, rain also introduces scattering processes (Spillard et al. 2002; Michailidis and Kanatas 2009). This scattering results in redirection of the radio waves in various directions, so that only a fraction of the incident energy is transmitted onward in the direction of the receiver. The scattering process is strongly frequency dependent, since wavelengths that are long compared to the particle size will be only weakly scattered.

10.4   MODELING OF THE STRATOSPHERIC RADIO CHANNEL

The propagation effects suffered by a signal transmitted over a stratospheric communication radio link are similar to those present over a terrestrial or satellite system, mainly in terms of shadowing and multipath. Besides, tropospheric (rain) effects substantially control the quality of the link of both satellite and stratospheric systems. Therefore, models for stratospheric channels could be based on models applied to terrestrial and satellite systems. However, classic terrestrial modeling artificially separates slow and fast variations due to shadowing and multipath, respectively, and models them independently. Conversely, in satellite systems, these two processes are usually treated statistically in a combined manner. The reason is that terrestrial propagation rarely exhibits LoS conditions and the propagation environment contributes to both small- and large-scale fading. Nevertheless, direct signal is usually present when the transmitter is located above Earth due to the higher elevation angles and impairments of the signal are mainly caused only by the local environment. Indeed, although terrestrial, satellite, and stratospheric channels exhibit similar multipath fading, the intensity of this small-scale effect is not the same, due to the different position of the effective scatterers. Moreover, the stratospheric systems exhibit distinct characteristics compared to satellite and terrestrial systems, with regard to the size of the coverage area, the length of the radio path between transmitter and receiver, the link geometry, and the propagation time delay. Note that the distance between the receiver and transmitter is shorter in the case of stratospheric systems compared with satellite systems resulting in smaller propagation delay and lower FSL, while the rate of elevation angle variation is different. In the case of geostationary Earth orbit (GEO) satellites, the elevation angle does not significantly vary when the mobile terminal changes its position. In low Earth orbit satellite systems, the rate of elevation angle variation is comparable to that in stratospheric systems, but it is mainly caused by satellite motion and hence it is easily predictable. Furthermore, in satellite systems, additional distortions of the signal occur when passing the ionosphere. Consequently, terrestrial and satellite channel models should not be directly used for describing propagation conditions in stratospheric communications and some modifications are essential.

The objective of this section is to provide an overview on recently used design methodologies enabling the development of channel models for stratospheric communication systems and discuss some open issues related to channel features not sufficiently reproduced by these models.

A classification of the different channel models is initially provided and the characteristics of the multiple-antenna channel models are described. The configurations under investigation range from very simple single-antenna stratospheric systems to quite complex and challenging multiantenna stratospheric systems. Emphasis is given on the presentation of cutting-edge research on the channel modeling of the latter. For the multiantenna stratospheric systems, the particular propagation characteristics crucially determine the viability of MIMO technologies and control the performance metrics. The spotlight is on point-to-point and relay-based stratospheric systems in outdoor radio propagation environments. However, stratospheric-to-indoor reception is also included. The presentation closes with a discussion of open research problems in this area.

10.4.1 Classification of Channel Models

The channel models can be classified into categories using different criteria (Vázquez-Castro et al. 2002; Yang et al. 2010). Among the most frequently used techniques to describe signal-level variations in the terrestrial, satellite, and stratospheric channels are statistical models, which can be used for the evaluation of designed fade mitigation techniques, access, modulation, and coding schemes, and time and carrier synchronization approaches. When channel characteristics are not static and vary in time, different channel conditions, states, or state transitions (if two or more states are defined) can be defined. If these transitions are independent (memoryless), the channel can be modeled as a Markov (continuous or discrete) chain (Lutz et al. 1991; Fontan et al. 1997; Karasawa et al. 1997; Bråten and Tjelta 2002; Cuevas-Ruíz and Delgado-Penín 2004a,b) and the corresponding model is called statistical switched. A criterion utilized to distinguish the individual channel models is bandwidth. In particular, the models for stratospheric channels can be divided into narrowband models, that is, the multipath fading is frequency-flat and wideband models, that is, the multipath fading is frequency-selective. These models can also be separated into mobile or fixed depending on the existence or absence of mobility. In addition, these models can be classified into physical or analytical (nonphysical), based on the modeling philosophy (Yu and Ottersten 2002; Almers et al. 2007). Physical channel models use important physical parameters to provide reasonable description of the channel characteristics and the surrounding scattering environment. Depending on the chosen complexity, these models allow for an accurate reproduction of the real channel. Indeed, choosing a small number of physical parameters makes it difficult, if not impossible, to identify and validate the models. Physical models can further be classified into deterministic models, geometry-based stochastic models, and nongeometric stochastic models. Deterministic models characterize the physical propagation parameters in a completely deterministic manner, for example, using ray-tracing techniques (Iskandar and Shimamoto 2005). With geometry-based stochastic models, the impulse response is characterized by the entire system and the geometry of the scattering environment. Moreover, nongeometric stochastic models characterize physical parameters via probability distribution functions without assuming a particular geometry. On the contrary, analytical models give limited insight to the propagation characteristics of the radio channels and synthesize the radio channel matrices in a mathematical analytical way. Finally, another way to model the radio channel is through field measurements of the channel responses. Then, empirical models are obtained. The accuracy of these models depends on the amount of data achieved by measurements and the quality of statistical data processing.

10.4.1.1 Particular Characteristics of Multiple-Antenna Channel Models

Extending single-antenna models to the multiple-antenna case is not straightforward or even not applicable mainly due to the need of utilization of the space domain, which is the essence of MIMO technology. Specifically, it is necessary to incorporate new parameters, such as the angles of arrival and departure, the angle spread, and the utilization of multiple antennas at both the ends of the link. At this point, it should be mentioned that the correlation in space and/or time and/or frequency dramatically influences MIMO performance and controls MIMO applicability (Salz and Winters

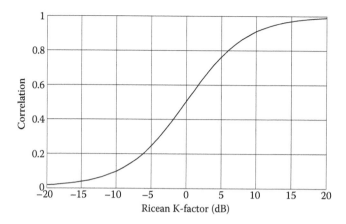

FIGURE 10.7 Absolute correlation of MIMO stratospheric channels for various values of the Ricean factor.

1994; Shiu et al. 2000). In the presence of a sufficiently large, theoretically infinite, number of non-coherent diffuse components in dense scattering propagation environment, the fading is described by a Rayleigh distribution (Pätzold 2002; Proakis and Salehi 2008), the channel matrix is full rank, and the elements of this matrix are uncorrelated, and are modeled as independent and identically distributed (i.i.d.). When a strong LoS signal also arrives at the receiver, the fading is described by a Rician distribution.

Figure 10.7 demonstrates the correlation as a function of the Ricean factor, in the general case. One observes that the absolute correlation easily exceeds 0.15, 0.5, and 0.7, as soon as the Ricean factor exceeds –7, 0, and 3 dB, respectively. Considering coherent diffuse components or sparse scattering and increased correlation in space and/or time and/or frequency, the rank of the MIMO channel matrix is deficient and the spectral efficiency is low. However, the degree of correlation is a complicated function of the degree of scattering and the antenna interelement spacing at both the transmitter and the receiver. Hence, an increase in this spacing is not sufficient to ensure decorrelation between the responses in the MIMO channel matrix. On the other hand, dense scattering in the propagation environment in combination with adequate antenna spacing ensures decorrelation. These two factors control the performance measures, that is, the diversity gain and the spatial multiplexing gain, and therefore the applicability of MIMO techniques to any communication system.

Consequently, the overwhelming majority of analytical and experimental work carried out on land mobile satellite (LMS) and stratospheric multiple-antenna channels at L and S frequency bands focuses on exploiting polarization diversity at both the transmitter and receiver to form a MIMO channel matrix (Erceg et al. 2006; Sellathurai et al. 2006; Horvath et al. 2007; Mohammed and Hult 2009; Hult et al. 2010; Arapoglou et al. 2011a). Then, multiple versions of a signal are transmitted and received via antennas with different polarization. Polarization diversity can overcome possible space limitation and still achieve the advantages predicted by MIMO theory. This approach seems beneficial, since a single satellite cannot provide the necessary antenna spacings required by MIMO theory to provide a high degree of channel decorrelation. Spatial, temporal, and/or polarization SIMO and MISO measurements at S and C bands for mobile satellite systems have been recently carried out by ESA (2015) employing existing satellites (MiLADY) and by Centre National d'Etudes Spatiales employing a helicopter (Lacoste et al. 2009). A relevant SIMO channel modeling approach is presented in Liolis et al. (2008). With regard to MIMO measurement campaigns, the relevant attempts are extremely scarce and have been conducted mainly in the frame of King (2007) in Guildford, UK, at 2.45 GHz. In order to obtain benefit from polarization dimension, the cross-polar transmissions, for example, transmission from vertically polarized antenna to horizontally polarized antenna, should be zero. However, in real scenarios, there is always some polarization mismatch since linearly polarized antenna arrays have nonzero patterns for cross-polar fields.

In addition, multipath effects, for example, diffraction, scattering, and reflection, may change the plane of polarization of incident EM waves at the receiver.

10.4.2 Statistical Channel Models

The statistical channel models express the distribution of the received signal by means of first-order statistics, such as the probability density function (PDF) or the cumulative distribution function and the second-order statistics, such the level crossing rate (LCR) and the AFD (Loo 1985; Vázquez-Castro et al. 2002; Yang et al. 2010). The LCR is basically a measure to describe the average number of times the signal envelope crosses a certain threshold level per second, while the envelope AFD is defined as the expected value of the time interval over which the fading signal envelope remains below a certain threshold level. Since multipath and shadowing effects are important in the signal propagation, the statistical models usually assume that the received signal consists of two components, the LoS and the NLoS. Then, the relative power of the direct and multipath components of the received signal is controlled by the Rician factor and the distributions of these two components are usually studied separately. A stratospheric communication channel is expected to be Rician in its general form. The signal envelope can be expressed as (Aragón-Zavala et al. 2008)

$$S(t)e^{j\theta(t)} = u(t)e^{j\alpha(t)} + v(t)e^{j\beta(t)}, \tag{10.6}$$

where $u(t)$ is a random variable that follows a Rayleigh distribution and $\alpha(t)$ is uniformly distributed within the range $(0, 2\pi)$, whereas the $v(t)$ and $\beta(t)$ are deterministic signals and are the magnitude and phase of the direct component, respectively. Then, the Rician factor is the average power ratio of the direct signal component to the multipath components and is given by

$$k = v^2/2\sigma^2, \tag{10.7}$$

where v is the direct component envelope of the received signal and $2\sigma^2$ is the average power of the multipath components. When the Rician factor is equal to 0, that is, rich multipath exists, the channel is described by a Rayleigh distribution, whereas a very large value of the Rician factor implies the presence of a Gaussian channel. Several values of the Rician factor have been reported in the literature from measurement campaigns and studies performed in the L and S frequency bands, for satellite (Jahn 2001) and stratospheric (Iskandar and Shimamoto 2006a) communication systems. According to these measurements, the value of the Rician factor depends on the elevation angle of the satellite/platform and the operating frequency. Note that HAP and/or user movement and HAP displacement cause a continuous change in the subplatform point and in the elevation angle of the platform. Thus, the Rician factor also varies with the elevation angle variation. Note that a lognormal distribution can characterize the shadowing, when the signal is subject to blocking of clutter and obstructions on terrain depending on the propagation area. Based on Loo model (Loo 1985), a statistical model for stratospheric channels was proposed in Bo et al. (2007). The key parameters of this model are the PDF of the amplitude of the received signal, the LCR, and the AFD.

10.4.3 Statistical-Switched-Channel Models

The switched-channel model in Cuevas-Ruíz and Delgado-Penín (2004b) allows for convenient and time-efficient system analysis of the links between HAPs and terrestrial stations, when the channel characteristics are dynamic and vary in time. In this model, Markov chains are used to describe blockage or shadowing effects due to buildings or vegetation close to the terminal. According to the definition of chain state and transition matrix between states, the transient behavior of a stochastic process can be described. The changes in shadowing and multipath are typically modeled by different propagation states, for example, "good" and "bad" states corresponding to

LoS/open/light shadowing areas and NLoS/blocked/heavy shadowing areas, respectively. Then, the goal of this model is to properly characterize a time-variant channel, which switches from one state to another at any time. These propagation states can be described by a first-order Markov chain with specific state and transition probabilities. The differences between the states are related to the type of fading that is affecting the channel. The model in Cuevas-Ruíz and Delgado-Penín (2004b) is based on the classic narrowband switched-channel model in Lutz et al. (1991) for the L band, which characterizes the process of fading through a switch between two states, one defined as good (Rice distribution) and the other defined as bad (Rayleigh-lognormal distribution). There are also three-state channel models proposed in Karasawa et al. (1997) and Fontan et al. (1997) that suggest different statistical distributions for each state of the signal. Typically, each state lasts a few meters along the traveled route.

The characteristics that define the channel, such as the finite number of states present in a discrete time, make the use of Markov chains possible. A Markov chain is fully defined by a state probability vector π and by a matrix $\mathbf{P}$ of transition probabilities between states matrix. Elements π_i of vector π show the time percentage in which the Markov chain belongs to state i, while the elements p_{ij} of the matrix $\mathbf{P}$ represent the probability that the chain changes from state i to state j. For a three-state chain, $\mathbf{P}$ is given by

$$\mathbf{P} = \begin{pmatrix} P_{AA} & P_{AB} & P_{AC} \\ P_{BA} & P_{BB} & P_{BC} \\ P_{CA} & P_{CB} & P_{CC} \end{pmatrix}. \tag{10.8}$$

A possible weakness of the Markov models is that they utilize the same distributions for all the channel states. Besides, the Markov chain can model very slow changes of the channel characteristics caused by large obstacles. Alternatively, a semi-Markov model for stratospheric broadband channels was proposed in Cuevas-Ruíz and Delgado-Penín (2004a) and considers that the time duration between states transition is random and can be characterized by some type of probability distribution. One of the greatest advantages of the semi-Markovian process is the possibility to make a clear distinction between the probability distribution of the duration of the fading present in a state and the type of fading present in that state, which helps to get a better approximation of the link channel conditions (Bråten and Tjelta 2002). The parameters for the durations of the channel states and the distributions of the durations were established in the ITU-R Recommendation P.681-6. A general feature of the Markov and semi-Markov approaches is the memoryless property of the channel, where one state is uncorrelated to other instances of the same state at different times.

10.4.4 Physical–Statistical Models

Physical or deterministic channel models based on ray-tracing algorithms can provide accurate results for a particular scenario. However, this approach is not often used due to increased computational complexity. On the other hand, statistical models are built around measurement data and provide reasonable reproduction of the real channel. However, they provide little insight into the propagation mechanisms and depend on the accuracy of the measurements. An intermediate approach between these models is the physical–statistical model. This type of modeling is the most appropriate in predicting the "ON/OFF" nature and investigating the small-scale fading effects over large coverage areas applicable to satellite/stratospheric communication systems (Tzaras et al. 1998).

King et al. (2005) proposed a physical–statistical model for LMS and stratospheric multiantenna channels for the L and S frequency bands. This model generates high-resolution time series data and power delay profile for communication links between satellite (or HAP) and terrestrial terminal antennas and also predicts the correlation between these links. In this model, the obstacles, for example, buildings and trees, are grouped into clusters of spherical shapes and the cluster centers are randomly positioned. Multiple scatterers are placed randomly around the cluster center and their position follows the Laplacian distribution, while the building heights follow the lognormal distribution. According to this model, the scatterers are nonuniformly distributed, that is, the user receives the signal only from particular directions. Three paths between a satellite and a mobile terminal are considered: an LoS path, a blocked LoS path, and an attenuated path by trees. The parameters obtained from experimental data collected in Munich, Germany at L band (1.54 GHz) for urban and highway environments were used to validate the model. The small-scale fading and the wideband parameters can be approximated using the output time series and spatial power delay profile data of the model. By considering two distinct HAPs and denoting the fast fading from each satellite/HAP as α_A and α_B, the correlation can be defined as (Saunders and Aragón-Zavala 2007)

$$\rho_{AB} = \frac{E[(a_A - \mu_A)(a_B - \mu_B)^*]}{\sigma_A \sigma_B}, \tag{10.9}$$

where E[.] is the statistical expectation operator, $(\cdot)^*$ denotes complex conjugate operation, and μ_A, μ_B are the means and σ_A, σ_B are the standard deviations of the fast-fading data from HAP antennas A and B, respectively. The results suggested that the HAP antennas require a separation of around 18 m at 1.54 GHz to ensure low correlation between each channel matrix coefficient. Hence, utilizing a single HAP is a viable solution.

As the maximum MIMO gain can be achieved with low correlation between antenna elements at both ends of a MIMO communication system, a fundamental way of achieving low antenna correlation is to use antenna elements with adequate separation. However, owing to size constraints, the physical–statistical model suggests that multiple antennas with large separation cannot be deployed at a single satellite. Hence, two satellites are required to achieve diversity. Nevertheless, employing two satellites gives rise to new challenges, such as waste of the limited satellite bandwidth for the transmission of the same signal, lack of synchronization in reception, and high implementation cost. The synchronization issues can be dealt with by employing cooperative satellite diversity concept (Ahn et al. 2010; Zang et al. 2010) or by using compact antennas (Horvath and Frigyes 2006; Mohammed and Hult 2009), in which the problem of synchronization does not exist. Multiple-HAP constellations employing compact MIMO antenna arrays were utilized in Mohammed and Hult (2009) and Hult et al. (2010) and the capacity performance was investigated. In multiple-HAP systems, virtual MIMO (V-MIMO) space-polarization channels can be created using HAP diversity in combination with the polarization and pattern diversity of a special type of compact MIMO antenna arrays, like a vector element antenna array (Andrews et al. 2001; Mohammed and Hult 2009), the MIMO-cube (Getu and Andersen 2005), or the MIMO-octahedron antenna (Mohammed and Hult 2009), which differ in complexity and design. Figure 10.8 illustrates the diversity setup for the case of vector element antennas and three HAPs separated by the angles $\theta_{1,2}$ and $\theta_{2,3}$. Although the multiple-HAP system outperforms the single-HAP system, the performance may be limited due to spatial correlation and mutual coupling between the separate antenna array elements. From the results, an optimal separation angle between HAPs that maximizes the total capacity of the system was also determined. Nevertheless, from this physical–statistical model in King et al. (2005), one also obtains that the application of multiple antennas at a single HAP may be viable. Nevertheless, analytical expressions for the space–time correlation were not derived and the effect of the array configuration was not studied.

10.4.5 Geometry-Based Single-Antenna Channel Models

In general, the channel characterization strongly depends on the location of the transmitter and the receiver. From Figure 10.9, the fundamental parameters, which describe the geometry of a basic

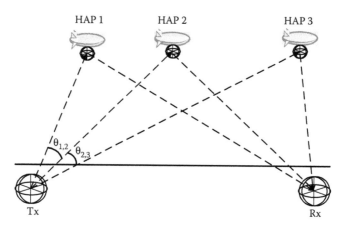

FIGURE 10.8 Multiple-HAP diversity system with three HAPs equipped with vector element antennas and the channel paths from the transmitter to the receiver. (Adapted from Mohammed, A. and T. Hult, *International Journal of Recent Trends in Engineering*, 1(3), 244–247, 2009.)

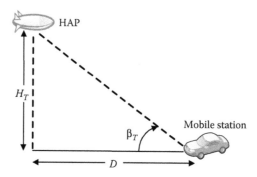

FIGURE 10.9 Typical geometry of a stratospheric system.

stratospheric system, are the elevation angle of the platform β_T, the height of the platform H_T, and the distance D between the mobile station and the subplatform point. Since impairments of the signal are mainly caused by the environment near to the user, a realistic positioning of the scatterers is essential for an accurate channel model. Although several channel models have been proposed for terrestrial and satellite communications channels, newer models are required to accurately characterize specific issues raised in stratospheric channels. For these channels, the physical–geometrical characteristics are critical. The establishment of a particular geometry allows for an accurate characterization of the multipath effects. Most of the physical/geometrical channel models postulate a scattering environment and attempt to capture the channel characteristics by involving scattering parameters. Such models can often illustrate the essential characteristics of the radio channel, as long as the constructed scattering environment is reasonable. The following subsections describe two different geometries for single-antenna stratospheric channels.

10.4.5.1 Geometry-Based Ellipsoid Model

Dovis et al. (2002) considered an ellipsoid as the volume containing all the scatterers in the terrain with transmitter and receiver as foci (see Figure 10.10). This model is based on the theoretical model proposed in Rappaport and Liberti (1996) for a terrestrial station and is extended to the case of a stratospheric station. In this extension, the receiver and the transmitter are no longer on the horizontal plane and the height of the transmitter is considered. This model is applicable to the

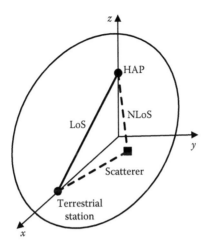

FIGURE 10.10 Geometry-based ellipsoid model for stratospheric channels. (Adapted from Dovis, F. et al., *IEEE Journal on Selected Areas in Communications*, 20(3), 641–647, 2002.)

propagation environments, where multipath effects are primary, for example, urban areas, and provides a really convenient method for estimating the small-scale fading of the communication links between HAPs and terrestrial stations and characterizing the time and frequency domain of fading channels through the power delay profile and the Doppler spectrum. Moreover, this model does not consider the Rayleigh or Rice distributions for the signal amplitude. Instead, it extracts the power distribution through the power delay profile. The results depicted that channel models neglecting the presence of scatterers close to the ground can yield too optimistic results. However, this model assumes that the scatterers are uniformly distributed in space, that is, the user receives signals from all directions with equal probabilities, an assumption that deviates from practical situations. Moreover, this model overestimates the effects of large delay components. This deficiency is due to the assumption of uniformly distributed single-bounce echoes. Therefore, a blockage-based channel model for stratospheric channels that was proposed in Liu et al. (2003), quantified the probability of the single-bounce echoes from different scatterers, obtained an improved distribution function of the excess delay through numerical integration, and estimated the joint time-angle spread of the underlying multipath channel. Although the blockage-based model seems more realistic and feasible than the ellipsoid model, it greatly complicates the derived process.

10.4.5.2　Geometry-Based Circular Cone Model

A three-dimensional (3D) multipath model based on circular straight cone geometry (see Figure 10.11) was also proposed in Cuevas-Ruíz et al. (2009). This geometry provides a better approximation to simulate multipath propagation, since it more accurately represents the coverage area of a wireless communication system for a link between a HAP station and terrestrial users. Specifically, this model assumes that the scatterers are not considered significant close to the platform, as Dovis et al. (2002) suggest, but they are more concentrated close to the base of the cone. This geometrical model can better resemble the scenario of a directive antenna on board the HAP illuminating a specific coverage area on the ground. From this model, the power delay profile was generated.

10.4.6　Geometry-Based Multiple-Antenna Channel Models

As mentioned in the previous section, the prerequisite so that single-HAP configurations fully exploit the spatial diversity and spatial multiplexing advantages of MIMO technology predicted by information theory is the existence of sufficient antenna spacing at the transmitter and receiver,

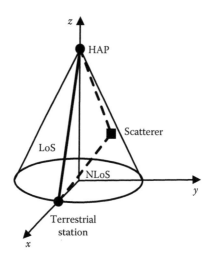

FIGURE 10.11 Geometry-based circular cone model for stratospheric channels. (Adapted from Cuevas-Ruíz, J.L. et al., *International Workshop on Satellite and Space Communications (IWSSC)*, pp. 235–239, 2009.)

as well as a rich scattering environment, which renders the fading paths between the antenna elements of the transmitter and the receiver independent. Therefore, the analysis and design of MIMO stratospheric communication systems require the development of space–time channel models, which enable us to properly characterize the fading channel and thoroughly study the channel statistics. The MIMO stratospheric channel models should take the distribution of the scatterers and the correlations among signal carriers into account. In practice, owing to the length of the radio path between HAPs and terrestrial stations, the transmit and/or receive antennas should be placed at significant distances from each other to ensure that the paths are really diverse. The application of multiple antennas to single HAPs may be plausible to be pursued in the frame of MIMO stratospheric systems, since spatial limitations on board are not as stringent as in satellites. Moreover, the length of the radio path between HAPs and terrestrial stations is significantly smaller than the corresponding one associated with satellites. King et al. (2005) provided a general estimation of the required antenna element separation at a HAP to achieve uncorrelated responses in the MIMO channel matrix. However, whether the size of a single HAP can support spatial diversity was thoroughly investigated through the models presented in the following subsections.

10.4.6.1 Geometry-Based Cylindrical Model for Narrowband Stratospheric Channels

Michailidis and Kanatas (2010) proposed a geometry-based model for MIMO stratospheric channels. This model utilizes L (1/2 GHz) and S (2/4 GHz) frequency bands and provides in-depth understanding and description of the statistical properties of MIMO stratospheric channels. A downlink narrowband MIMO stratospheric communication channel was considered with n_T transmit and n_R receive antenna elements. All antennas are fixed, omnidirectional, and are numbered as $1 \leq p \leq q \leq n_T$ and $1 \leq l \leq m \leq n_R$, respectively. The n_T antenna elements of the free of local scattering stratospheric base station (SBS) are situated approximately 20 km above the ground and it is assumed that the n_R antenna elements of the TMS are in motion. Considering slowly varying and frequency-flat-fading channels, the link between the SBS and TMS antenna arrays is represented using the following complex baseband vector equation:

$$\mathbf{r} = \mathbf{Hs} + \mathbf{n}, \tag{10.10}$$

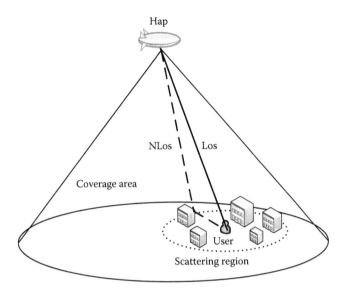

FIGURE 10.12 Basic concept of a stratospheric communication system and a cylindrical scattering region inside the coverage area.

where $\mathbf{s} \in \mathbb{C}^{n_T \times 1}$ is the transmitted signal vector, $\mathbf{r} \in \mathbb{C}^{n_R \times 1}$ is the received signal vector, and $\mathbf{n} \in \mathbb{C}^{n_R \times 1}$ is the noise vector, which denotes the additive white Gaussian noise at the receiver branches. The entries of the noise vector are i.i.d. complex Gaussian random variables with zero mean and variance N_0, where N_0 is the noise power spectral density (PSD). Finally, $\mathbf{H} = \left[h_{ij} \right]_{n_R \times n_T} \in \mathbb{C}^{n_R \times n_T}$ is the matrix of complex faded channel gains. As shown in Figure 10.12, this model constructs the received complex faded envelope as a superposition of the LoS and the NLoS rays and assumes that the local scatterers in the vicinity of the mobile user are nonuniformly distributed within a cylinder, that is, the cylinder is considered as the volume containing all the scatterers. According to this model, the waves may travel in both horizontal and vertical planes, and the propagation environment is characterized by 3D nonisotropic scattering conditions. Several parameters related to the physical properties of the stratospheric communication system were considered, for example, the elevation angle of the platform, the antenna array orientation and elevation, the degree and spread of scattering for the terrestrial user, the height of the scatterers, and the HAP displacement due to stratospheric winds. An alternative version of this model was presented in Michailidis and Kanatas (2008) and Michailidis et al. (2008), but only the first tier of scatterers lying on the surface of a cylinder was taken into account. These studies were based on existing terrestrial 3D MIMO channel models that deal with fixed-to-mobile (F-to-M) (Leong et al. 2004) or mobile-to-mobile (M-to-M) (Zajić and Stüber 2008) cases.

Based on the model in Michailidis and Kanatas (2010), the space–time correlation function (STCF) was derived as a function of the model parameters. In particular, the STCF between two arbitrary subchannels $h_{pl}(t)$ and $h_{qm}(t)$ is defined as

$$R_{pl,qm}(\delta_T, \delta_R, \tau, t) = \mathrm{E}\left[h_{pl}(t) h_{qm}^*(t + \tau) \right], \tag{10.11}$$

where δ_T and δ_R denote the spacing between two adjacent antenna elements at the SBS and TMS, respectively. The distributions for the azimuth AoA of the scattered waves, the distance between the effective scatterers and the user, and the height of the effective scatterers were modeled by the von Mises (Abdi and Kaveh 2002), the hyperbolic (Mahmoud et al. 2002), and the lognormal

(Tzaras et al. 1998) distributions, respectively, which all have previously shown to be successful in describing measured data. The von Mises PDF is defined as

$$f(a_R) = \frac{e^{k\cos(a_R - \mu)}}{2\pi I_0(k)}, \quad -\pi \leq a_R \leq \pi, \tag{10.12}$$

where a_R stands for the azimuth AoA of the scattered waves, $I_0(\cdot)$ is the zeroth-order modified Bessel function of the first kind, $\mu \in [-\pi, \pi]$ is the mean angle at which the scatterers are distributed in the x–y plane, and $k \geq 0$ controls the spread around the mean. Note that the scattering becomes increasingly nonisotropic, as k increases, whereas setting $k = 0$, that is, $f(a_R) = 1/2\pi$, incurs isotropic scattering. The hyperbolic PDF is defined as

$$f(R_S) = \frac{a}{\tanh(aR_{S,\max})\cosh^2(aR_S)}, \quad 0 < R_S \leq R_{S,\max}, \tag{10.13}$$

where R_S and $R_{S,\max}$ denote the distance and the maximum distance, respectively, between the effective scatterers and the user and the parameter $a \in (0, 1)$ controls the spread (standard deviation) of the scatterers around the TMS. Decreasing a increases the spread of the PDF of R_S and increases the mean distance between the scatterers and the terrestrial user. Considering a HAP-based communications system, as the elevation angle of the platform decreases, the scatterers are expected to be more widely distributed, which corresponds to a possible decrease of a. Nevertheless, a can be accurately obtained through measurements in different propagation environments, that is, urban, suburban, or rural environments. Figure 10.13 shows the mean distance between TMS and the scatterers for several values of the parameter a and $R_{S,\max} = 200$ m. The lognormal PDF is defined as

$$f(H_S) = \frac{e^{-([\ln(H_S) - \ln(H_{S,\text{mean}})]^2/2\sigma^2)}}{H_S\sigma\sqrt{2\pi}}, \quad 0 < H_S \leq H_{S,\max}, \tag{10.14}$$

where H_S denotes the height of the effective scatterers and $H_{S,\text{mean}}$ and σ are the mean and standard deviation of H_S, respectively.

The results indicated the required HAP antenna separation to attain uncorrelated MIMO stratospheric channels in different propagation environments. For instance, assuming that an absolute

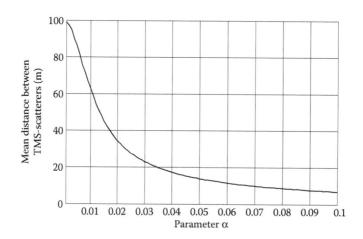

FIGURE 10.13 Mean distance between the TMS and the scatterers for different values of the parameter α.

correlation of 0.15 or less amounts to nearly uncorrelated links and considering 2.1 GHz carrier frequency (ITU 2007c) and 60° elevation angle, HAP antennas require a separation of around 10 and 18 m in isotropic and nonisotropic scattering environments, respectively. This suggests that utilizing MIMO techniques on a single aircraft or airship is a viable solution, as long as the Ricean factor is small and the multipath is rich, for example, considering propagation in dense urban areas, where the scatterers are usually dense and tall.

From this model, the channel capacity of a stratospheric communication system equipped with multielement antennas at both sides was defined and investigated in Michailidis and Kanatas (2011). Assuming that the channel is known to the TMS and unknown to the SBS, the available capacity can be obtained as follows (Paulraj et al. 2003):

$$C = \log_2 \det\left(\mathbf{I}_{n_R} + \left(\frac{\text{SNR}}{n_T} \right) \mathbf{H}\mathbf{H}^H \right) \text{ bps/Hz}, \qquad (10.15)$$

where $\mathbf{I}_{n_R}$ is an identity matrix of size n_R, $(\cdot)^H$ denotes the complex conjugate (Hermitian) transpose operator, and $\det(\cdot)$ denotes the matrix determinant. The results demonstrated that the multiple-antenna architecture outperforms the conventional single-antenna architecture in terms of the channel capacity. These results also revealed that the capacity depends on the strength of the LoS signal, and the received SNR. In particular, increasing the Rician factor decreases the capacity, while increasing the SNR increases the capacity. These results also demonstrated the influence of the spatial and temporal correlation on the capacity. Specifically, it has been shown that increasing the elevation angle of the platform increases the capacity, while increasing the density of the scatterers in the vicinity of the user and the spacing between the antennas increases the capacity. Moreover, broadside HAP antennas maximize the capacity, while vertically placed antennas at the mobile terminal provide considerable capacity gain in highly urbanized areas. Changing the velocity and the moving direction of the user significantly affects the capacity. Furthermore, the results underlined the effects of spatial and temporal correlation on the capacity of uniform linear arrays (ULAs) and suggested that applying MIMO techniques to a single HAP can effectively enhance the capacity, as soon as the Rician factor is small.

The first-order statistics of the channel impulse response are not sufficient to assess system characteristics, such as the handover, the velocities of the transmitter and receiver, the fading rate, and design effective error control mechanisms for optimal packet radio transmission over burst error correlated fading channels. Hence, accurate characterization of the second-order statistics, that is, the LCR and the AFD, is necessary. Based on Zajić et al. (2008, 2009) and using the model for narrowband MIMO stratospheric channels in Michailidis and Kanatas (2010), the corresponding analytical expressions for the envelope LCR and AFD for a 3D nonisotropic scattering environment were derived in Eldowek et al. (2014). Note that a 3D cylindrical model was also proposed for UAV-MIMO systems (Gao et al. 2012; Xi Jun et al. 2014). This model adopts the method of channel matrix decomposition and normalization to deduce the UAV-MIMO average channel correlation matrix and directly analyzes the effect of UAV multiple-antenna layout, flight distance, and the position of scatterers and other parameters on the UAV-MIMO channel.

10.4.6.2 Geometry-Based Concentric-Cylinder Model for Wideband Stratospheric Channels

Currently, a number of standardization bodies supported by industries and research institutes are trying to establish new system standards for future high-speed wireless systems employing MIMO techniques. One important feature of future communication systems is that they demand considerably larger bandwidths than today's systems to support advanced multimedia and broadcasting services. Thus, realistic wideband MIMO stratospheric channel models are essential for the design and concise evaluation of future stratospheric communication systems. Wideband effects usually impose a frequency-selective transfer function, which can be modeled by a TDL with each tap defined using different weights and distributions (Bello 1963). Thus, provided that wideband transmissions are present, a statistical TDL model can be used, where each tap is described by a corresponding narrowband model. Nevertheless, the propagation environment necessitates fundamental limitations on the performance. Hence, when terrain and scattering distributions are available, physical–geometrical channel modeling is preferred to ensure model accuracy and versatility. The realization of the model in Michailidis and Kanatas (2010) is limited to narrowband, that is, frequency-nonselective, communications. In particular, it was assumed that the propagation delays of all incoming scattered waves are approximately equal and small in comparison to the data symbol duration. In wideband communications, the data symbol duration is small and multipath delay spread is introduced. Therefore, the propagation delay differences cannot be neglected.

Michailidis and Kanatas (2014) extended the aforementioned model for narrowband MIMO stratospheric channels with respect to frequency selectivity and a 3D reference model for wideband MIMO stratospheric channels was proposed. The proposed model utilizes carrier frequencies well below 10 GHz. Hence, both LoS and NLoS links should be considered, while rain effects are insignificant. Specifically, a 3D geometrical model for wideband MIMO stratospheric channels was introduced, referred to as the "two concentric-cylinders" model. Based on the modified model geometry, the scatterers occupy the volume between two concentric-cylinders. Several parameters related to the physical properties of wideband stratospheric communications were considered, in order to properly and thoroughly characterize the wideband MIMO stratospheric channel.

From the reference model, the space–time–frequency correlation function (STFCF), the space-Doppler power spectrum (SDPS), and the power space-delay spectrum (PSDS) were derived for a 3D nonisotropic scattering environment and a wide-sense stationary uncorrelated scattering MIMO stratospheric channel. The normalized STFCF between two time-variant transfer functions $T_{pl}(t, f)$ and $T_{qm}(t, f)$ is defined as (Bello 1963; Fleury 2000)

$$R_{pl,qm}(\delta_T, \delta_R, \Delta t, \Delta f) = \mathrm{E}\left[T_{pl}^*(t, f) T_{qm}(t + \Delta t, f + \Delta f) \right]. \tag{10.16}$$

The results revealed that that the frequency correlation decreases as the elevation angle of the platform increases. The SDPS was obtained by calculating the FT of the STCF, that is, $R_{pl,qm}(\delta_T, \delta_R, \Delta t, \Delta f = 0)$, while the PSDS was obtained by calculating the imaging Fourier transform (IFT) of the space–frequency correlation function, that is, $R_{pl,qm}(\delta_T, \delta_R, \Delta t = 0, \Delta f)$. Future empirical results could be easily compared with theoretical results, since this model is flexible and applicable to a wide range of propagation environments, that is, one may choose proper values for the model parameters to fit a particular environment.

10.4.6.3 Geometry-Based Simulation Models for Stratospheric Channels

The theoretical narrowband and wideband channel models in Michailidis and Kanatas (2010, 2014) assume an infinite number of scatterers, which prevents practical realization, that is, software/hardware implementation. Although these models can be ideally verified through experimental real-time field trials, simulation of the radio propagation environment is commonly used as an alternative, cost-effective, and time-saving approach to the test, optimization, and performance evaluation of wireless communication systems. Hence, the development and design of accurate and efficient simulation models for MIMO stratospheric channels is essential to reproduce their statistical properties. The prime requirement of a simulation setup is to capture the fading effects created by the radio channel and the goal of any simulation model is to properly reproduce the channel properties. Indeed, many different methods have been adopted for the simulation of fading channels. The most widely accepted methods are the filtered noise models (Fechtel 1993; Verdin and Tozer 1993; Young and Beaulieu 2000) and the sum-of-sinusoids (SoS) models (Clarke 1968; Pätzold et al. 1998; Pop and Beaulieu 2001; Xiao et al. 2006; Wang et al. 2007; Zajić and Stüber 2008). The filtered noise models intend to simulate the channel properties by means of signal processing techniques, without considering the underlying propagation mechanisms. These models filter Gaussian noise through appropriately designed linear time-invariant filters to generate the channel waveform with the desired channel PSD and capture the important first- and second-order channel statistics. However, the efficiency of this approach is limited by the utilized filter. The sum-of-sinusoids (SoS) principle introduced by Rice (1944) has been widely accepted by academia and industry as an adequate basis for the design of simulation models due to its reasonably low computational costs. According to this principle, the overall channel waveform is the sum of several complex sinusoids having frequencies, amplitudes, and phases that are appropriately selected to accurately approximate the desired statistical properties.

Two main categories of SoS-based simulation models are reported in the literature, the deterministic simulation models (Pätzold et al. 1998) and the statistical simulation models (Patel et al. 2005). The deterministic models are easy to implement and have short simulation times. Specifically, they have fixed parameters for all simulation trials and converge to the desired properties in a single simulation trial leading to deterministic statistical properties. On the contrary, the statistical (Monte Carlo) models have at least one of the parameters as random variables that vary with each simulation trial. Hence, their statistical properties also vary for each simulation trial and converge to the desired ones in the statistical sense, that is, when averaged over a sufficiently large number of simulation trials. In contrast to filtered noise models, SoS-based models produce channel waveforms that have high accuracy and a perfectly band-limited spectrum. In addition, their complexity is typically reduced by cleverly choosing the model parameters to reduce the computation load. Furthermore, SoS-based models can be easily extended to develop simulation channel models for MIMO communication systems due to the explicit inclusion of spatial information, such as the multipath angles of arrival and departure.

Owing to these advantages, SoS-based models with a finite number of scatterers for narrowband (Michailidis and Kanatas 2011, 2012) and wideband (Michailidis and Kanatas 2014) MIMO stratospheric channels were proposed, under the framework of the reference models in Michailidis and Kanatas (2010, 2014). The statistical properties of these simulation models were verified by comparison with the corresponding statistical properties of the reference model. Although the deterministic simulation model (Michailidis and Kanatas 2011) has the potential of becoming a standard procedure due to its simplicity, efficiency, and reproducibility, the slightly complex statistical simulation model in Michailidis and Kanatas (2012) provides the highest performance with a relatively small number of simulation trials. In most applications, software-based simulation is performed on a workstation or a personal computer. Nevertheless, the feasibility of porting these simulation models into hardware by using digital signal processors is also of interest. In particular, hardware channel simulations can significantly increase the usefulness of the simulation models by enabling simulations in real time. Table 10.3 reviews and compares the relative complexity of the proposed simulation models, which generally depends on the number of the utilized scatterers, the number of the performed simulation trials, and the required number of random variables.

TABLE 10.3
Complexity of Simulation Models

Simulation Model	Number of Scatterers	Number of Simulation Trials	Relative Number of Calculations	Number of Random Variables
Stochastic	N_{Stoc}	∞	∞	3
Deterministic	N_{Det}	1	N_{Det}	0
Statistical	N_{Stat}	N_{trials}	$N_{Stat}N_{trials}$	3

10.4.6.4   Geometry-Based Two-Cylinder Model for Relay-Based Stratospheric Channels

A key objective in the development of next-generation systems is the seamless integration of wireless terrestrial and aerospace infrastructures over heterogeneous networks (Evans et al. 2005; Giuliano et al. 2008; Paillassa et al. 2011). Hybrid satellite/terrestrial networks are a typical example of cooperation between different architectures. Motivated by this observation, the use of a radio-relay installed on a HAP, which transfers information between two TMSs was investigated in Michailidis et al. (2013b) and a model for MIMO M-to-M (mobile-to-mobile) via stratospheric relay (MMSR) fading channels in single-relay dual-hop amplify-and-forward (AF) networks was proposed. The idea of using stratospheric relay nodes to provide surveillance, monitoring, maritime, and 3G services was initially conceived in Jull et al. (1985), Antonini et al. (2003), Oodo et al. (2005), and Giuliano et al. (2008), while a relay system based on UAVs was presented in Zhan et al. (2011). Compared to other relaying methods, AF relaying leads to low-complexity relay transceivers and to lower power consumption, since signal processing and decoding procedures are not required. This transmission scheme intends to improve the link reliability and extend the network range of point-to-point M-to-M communication networks by preserving the end-to-end communication between the source (S) and the destination (D) via the intermediate stratospheric relay (R). Indeed, harsh multipath fading effects usually degrade the transmission link quality of M-to-M systems, while a high attenuation in the propagation medium could preclude the link from the transmitter to the receiver. Since the channel capacity gain is larger for MIMO channels by effectively exploiting multipath propagation environments, it is considered that multiple antennas are used at the transmitting, the relaying, and the receiving end. It is assumed that the local scatterers in the vicinity of the source and the destination are nonuniformly distributed within two separate cylinders, which reflect the influence of two heterogeneous 3D scattering environments. The uplink (S-R link) and the downlink (R-D link) may experience either symmetric, that is, Rayleigh/Rayleigh or Rician/Rician, or asymmetric (Suraweera et al. 2009), that is, Rayleigh/Rician or Rician/Rayleigh, fading phenomena depending on the strength of the LoS component. According to the propagation scenario in Figure 10.14, the waves emitted from the source antennas travel over paths with different lengths and impinge the relay antennas from different directions due to the 3D nonisotropic scattering conditions within the cylinder 1. Similarly, the waves emitted

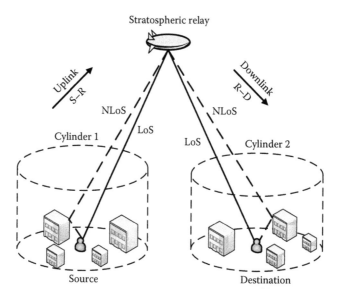

FIGURE 10.14 Simple representation of an MMSR fading channel. (Adapted from Michailidis, E.T., P. Theofilakos, and A.G. Kanatas, *IEEE Transactions on Vehicular Technology,* 62(5), 2014–2030, 2013b.)

from the relay antennas travel over paths with different lengths and impinge the destination antennas from different directions due to the 3D nonisotropic scattering conditions within the cylinder 2. The resulted model is parametric and adaptable to a wide variety of propagation environments by simply adjusting the model parameters.

The impulse response $h_{pl}(t) = [\mathbf{H}(t)]_{lp}$, corresponding to the transmission link from source antenna element p to destination antenna element l via the relay antenna element q, is given by

$$h_{pl}(t) = \sum_{q=1}^{L_R} h_{pq}^{SR}(t) h_{ql}^{RD}(t), \tag{10.17}$$

where $h_{pq}^{SR}(t)$ is the impulse response for the transmission link from the pth source antenna element to the qth relay antenna element and $h_{ql}^{RD}(t)$ is the impulse response for the transmission link from the qth relay antenna element to the lth destination antenna element. Based on the reference model, general analytical formulas for the corresponding STCF were derived. The results underlined that the correlation depends mainly on the interelement spacing at the source and the destination antennas, respectively. Providing that the spacing is adequate, that is, larger than 1λ, where λ is the carrier wavelength, low spatial correlation can be maintained regardless of the interelement spacing at the relay.

Nomikos et al. (2013) analyzed the performance evaluation of an MMSR system through rigorous simulations in terms of the BER by employing a hierarchical broadcast technique and minimum mean square error receivers. The results revealed the relationship between channel correlation and BER for varying fading condition and distribution of the scatterers, thus illustrating valuable information for a stratospheric relay system implementation.

10.4.6.5 Geometry-Based Cylindrical Model for Stratospheric Dual-Polarized Channel

Although the size of a single HAP possibly allows for reasonably low correlation, when two antennas are accommodated, as shown in Michailidis and Kanatas (2010), other ways of decorrelating MIMO branches are desired, in case of environments dominated by a strong LoS component. A promising, attractive, and potential strategy due to the recent advances in MIMO compact antennas (Getu and Andersen 2005) to achieve low correlation and increased channel capacity in free-space communications is to exploit the benefits of polarization diversity via dual-polarized (DP) antennas. Then, the two spatially separated single-polarized antennas are replaced by a single-antenna structure employing two orthogonal polarizations. For DP systems, the cross-polarization discrimination (XPD) factor is the usual evaluation parameter (Quitin et al. 2009) and estimates the depolarization effects that arise due to the scattering mechanisms. Specifically, the XPD is defined as the ratio of the copolarized average received power to the cross-polarized average received power.

To comprehensively understand stratospheric DP systems, modeling of the underlying dispersive channel is important. Since the waves may travel in both horizontal and vertical planes, 3D channel modeling is required to ideally characterize the propagation environment and accurately represent important aspects of polarized stratospheric DP channels. Kwon and Stüber (2011) proposed 3D models for XPD in F-to-M (fixed-to-mobile) and M-to-M channels. Nevertheless, these models consider only the first tier of scatterers lying on the surface of one (F-to-M) or two (M-to-M) cylinders, which is unrealistic for stratospheric communication systems (Michailidis and Kanatas 2010). Based on the geometrical theory of channel depolarization introduced in Kwon and Stüber (2011) and the model in Michailidis and Kanatas (2010), a 3D geometry-based model for narrowband stratospheric DP channels was proposed in Michailidis et al. (2013a). In contrast to other research work that deals with compact MIMO antenna array configurations in conjunction with multiple-HAP constellations (Hult et al. 2010), the channel model in Michailidis et al. (2013a) utilizes the polarization diversity from a single HAP. According to this model, a vertically (horizontally) polarized wave emitted from the HAP gives rise to a horizontally (vertically) polarized wave being

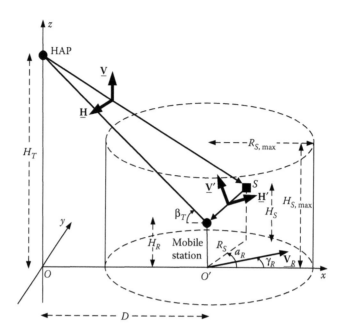

FIGURE 10.15 Simple representation of the cylindrical scattering model for XPD on Rician stratospheric channels. (From Michailidis, E.T., P. Petropoulou, and A.G. Kanatas, *7th European Conference on Antennas and Propagation (EuCAP)*, 2013a.)

received at the mobile station. Figure 10.15 demonstrates the vertical polarization vector **V** and the horizontal polarization vector **H** of a wave transmitted from the HAP and the vertical polarization vector **V′** and the horizontal polarization vector **H′** of the wave reflected by the scatterer S. Note that **V′** and **H′** are not entirely vertically or horizontally polarized, but have also cross-polarization components. This effect is called depolarization and is significantly affected by the distribution and the electrical characteristics of the scatterers within the cylinder. One observes that the HAP, the mobile station, and one scatterer define a plane for each wave that is emitted from the HAP and received at the mobile station.

From this model, the XPD can be derived, under 3D nonisotropic scattering, as a function of the model parameters, that is, the distribution of the scatterers and the elevation angle of the platform. The XPD is defined as the ratio of the copolarized average received power to the cross-polarized average received power and is given by

$$\overline{\text{XPD}}_{VV/HV} = \frac{P_{VV}}{P_{HV}}, \tag{10.18}$$

$$\overline{\text{XPD}}_{HH/VH} = \frac{P_{HH}}{P_{VH}}, \tag{10.19}$$

where P_{VV}, P_{HV}, P_{VH}, and P_{HH} are the received power with vertically polarized TMS and HAP antennas (VV channel), with a horizontally polarized TMS antenna and a vertically polarized HAP antenna (HV channel), with a vertically polarized TMS antenna and a horizontally polarized HAP antenna (VH channel), and with horizontally polarized TMS and HAP antennas (HH channel), respectively. By taking into account the statistical distribution of scatterers, the total power of the vertical and horizontal polarization components can be derived via the superposition of the LoS and NLoS components by averaging over the PDFs defined in Equations 10.12 through 10.14. The results showed the relation between XPD and the 3D nonisotropic distribution of scatterers in the vicinity of the mobile station. In addition, the XPD strongly depends on the elevation angle of the platform with respect to the mobile station.

10.4.6.6 Modeling of LoS MIMO Stratospheric Channels

Nowadays, as the telecommunication market is driven by the increasing demand for broadcast services and high-speed ubiquitous Internet access, BFWA is gaining increased popularity. Nevertheless, frequency bandwidths wide enough to carry such services are only available at mmwave frequency bands. Apart from providing mobile communications services, HAPs can also serve as base stations for the provision of BFWA services at the licensed Ka and V frequency bands. As MIMO technology is deemed a necessity due to its potentially high-bandwidth efficiency, new wireless system design approaches at the physical layer and realistic MIMO stratospheric channel models are required. Chen et al. (2005), Grace et al. (2005b), and Celcer et al. (2006) investigated the performance of constellations of multiple HAPs that intend to provide BFWA services at mmwave frequency bands. These constellations enhance the capacity by exploiting antenna user directionality, when using shared spectrum in colocated coverage areas. The capacity is controlled by the minimum angular separation of the HAPs and the sidelobe level of the terrestrial antenna.

Although applying MIMO to BFWA networks may be beneficial, rain effects severely affect signal propagation in mm-wave frequencies. Specifically, at these frequencies, the propagation characteristics are drastically different compared to those observed in frequencies well below 10 GHz, that is, L and S bands. Besides, the MIMO gain strongly depends on the channel characteristics, which are mainly determined by the antenna configuration and the richness of scattering at frequencies well below 10 GHz. Propagation at mm-wave bands requires a strong, dominant LoS signal for sufficient coverage due to the severe attenuation of the NLoS links (Grace et al. 2001; Aragón-Zavala et al. 2008). Conceptually, this corresponds to a rank-deficient MIMO channel matrix and low spectral efficiency due to the increased spatial correlation introduced by the linear relationship of the phases of the received signals (Cottatellucci and Debbah 2004; Sakaguchi et al. 2005). Contrary to these observations, a number of studies have suggested that the LoS response is not inherently correlated and have shown the possibility of getting high MIMO gain for LoS channels by preserving the orthogonality of the received signals (Driessen and Foschini 1999; Bohagen et al. 2005, 2007; Sarris and Nix 2007). In detail, using specifically designed antenna arrays, that is, placing the antenna elements in positions, where the LoS rays are orthogonal, a full-rank MIMO channel may be achieved.

The application of MIMO techniques to fixed stratospheric systems at Ka and V bands was examined in Michailidis and Kanatas (2009) and geometrical design recommendations were introduced, in order to construct a full-rank MIMO channel matrix. The SBS and the terrestrial fixed station employ ULAs with n_T transmit and n_R receive antenna elements, respectively. All antennas are fixed and highly directional. Since it is more difficult to generate power at the susceptible to propagation impairments Ka/V bands than at the lower L/S bands, omnidirectional antennas are essentially not applicable. Thus, it is easier to realize highly directive antennas, which can increase the fade margin by adding more gain, and improve the link availability. The maximum capacity was obtained for

$$\mathbf{H}_{LoS}\mathbf{H}_{LoS}^{H} = n_T \mathbf{I}_{n_R}, \tag{10.20}$$

where $\mathbf{H}_{LoS}$ is a deterministic $n_R \times n_T$ matrix containing the nonfading LoS responses between all array elements and $\mathbf{I}_{n_R}$ is an $n_R \times n_R$ identity matrix. The parameters of interest were the height and the elevation angle of the platform, the carrier frequency, and the array configuration. Since rain has an important impact on the quality of the link at mm-wave frequencies, both clear sky and rain conditions were considered. In particular, rain introduces not only severe attenuation but also a short-term variation to the received signal. These channel dynamics is a result of the radiowave propagation through the rainfall medium. Then, the scattering introduced by rain can be modeled as a stochastic process (Bohagen et al. 2005). As experimentally shown in Xu et al. (2000), multipath may be observed even in unobstructed LoS links during rain but not during clear sky conditions. The results in Michailidis and Kanatas (2009) showed that the proposed design method overcomes the problem of reduced capacity in LoS conditions and outperforms the conventional rank-one LoS architecture in terms of channel capacity. The sensitivity to possible deviation of the parameters values chosen to satisfy the optimal design constraints was also investigated. Thus, a deviation factor was introduced and an analysis of the sensitivity to nonoptimal design with regard to channel capacity was performed.

10.4.6.7 Modeling of LoS MIMO HAP-to-Train Architectures

The need for efficient high-speed Internet access and audio, video, and file transfer services on commercial train routes have prompted the use and the development of new satellite and wireless terrestrial network services (Karimi et al. 2012). GEO satellites intend to exploit LoS connections, whereas the terrestrial cellular infrastructure offers link availability in propagation environments where the direct communication to the satellite might not be feasible, for example, in tunnels and train stations. Although the Ka and V frequency bands were licensed for BFWA services through HAPs, their application may be extended to mobile scenarios, such as vehicular, maritime, aeronautical, and railway scenarios (Morlet et al. 2007). As the demand for uninterrupted quality broadband wireless services in the area of passenger transport in railways and high-data rate communications to multiple moving trains grow, the railroad scenario seems a promising and commercially attractive field for stratospheric systems. To meet the LTE for railway (LTE-R) requirements (Guan et al. 2011) imposed by the International Union of Railways (UIC) and enhance the achievable data throughput, the application of MIMO technology seems necessary. Although the propagation at mm-wave frequencies requires a strong, dominant LoS signal for sufficient coverage, mobility effects, multipath, shadowing, and blockage, which are also encountered at lower frequency bands, may exist due to the local environment in the vicinity of the trains, for example, due to the presence of various metallic obstacles along the train trajectory for electrical power supply, bridges, etc. Nevertheless, the railway environment is generally characterized by sparse insignificant multipath (Gao et al. 2008) and the elevation angles of the HAPs with respect to the trains are usually high, which in turn implies nearly LoS (open) propagation. Then, the MIMO channel is rank-deficient and the spectral efficiency is low. Although geometrical design guidelines to construct full-rank capacity optimized MIMO stratospheric channels were suggested in Michailidis and Kanatas (2009), there are some technical challenges for LTE-R due to train speed (up to 500 km/h), which leads to a time-varying and nonstationary radio channel. In addition, the stratospheric winds may vary the position of the HAP, which in turn influences the stabilization of HAP antenna beam pointing angles. The viability of reliable high-data rate communications for trains through HAPs at mm-wave frequencies was examined in Michailidis et al. (2013c). Clear sky conditions were considered without incorporating atmospheric fading, which leads to different channel characteristics (Arapoglou et al. 2012). The parameters of interest are the height and the elevation angle of the platform with respect to the train, the carrier frequency, and the antenna array configuration. The system model considered multiple transmit and receive antenna elements installed on a HAP and the top of a train, respectively. It was assumed that there are many trains within the coverage area at a given time, which may be on the adjacent track and may travel toward or away from each other. These trains simultaneously transmit uncorrelated signals within the same frequency band. The maximum speed of the trains was considered to be 300 km/h. Therefore, a multiple-HAP constellation should be used for sufficient coverage. Single-carrier quaternary phase-shift keying transmission is assumed based on the IEEE 802.16 standard for BFWA (IEEE 2001). An analysis of the sensitivity to imperfect positioning and orientation of the antenna arrays was performed with regard to the channel capacity. The results revealed relatively low sensitivity of the underlying system to displacements of antenna arrays from the optimal point. These results also depicted that the orientation of the antenna arrays and the elevation angle of the platform significantly affect the channel capacity.

10.4.7 ANALYTICAL MODELS

The application of multiantenna technology to stratospheric systems operating at the Ka frequency band was discussed in Falletti et al. (2006a) and an analytical vector-based spatial–temporal channel model was proposed for point-to-point HAP-to-ground communication links. This model extends (Stephenne and Champagne 2000) by adding antenna arrays at both ends of the underlying communication system. Contrary to long-term models, which predict changes from a channel state to another by means of Markov chain approaches, this model provides the short-term time series of the fading processes affecting the signal propagation from each transmit antenna to each receive antenna, for each channel state. Both temporal and spatial correlations among these processes were considered as well as possible impairments due to scattering objects in the vicinity of the ground station and Doppler effects were taken into account. From the analytical model, a link-level channel simulator with a relatively low computational complexity was implemented. Results regarding the distribution of the fading coefficients and the time series of the fading processes in both LoS (clear sky) and non-LoS (dense trees or buildings) propagation conditions were demonstrated, when a four-elements antenna array and a nine-elements antenna array is employed at the transmitter and the receiver, respectively, and the relative velocity of the HAP and the ground terminal is either 200 km/h (for HAP-to-train transmissions) or 60 km/h (for quasi-still ground terminals).

An analytical channel modeling approach for MIMO LMS systems operating at the Ku frequency band and above was described in Liolis et al. (2007). In this model, which may be also used for stratospheric systems, two schemes of MIMO techniques are presented: (i) a 2×2 MIMO spatial multiplexing system is used to achieve capacity improvements and a closed form expression for the outage capacity is derived and (ii) a MIMO spatial diversity scheme with receive antenna selection is applied in order to reduce interference in LMS communication links. In addition, an analytical closed-form expression for interference mitigation on forward link of a satellite 2×2 MIMO diversity system with antenna selection is also obtained. In order to discuss the features of MIMO techniques, the model assumes high antenna directivity and propagation phenomena, such as clear LoS, rain fading, and rainfall spatial homogeneity. The propagation delay offset (synchronization problem) in LMS communications is also considered and a practical solution to this problem is suggested by first applying matched filters to the received signals for the detection of propagation delay offset and then by feeding the resulting signals to a timing aligner. The effect of the system parameters on the performance of a 2×2 MIMO spatial multiplexing system and a MIMO spatial diversity scheme with receive antenna selection was analyzed and a comparison between each of these systems and a conventional single-antenna system was performed. The results revealed that the channel outage capacity gain of the MIMO spatial multiplexing system is significant for moderate and high SNR levels. This capacity also increases, as the rain fading becomes weaker. Moreover, increasing the angular separation between satellites decreases the spatial correlation coefficient due to rainfall medium and increases the outage capacity. The results regarding the MIMO spatial diversity system with receive antenna selection suggest that this system significantly improves the signal-to-interference ratio with respect to the single-antenna system, especially for high system availability and high frequency.

10.4.8 EMPIRICAL CHANNEL MODELS

Empirical channel models are based on experimental measurements in a specific environment. From the empirical models, efficient cell planning, link budget calculations, system performance simulations, and coverage planning can be attained. Iskandar and Shimamoto (2006a) defined and analyzed an empirical model for the link between HAPs and terrestrial users, based on experiments carried out in a "semiurban" environment. From this model, narrowband channel characteristics were presented in terms of the Rician factor and the local mean received power over a wide range of elevation angles. In particular, the Rician factor varies between 0.9 and 18.6 dB at a frequency of 1.2 GHz, and between 1.4 and 16.8 dB at a frequency of 2.4 GHz, for the elevation angle ranging from 10° to 90°. Nevertheless, the value of the Rician factor also depends on the propagation area and the degree of urbanization. Thus, this value is expected to be lower in highly urbanized areas, where the scatterers are usually dense and tall. However, this model treated only the total received power, without differentiating the LoS power and the multipath scattered power during measurements. Axiotis and Theologou (2003), Fontan et al. (2008), and Holis and Pechac (2008c) presented results from measurement campaigns emulating the HAP-to-indoor channel at S band and developed empirical models to estimate the penetration loss into buildings for several high elevation angles. The results demonstrated that the penetration loss depends on the elevation angle, the position of the user within the building, and the type of building. These results were in good agreement with the results obtained by a 3D ray-tracing simulation. However, the aforementioned empirical models were established based on experiments performed in a specific area, which limit their merit. Indeed, it is questionable whether they can be applied to other propagation environments. Note that empirical models based on MIMO channel measurement campaigns are only reported for terrestrial (Molisch et al. 2002) and satellite (King 2007) systems.

10.5 FUTURE RESEARCH DIRECTIONS

With their unique characteristics, stratospheric systems seem to represent an efficacious alternative infrastructure with regard to terrestrial and satellite systems, which can revolutionize the telecommunication industry. It is envisaged that stratospheric systems will be potentially capable of providing and delivering a compelling range of current and next-generation mobile/fixed services. Hence, research on stratospheric communications will gain much interest from academia, research centers, and industry worldwide. Although current modeling approaches constitute a robust basis for the characterization of single- and multiple-antenna stratospheric radio channels, they could be further improved or extended into different areas.

Future research efforts may be devoted to collecting measured channel data and developing empirical models. Although measurement campaigns are expensive, time consuming, and difficult to carry out, conducting real-world measurements and collecting measured channel data are preconditions for the successful validation of the results of preliminary theoretical efforts and ascertains the benefits of employing different configurations of stratospheric systems. Owing to the lack of channel-sounding measurement campaigns, it is important to verify the channel models in real-world propagation conditions. Specifically, HAPs and UAVs may be emulated by using a helicopter, a small plane, or a balloon containing two or more antennas sufficiently separated. However, access to a real HAP or UAV would be even more ideal.

Moreover, the analysis in Michailidis and Kanatas (2010, 2014) and Michailidis et al. (2013b) is restricted to MIMO stratospheric systems employing ULAs on both sides of the communication link. However, the proposed channel models can be modified to employ other antenna array geometries, such as uniform planar arrays, uniform circular arrays, and spherical antenna arrays, or a combination of them. Furthermore, apart from considering only single-bounce rays, the models in Michailidis and Kanatas (2010, 2014) and Michailidis et al. (2013b), can be extended to additionally support double- or multiple-bounce rays, due to multiple scattering, reflection, or diffraction of the radiated energy, which individually contribute in the total transmitted power of the link. Then, more sophisticated and realistic channel models can be obtained. These models also assume that all scattering objects are stationary. However, this assumption does not hold in some cases. Therefore, these models can be extended to include nonstationary scattering objects, such as vehicles and people.

Although signal transmissions at mm-wave frequencies are usually associated with fixed broadband services, currently there is a clear trend to extend the applicability of these services to mobile scenarios in order to benefit from existing air interfaces and accelerate the development of new applications, such as the provision of high-speed Internet access, audio, and video on demand and file transfer to vehicles, airplanes, trains, and ships (Morlet et al. 2007). Modeling the propagation effects associated with the licensed Ka and V frequency bands is an essential precondition. The propagation effects are usually classified into two main categories, the local environment propagation effects, for example, multipath, shadowing, and blockage and the tropospheric effects, for example, rainfall, oxygen absorption, water vapor, clouds, and precipitation. The aforementioned models independently treat the local environment propagation effects and the tropospheric effects depending on the operation frequency. Hence, their possible correlation has not been yet investigated. However, an accurate and realistic MIMO stratospheric channel modeling of broadband mobile communication systems operating at mm-wave frequencies should treat these effects together. Under such conditions, it could be interesting to theoretically and experimentally investigate the relation between the multipath introduced by the local scatterers and rainfall effects. The interested reader is referred to Liolis et al. (2010) for a first theoretical approach to this open research problem, where a novel statistical analysis for single-antenna mobile satellite systems is presented.

Future research efforts include the channel modeling and performance analysis of stratospheric DP systems accommodating two spatially separated DP antennas, which form four-antenna arrays. Future directions include the use of the 3D channel model in more complex scenarios, where integrated satellite/HAP/terrestrial architectures are considered. In addition, other relaying protocols such as decode-and-forward could be investigated in order to examine the effect of channel correlation in this type of operation. Finally, employing opportunistic relaying either in stratospheric relay selection or in relay selection on the ground is an attractive research field.

REFERENCES

Abdi, A. and M. Kaveh, A space–time correlation model for multielement antenna systems in mobile fading channels, *IEEE Journal on Selected Areas in Communications*, 20(3), 550–560, 2002.

Agrawal, S.K. and P. Garg, Calculation of channel capacity and Rician factor in the presence of vegetation in higher altitude platforms communication systems, in *15th International Conference on Advanced Computing and Communications (ADCOM 2007)*, India, pp. 243–248, 2007.

Agrawal, S.K. and P. Garg, Effect of urban-site and vegetation on channel capacity in higher altitude platform communication system, *IET Microwaves, Antennas & Propagation*, 3(4), 703–713, 2009.

Ahmed, B.T.T., M.C. Ramon, and L.H. Ariet, On the UMTS-HSDPA in high altitude platforms (HAPs) communications, in *3rd International Symposium on Wireless Communication Systems (ISWCS)*, Valencia, Spain, pp. 704–708, 2006.

Ahn, D.S., S. Kim, H.W. Kim, and D.-C. Park, A cooperative transmit diversity scheme for mobile satellite broadcasting systems, *International Journal of Satellite Communications and Networking*, 28(5–6), 352–368, 2010.

Akalestos, K., T.C. Tozer, and D. Grace, Emergency communications from high altitude platforms, in *International Workshop on High Altitude Platform Systems*, Athens, Greece, 2005.

Almers, P. et al., Survey of channel and radio propagation models for wireless MIMO systems, *EURASIP Journal on Wireless Communications and Networking*, 2007, 019070, 2007.

Andrews, J.G., A. Ghosh, and R. Muhamed. *Fundamentals of WiMAX: Understanding Broadband Wireless Networking*, Upper Saddle River, New Jersey: Prentice Hall, 2007.

Andrews, M.R., P.P. Mitra, and R. de Carvalho, Tripling the capacity of wireless communications using electromagnetic polarization, *Nature*, 409, 316–318, 2001.

Antonini, M., E. Cianca, A. De Luise, M. Pratesi, and M. Ruggieri, Stratospheric relay: Potentialities of new satellite-high altitude platforms integrated scenarios, in *IEEE Aerospace Conference Proceedings (Cat. No.03TH8652)*, Montana, USA, Vol. 3, pp. 3-1211–3-1219, 2003.

Aragón-Zavala, A., J.L. Cuevas-Ruíz, and J.A. Delgado-Penín, *High-Altitude Platforms for Wireless Communications*, Chichester, UK: Wiley, 2008.

Arapoglou, P.-D., K.P. Liolis, M. Bertinelli, A. Panagopoulos, P. Cottis, and R. De Gaudenzi, MIMO over satellite: A review, *IEEE Communications Surveys & Tutorials*, 13(1), 27–51, 2011b.

Arapoglou, P.-D., K.P. Liolis, and A.D. Panagopoulos, Railway satellite channel at Ku band and above: Composite dynamic modeling for the design of fade mitigation techniques, *International Journal of Satellite Communications and Networking*, 30(1), 1–17, 2012.

Arapoglou, P.-D., E.T. Michailidis, A.D. Panagopoulos, A.G. Kanatas, and R. Prieto-Cerdeira, The land mobile earth–space channel: SISO to MIMO modeling from L- to Ka-bands, *IEEE Vehicular Technology Magazine, Special Issue on Trends in Mobile Radio Channels: Modeling, Analysis, and Simulation*, 6(2), 44–53, 2011c.

Arapoglou, P.-D., M. Zamkotsian, and P. Cottis, Dual polarization MIMO in LMS broadcasting systems: Possible benefits and challenges, *International Journal of Satellite Communications and Networking*, 29(4), 349–366, 2011a.

Axiotis, D.I. and M.E. Theologou, An empirical model for predicting building penetration loss at 2 GHz for high elevation angles, *IEEE Antennas and Wireless Propagation Letters*, 2(1), 234–237, 2003.

Axiotis, D.I., M.E. Theologou, and E.D. Sykas, The effect of platform instability on the system level performance of HAPS UMTS, *IEEE Communications Letters*, 8(2), 111–113, 2004.

Bello, P., Characterization of randomly time-variant linear channels, *IEEE Transactions on Communications*, 11, 360–393, 1963.

Biglieri, E., R. Calderbank, A. Constantinides, A. Goldsmith, A. Paulraj, and H. Vincent Poor, *MIMO Wireless Communications*, Cambridge: Cambridge University Press, 2007.

Bo, Z., R. Qinghua, L. Yunjiang, C. Zhenyong, and Z. Feng, Characteristic and simulation of the near space communication channel, in *International Symposium on Microwave, Antenna, Propagation and EMC Technologies for Wireless Communications*, Hangzhou, China, pp. 769–773, 2007.

Bohagen, F., P. Orten, and G.E. Oien, Modeling and analysis of a 40 GHz MIMO system for fixed wireless access, in *IEEE 61st Vehicular Technology Conference*, Stockholm, Sweden, Vol. 3, pp. 1691–1695, 2005.

Bohagen, F., P. Orten, and G.E. Oien, Design of optimal high-rank line-of-sight MIMO channels, *IEEE Transactions on Wireless Communications*, 6(4), 1420–1425, 2007.

Bråten, L.E. and T. Tjelta, Semi-Markov multistate modeling of the land mobile propagation channel for geostationary satellites, *IEEE Transactions on Antennas and Propagation*, 50(12), 1795–1802, 2002.

Bria, A. et al., 4th-generation wireless infrastructures: Scenarios and research challenges, *IEEE Personal Communications*, 8(6), 25–31, 2001.

Celcer, T., G. Kandus, T. Javornik, M. Mohorcic, and S. Plevel, Evaluation of diversity gain and system capacity increase in a multiple HAP system, in *International Workshop on Satellite and Space Communications*, Leganes-Madrid, Spain, pp. 114–118, 2006.

Chen, G., D. Grace, and T.C. Tozer, Performance of multiple high altitude platforms using directive HAP and user antennas, *Wireless Personal Communications*, 32(3–4), 275–299, 2005.

Chen, H. and M. Guizani, *Next Generation Wireless Systems and Networks*, Chichester: John Wiley, 2006.

Clarke, R.H., A statistical theory of mobile-radio reception, *Bell System Technical Journal*, 47, 957–1000, 1968.

Colella, N., J. Martin, and I. Akyildiz, The HALO network, *IEEE Communications Magazine*, 38(6), 142–148, 2000.

Cottatellucci, L. and M. Debbah, On the capacity of MIMO rice channels, in *42nd Allerton Conference on Communication, Control, and Computing*, Monticello-Illinois, USA, pp. 1506–1516, 2004.

Cuevas-Ruíz, J.L., A. Aragón-Zavala, G.A. Medina-Acosta, and J.A. Delgado-Penín, Multipath propagation model for high altitude platform (HAP) based on circular straight cone geometry, in *International Workshop on Satellite and Space Communications (IWSSC)*, Siena-Tuscany, Italy, pp. 235–239, 2009.

Cuevas-Ruíz, J.L. and J.A. Delgado-Penín, Channel model based on semi-Markovian processes, an approach for HAPS systems, in *XIV International Conference on Electronics, Communications, and Computers (CONIELECOMP)*, Veracruz, Mexico, pp. 52–56, 2004a.

Cuevas-Ruíz, J.L. and J.A. Delgado-Penín, A statistical switched broadband channel model for HAPS links, in *IEEE Wireless Communications and Networking Conference (WCNC)*, Atlanta, USA, 2004b.

Cuevas-Ruíz, J.L. and J.A. Delgado-Penín, HAPS systems performance using a Ka-band channel model based on a time series generator, in *15th International Conference on Electronics, Communications and Computers (CONIELECOMP)*, Puebla, Mexico, pp. 10–15, 2005.

Deaton, J.D., High altitude platforms for disaster recovery: Capabilities, strategies, and techniques for emergency telecommunications, *EURASIP Journal on Wireless Communications and Networking*, 2008, 153469, 2008.

Demestichas, P., G. Vivier, K. El-Khazen, and M. Theologou, Evolution in wireless systems management concepts: From composite radio to reconfigurability, *IEEE Communications Magazine*, 42(5), 90–98, 2004.

Dovis, F., R. Fantini, M. Mondin, and P. Savi, Small-scale fading for high-altitude platform (HAP) propagation channels, *IEEE Journal on Selected Areas in Communications*, 20(3), 641–647, 2002.

Driessen, P.F. and G.J. Foschini, On the capacity formula for multiple input-multiple output wireless channels: A geometric interpretation, *IEEE Transactions on Communications*, 47(2), 173–176, 1999.

Eldowek, B.M. et al., Complex envelope second-order statistics in high-altitude platforms communication channels, *Wireless Personal Communications*, 77(4), 2517–2535, 2014.

Erceg, P.V., H. Sampath, and S. Catreux-Erceg, Dual-polarization versus single-polarization MIMO channel measurement results and modeling, *IEEE Transactions on Wireless Communications*, 5(1), 28–33, 2006.

ESA. MiLADY (Mobile satellite channeL with Angle DiversitY) project. https://artes.esa.int/projects/milady-mobile-satellite-channel-angle-diversity, Accessed August 21, 2015.

ESA-ESTEC Contract 162372/02/NL/US. *STRATOS: Stratospheric Platforms, A Definition Study for an ESA Platform,* Final Report, pp. 1–34, 2005.

Evans, B.G. and K. Baughan, Visions of 4G, *IEE Electronics & Communication Engineering Journal*, 12(6), 293–303, 2000.

Evans, B.G. et al., Integration of satellite and terrestrial systems in future multimedia communications, *IEEE Wireless Communications*, 12(5), 72–80, 2005.

Falletti, E., M. Laddomada, M. Mondin, and F. Sellone, Integrated services from high-altitude platforms: A flexible communication system, *IEEE Communications Magazine*, 44(2), 85–94, 2006b.

Falletti, E., F. Sellone, C. Spillard, and D. Grace, A transmit and receive multi-antenna channel model and simulator for communications from high altitude platforms, *International Journal of Wireless Information Networks*, 13(1), 59–75, 2006a.

Fechtel, S.A., A novel approach to modeling and efficient simulation of frequency-selective fading radio channels, *IEEE Journal on Selected Areas in Communications*, 11(3), 422–431, 1993.

Feng, Q., J. McGeehan, E.K. Tameh, and A.R. Nix, Path loss models for air-to-ground radio channels in urban environments, in *IEEE 63rd Vehicular Technology Conference (VTC Spring)*, Melbourne, Australia, Vol. 6, pp. 2901–2905, 2006.

Fleury, B.H, First- and second-order characterization of direction dispersion and space selectivity in the radio channel, *IEEE Transactions on Information Theory*, 46(6), 2027–2044, 2000.

Fontan, F.P., J.P. González, M.J.S. Ferreiro, M.A.V. Castro, S. Buonomo, and J.P. Baptista, Complex envelope three-state Markov model based simulator for the narrow-band LMS channel, *International Journal of Satellite Communications*, 15(1), 1–15, 1997.

Fontan, F.P. et al., Building entry loss and delay spread measurements on a simulated HAP-to-indoor link at S-band, *EURASIP Journal on Wireless Communications and Networking*, 2008, 1–6, 2008.

Gao, L., Z. Zhong, B. Ai, and L. Xiong, Estimation of the Ricean K-factor in the high speed railway scenarios, in *IEEE Wireless Communications and Networking Conference (WCNC'08)*, Las Vegas, USA, pp. 775–779, 2008.

Gao, X., Z. Chen, J. Lv, and Y. Li, The correlation matrix model of capacity analysis in unmanned aerial vehicle MIMO channel, *WSEAS Transactions on Communications*, 11(12), 476–485, 2012.

Getu, B.N. and J.B. Andersen, The MIMO cube—A compact MIMO antenna, *IEEE Transactions on Wireless Communications*, 4(3), 1136–1141, 2005.

Giuliano, R., M. Luglio, and F. Mazzenga, Interoperability between WiMAX and broadband mobile space networks, *IEEE Communication Magazine*, 46(3), 50–57, 2008.

Google, Project Loon. http://www.google.com/loon/, Accessed August 21, 2015.

Grace, D., N.E. Daly, T.C. Tozer, A.G. Burr, and D.A.J. Pearce, Providing multimedia communications services from high altitude platforms, *International Journal of Satellite Communications*, 19(6), 559–580, 2001.

Grace, D., M. Mohorcic, M.H. Capstick, M.B. Pallavicini, and M. Fitch, Integrating users into the wider broadband network via high altitude platforms, *IEEE Transactions on Wireless Communications*, 12(5), 98–105, 2005a.

Grace, D., J. Thornton, C. Guanhua, G.P. White, and T.C. Tozer, Improving the system capacity of broadband services using multiple high-altitude platforms, *IEEE Transactions on Wireless Communications*, 4(2), 700–709, 2005b.

Guan, K., Z. Zhong, and B. Ai, Assessment of LTE-R using high speed railway channel model, in *Third International Conference on Communications and Mobile Computing (CMC)*, Qingdao, China, pp. 461–464, 2011.

Holis, J. and P. Pechac, Coexistence of terrestrial and HAP 3G networks during disaster scenarios, *Radioengineering*, 17(4), 1–7, 2008a.

Holis, J. and P. Pechac, Elevation dependent shadowing model for mobile communications via high altitude platforms in built-up areas, *IEEE Transactions on Antennas and Propagation*, 56(4), 1078–1084, 2008b.

Holis, J. and P. Pechac, Penetration loss measurement and modeling for HAP mobile systems in urban environment, *EURASIP Journal on Wireless Communications and Networking*, 2008, 543290, 2008c.

Holis, J. and P. Pechac, Provision of 3G mobile services in sparsely populated areas using high altitude platforms, *Radioengineering*, 17(1), 43–49, 2008d.

Horvath, P. and I. Frigyes, Application of the 3D polarization concept in satellite MIMO systems, in *IEEE Global Communications Conference (Globecom)*, San Francisco, USA, pp. 1–5, 2006.

Horvath, P., G. Karagiannidis, P.R. King, S. Stavrou, and I. Frigyes, Investigations in satellite MIMO channel modeling: Accent on polarization, *EURASIP Journal on Wireless Communications and Networking*, 2007, 098942, 2007.

Hult, T., A. Mohammed, Z. Yang, and D. Grace, Performance of a multiple HAP system employing multiple polarization, *Wireless Personal Communications*, 52(1), 105–117, 2010.

IEEE Std 802.16-2001, *IEEE Standard for Local and Metropolitan Area Networks, Part 16: Air Interface for Fixed Broadband Wireless Access Systems*, 2001.

IEEE Std. 802.16e-2005, *Air Interface for Fixed and Mobile Broadband Wireless Access Systems—Amendment for Physical and Medium Access Control Layers for Combined Fixed and Mobile Operation in Licensed Band*, 2005.

Iskandar and S. Shimamoto, Ray tracing for urban site propagation in stratospheric platform mobile communications, in *Asia-Pacific Conference on Communications*, Perth, Australia, pp. 212–216, 2005.

Iskandar and S. Shimamoto, Channel characterization and performance evaluation of mobile communication employing stratospheric platforms, *IEICE Transactions on Communications*, E89-B(3), 937–944, 2006a.

Iskandar and S. Shimamoto, Prediction of propagation path loss for stratospheric platforms mobile communications in urban site LOS/NLOS environment, in *IEEE International Conference on Communications (ICC)*, Istanbul, Turkey, Vol. 12, pp. 5643–5648, 2006b.

ITU. RESOLUTION 122 (Rev.WRC-07): *Use of the Bands 47.2–47.5 GHz and 47.9–48.2 GHz by High-Altitude Platform Stations in the Fixed Service and by Other Services*, 2007a.

ITU. RESOLUTION 145 (Rev.WRC-07): *Use of the Bands 27.9–28.2 GHz and 31–31.3 GHz by High-Altitude Platform Stations in the Fixed Service*, 2007b.

ITU. RESOLUTION 221 (Rev.WRC-07): *Use of High-Altitude Platform Stations Providing IMT-2000 in the Bands 1885–1980 MHz, 2010–2025 MHz and 2110–2170 MHz in Regions 1 and 3 and 1885–1980 MHz and 2110–2160 MHz in Region 2*, 2007c.

ITU. RESOLUTION 734 (Rev.WRC-07): *Studies for Spectrum Identification for Gateway Links for High-Altitude Platform Stations in the Range from 5850 to 7075 MHz*, 2007d.

ITU-R P.838-1. *Specific Attenuation Model for Rain for Use in Prediction Methods*, 1997.

ITU-R P.837-4. *Characteristics of Precipitation for Propagation Modeling*, 2003.

ITU-T. *Vocabulary of Terms for Broadband Aspects of ISDN, I.113 Recommendation*, 1992.

Jahn, A. Propagation considerations and fading countermeasures for mobile multimedia services, *International Journal on Satellite Communications*, 19(3), 223–250, 2001.

Jamison, L., G.S. Sommer, and I.R. Porche, *High-Altitude Airships for the Future Force Army*. Technical Report 234, 2005.

Jull, G.W., A. Lillemark, and R.M. Turner, SHARP (stationary high altitude relay platform) telecommunications missions and systems, in *IEEE Global Telecommunications Conference (Globecom)*, New Orleans, USA, pp. 955–959, 1985.

Kandus, G., M. Mohorcic, E. Leitgeb, and T. Javornik, Modelling of atmospheric impairments in stratospheric communications, in *2nd WSEAS International Conference on Circuits, Systems, Signal and Telecommunications*, Acapulco, Mexico, pp. 86–91, 2008.

Karapantazis, S. and F.-N. Pavlidou, Broadband communications via high-altitude platforms: A survey, *IEEE Communications Surveys and Tutorials*, 7(1), 2–31, 2005a.

Karapantazis, S. and F.-N. Pavlidou, The role of high altitude platforms in beyond 3G networks, *IEEE Wireless Communications*, 12(6), 33–41, 2005b.

Karasawa, Y., K. Kimura, and K. Minamisono, Analysis of availability improvements in LMSS by means of satellite diversity based on three-state propagation channel model, *IEEE Transactions on Vehicular Technology*, 46(4), 1047–1056, 1997.

Karimi, O.B., L. Jiangchuan, and W. Chonggang, Seamless wireless connectivity for multimedia services in high speed trains, *IEEE Journal on Selected Areas in Communications*, 30(4), 729–739, 2012.

King, P.R., Modelling and measurement of the land mobile satellite MIMO radio propagation channel. PhD dissertation, University of Surrey, 2007.

King, P.R., B.G. Evans, and S. Stavrou, Physical–statistical model for the land mobile-satellite channel applied to satellite/HAP MIMO, in *11th European Wireless Conference*, Nicosia, Cyprus, Vol. 1, pp. 198–204, 2005.

Konefal, T., C.L. Spillard, and D. Grace, Site diversity for high-altitude platforms: A method for the prediction of joint site attenuation statistics, *IEE Proceedings of Microwaves, Antennas and Propagation*, 149(2), 124–128, 2002.

Kong, M., O. Yorkinov, and S. Shimamoto, Evaluations of urban shadowing characteristics for HAPS communications, in *IEEE 5th Consumer Communications and Networking Conference (CCNC)*, Las Vegas, USA, pp. 555–559, 2008.

Ku, B.-J., D.-S. Ahn, and N. Kim, An evaluation of interference mitigation schemes for HAPS systems, *EURASIP Journal on Wireless Communications and Networking*, 2008, 865393, 2008.

Kwon, S.-C. and G.L. Stüber, Geometrical theory of channel depolarization, *IEEE Transactions on Vehicular Technology*, 60(8), 3542–3556, 2011.

Lacoste, F., F. Carvalho, F.P. Fontan, A. Nunez Fernandez, V. Fabro, and G. Scot, Polarization and spatial diversity measurements of the land mobile satellite propagation channel at S-band, in *COST Action IC0802*, Toulouse, France, 2009.

Lee, Y. and H. Ye, Sky station stratospheric telecommunications system, a high speed low latency switched wireless network, in *17th AIAA International Communications Satellite Systems Conference*, Yokohama, Japan, pp. 25–32, 1998.

Leong, S.-Y., Y.R. Zheng, and C. Xiao, Space–time fading correlation functions of a 3-D MIMO channel model, in *IEEE Wireless Communications Networking Conference (WCNC)*, Atlanta, USA, Vol. 2, pp. 1127–1132, 2004.

Liolis, K.P., I. Andrikopoulos, and P. Cottis, On statistical modeling and performance evaluation of SIMO land mobile satellite channels, in *4th Advanced Satellite Mobile Systems (ASMS)*, Bologna, Italy, pp. 76–81, 2008.

Liolis, K.P., A.D. Panagopoulos, and P.G. Cottis, Multi-satellite MIMO communications at Ku band and above: Investigations on spatial multiplexing for capacity improvement and selection diversity for interference mitigation, *EURASIP Journal on Wireless Communications and Networking*, 2007, 059608, 2007.

Liolis, K.P., A.D. Panagopoulos, and S. Scalise, On the combination of tropospheric and local environment propagation effects for mobile satellite systems above 10 GHz, *IEEE Transactions on Vehicular Technology*, 59(3), 1109–1120, 2010.

Liu, S., Z. Niu, and Y. Wu, A blockage based channel model for high altitude platform communications, in *IEEE 57th IEEE Vehicular Technology Conference (VTC Spring)*, Seoul, Korea, Vol. 2, pp. 1051–1055, 2003.

Loo, C., A statistical model for a land mobile satellite link, *IEEE Transactions on Vehicular Technology*, 34, 122–127, 1985.

LTE-Advanced, http://www.3gpp.org/LTE-Advanced, Accessed August 21, 2015.

Lum, M.J.H. et al., Telesurgery via unmanned aerial vehicle (UAV) with a field deployable surgical robot, *Studies in Health Technology and Informatics*, 125, 313–315, 2006.

Lutz, E., D. Cygan, M. Dippold, F. Dolainsky, and W. Papke, The land mobile satellite communications channel—Recording, statistics, and channel model, *IEEE Transactions on Vehicular Technology*, 40(2), 375–386, 1991.

Mahmoud, S.S., Z.M. Hussain, and P. O'Shea, Space–time model for mobile radio channel with hyperbolically distributed scatterers, *IEEE Antennas and Wireless Propagation Letters*, 1, 211–214, 2002.

Michailidis, E.T., P.N. Daskalaki, and A.G. Kanatas, Performance of capacity optimized line-of-sight MIMO HAP-to-train architectures, in *34th Progress in Electromagnetics Research (PIERS) Symposium*, Stockholm, Sweden, 2013c.

Michailidis, E.T., G. Efthymoglou, and A.G. Kanatas, Spatially correlated 3-D HAP-MIMO fading channels, in *International Workshop on Aerial & Space Platforms: Research, Applications, Vision of IEEE Global Communications Conference (Globecom)*, New Orleans, USA, pp. 1–7, 2008.

Michailidis, E.T. and A.G. Kanatas, A three dimensional model for land mobile-HAP-MIMO fading channels, in *10th International Workshop on Signal Processing for Space Communications (SPSC)*, Rhodes Island, Greece, 2008.

Michailidis, E.T. and A.G. Kanatas, Capacity optimized line-of-sight HAP-MIMO channels for fixed wireless access, in *International Workshop on Satellite and Space Communications (IWSSC)*, Siena, Italy, pp. 73–77, 2009.

Michailidis, E.T. and A.G. Kanatas, Three-dimensional HAP-MIMO channels: Modeling and analysis of space–time correlation, *IEEE Transactions on Vehicular Technology*, 59(5), 2232–2242, 2010.

Michailidis, E.T. and A.G. Kanatas, Capacity analysis and simulation of 3-D space–time correlated HAP-MIMO channels, *International Journal on Advances in Telecommunications*, 4(1&2), 12–23, 2011.

Michailidis, E.T. and A.G. Kanatas, Statistical simulation modeling of 3-D HAP-MIMO channels, *Wireless Personal Communications*, 65(4), 833–841, 2012.

Michailidis, E.T. and A.G. Kanatas, Wideband HAP-MIMO channels: A 3-D modeling and simulation approach, *Wireless Personal Communications*, 74(2), 639–664, 2014.

Michailidis, E.T., P. Petropoulou, and A.G. Kanatas, Geometry-based modeling of cross-polarization discrimination in HAP propagation channels, in *7th European Conference on Antennas and Propagation (EuCAP)*, Gothenburg, Sweden, 2013a.

Michailidis, E.T., P. Theofilakos, and A.G. Kanatas, Three-dimensional modeling and simulation of MIMO mobile-to-mobile via stratospheric relay fading channels, *IEEE Transactions on Vehicular Technology*, 62(5), 2014–2030, 2013b.

Milas, V.F. and P. Constantinou, Interference environment between high altitude platform networks (HAPN) geostationary (GEO) satellite and wireless terrestrial systems, *Wireless Personal Communications*, 32(3–4), 257–274, 2005.

Mohammed, A. and T. Hult, Capacity evaluation of a high altitude platform diversity system equipped with compact MIMO antennas, *International Journal of Recent Trends in Engineering*, 1(3), 244–247, 2009.

Mohammed, A. and Z. Yang, Broadband communications and applications from high altitude platforms, *International Journal of Recent Trends in Engineering*, 1(3), 239–243, 2009.

Molisch, A.F., M. Steinbauer, M. Toeltsch, E. Bonek, and R.S. Thoma, Capacity of MIMO systems based on measured wireless channels, *IEEE Journal on Selected Areas in Communications*, 20(3), 539–549, 2002.

Mondin, M., F. Dovis, and P. Mulassano, On the use of HALE platforms as GSM base stations, *IEEE Personal Communications*, 8(2), 37–44, 2001.

Morlet, C., A. Bolea-Alamañac, G. Gallinaro, L. Erup, P. Takats, and A. Ginesi, Introduction of mobility aspects for DVB-S2/RCS broadband systems, *International Journal of Space Communications*, 21(1–2), 5–17, 2007.

Nomikos, N., E.T. Michailidis, D. Vouyioukas, and A.G. Kanatas, Performance analysis of a two-hop MIMO mobile-to-mobile via stratospheric-relay link employing hierarchical modulation, *International Journal of Antennas and Propagation*, 2013, 10, 2013.

Oodo, M. et al., Experiments on IMT-2000 using unmanned solar-powered aircraft at an altitude of 20 km, *IEEE Transactions on Vehicular Technology*, 54(4), 1278–1294, 2005.

Paillassa, B., B. Escrig, R. Dhaou, M.-L. Boucheret, and C. Bes, Improving satellite services with cooperative communications, *International Journal of Satellite Communications and Networking*, 29(6), 479–500, 2011.

Palma-Lázgare, I.R., J.A. Delgado-Penín, and F. Pérez-Fontán, An advance in wireless broadband communications based on a WiMAX-HAPS architecture in *26th International Communications Satellite Systems Conference (ICSSC)*, San Diego, USA, 2008.

Panagopoulos, A.D., E.M. Georgiadou, and J.D. Kanellopoulos, Selection combining site diversity performance in high altitude platform networks, *IEEE Communications Letters*, 11(10), 787–789, 2007.

Park, J.-M., B.-J. Ku, and D.-S. Oh, Technical and regulatory studies on HAPS, in *International Workshop on Aerial & Space Platforms: Research, Applications, Vision of IEEE Global Communications Conference (Globecom)*, New Orleans, USA, pp. 1–5, 2008.

Patel, C.S., G.L. Stüber, and T.G. Pratt, Comparative analysis of statistical models for the simulation of Rayleigh faded cellular channels, *IEEE Transactions on Communications*, 53(6), 1017–1026, 2005.

Pätzold, M., *Mobile Fading Channels*, Chichester, England: John Wiley, 2002.

Pätzold, M., U. Killat, F. Laue, and Y. Li, On the statistical properties of deterministic simulation models for mobile fading channels, *IEEE Transactions on Vehicular Technology*, 47(1), 254–269, 1998.

Paulraj, A., R. Nabar, and D. Gore, *Introduction to Space–Time Wireless Communications*, Cambridge: Cambridge University Press, 2003.

Pawlowski, W., Radio wave propagation effects in high-altitude platform systems, in *International Conference on Microwaves, Radar and Wireless Communications (MIKON)*, Wroclaw, Poland, Vol. 1, pp. 185–188, 2000.

Pop, M.F. and N.C. Beaulieu, Limitations of sum-of-sinusoids fading channel simulators, *IEEE Transactions on Communications*, 49(4), 699–708, 2001.

Proakis, J.G. and M. Salehi, *Digital Communications*, 5th Ed., New York: McGraw-Hill, 2008.

Quitin, F., C. Oestges, F. Horlin, and P. De Doncker, Multipolarized MIMO channel characteristics: Analytical study and experimental results, *IEEE Transactions on Antennas and Propagation*, 57(9), 2739–2745, 2009.

Rappaport, T.S., *Wireless Communications: Principles and Practice*, 2nd Ed., Upper Saddle River, New Jersey: Prentice Hall PTR, 2002.

Rappaport, T.S. and J.C. Liberti, A geometrical-based model for line-of sight multipath radio channel, in *IEEE 46th Vehicular Technology Conference (VTC)*, Atlanta, USA, pp. 844–848, 1996.

Rice, S.O., Mathematical analysis of random noise, *Bell Systems Technical Journal*, 23, 282–332, 1944.

Sakaguchi, K., H.Y.E. Chua, and K. Araki, MIMO channel capacity in an indoor line-of-sight (LoS) environment, *IEICE Transactions on Communications*, E88-B(7), 3010–3019, 2005.

Salz, J. and J.H. Winters, Effect of fading correlation on adaptive arrays in digital mobile radio, *IEEE Transactions on Vehicular Technology*, 43(4), 1049–1057, 1994.

Sarris, I. and A.R. Nix, Design and performance assessment of high-capacity MIMO architectures in the presence of a line-of-sight component, *IEEE Transactions on Vehicular Technology*, 56(4), 2194–2202, 2007.

Saunders, S.R. and A. Aragón-Zavala, *Antennas and Propagation for Wireless Communication Systems*, 2nd Ed., Chichester, England: John Wiley & Sons, 2007.

Sellathurai, M., P. Guinand, and J. Lodge, Space–time coding in mobile satellite communications using dual-polarized channels, *IEEE Transactions on Vehicular Technology*, 55(1), 188–199, 2006.

Shiu, D.-S., G.J. Foschini, M.J. Gans, and J.M. Kahn, Fading correlation and its effect on the capacity of multielement antenna systems, *IEEE Transactions on Communications*, 48(3), 502–513, 2000.

Simunek, M., P. Pechac, and F.P. Fontan, Excess loss model for low elevation links in urban areas for UAVs, *Radioengineering*, 20(3), 561–568, 2011.

Smolnikar, M., M. Mohorcic, T. Javornik, and D. Grace, Propagation impairment countermeasures in mobile stratospheric operating environment, in *IEEE 69th Vehicular Technology Conference (VTC Spring)*, Barcelona, Spain, pp. 1–5, 2009.

Spillard, C.L., D. Grace, J. Thornton, and T.C. Tozer, Effect of ground station antenna beamwidth on rain scatter interference in high-altitude platform links, *Electronics Letters*, 38(20), 1211–1213, 2002.

Spillard, C.L., T.C. Tozer, B. Gremont, and D. Grace, The performance of high-altitude platform networks in rainy conditions, in *22nd AIAA International Communications Satellite Systems Conference*, Monterey, USA, 2004.

Stavroulaki, V., S. Buljore, P. Roux, and E. Melin, Equipment management issues in B3G, end-to-end reconfigurable systems, *IEEE Wireless Communications*, 13(3), 24–32, 2006.

Stephenne, A. and B. Champagne, Effective multi-path vector channel simulator for antenna array systems, *IEEE Transactions on Vehicular Technology*, 49(6), 2370–2381, 2000.

Suraweera, H.A., G.K. Karagiannidis, and P.J. Smith, Performance analysis of the dual-hop asymmetric fading channel, *IEEE Transactions on Wireless Communications*, 8(6), 2783–2788, 2009.

Taha-Ahmed, B., M. Calvo-Ramon, and L. de Haro-Ariet, High altitude platforms (HAPs) W-CDMA system over cities, in *IEEE 61st Vehicular Technology Conference (VTC Spring)*, Stockholm, Sweden, Vol. 4, pp. 2673–2677, 2005.

Thornton, J. and D. Grace, Effect of lateral displacement of a high altitude platform on cellular interference and handover, *IEEE Transaction on Wireless Communications*, 4(4), 1483–1490, 2005.

Thornton, J., D. Grace, C. Spillard, T. Konefal, and T.C. Tozer, Broadband communications from high altitude platforms—The European Helinet programme, *IEE Electronics & Communication Engineering Journal*, 13(3), 138–144, 2001.

Thornton, J., A.D. White, and T.C. Tozer. A WiMAX payload for high altitude platform experimental trials, *EURASIP Journal on Wireless Communications and Networking*, 2008, 498517, 2008.

Toshiaki, T. and H. Masatoshi, Navigation and positioning system using high altitude platforms systems (HAPs), *Journal of the Japan Society for Aeronautical and Space Sciences*, 52(603), 175–185, 2004.

Tzaras, C., B.G. Evans, and S.R. Saunders, Physical–statistical analysis of land mobile-satellite channel, *Electronics Letters*, 34(13), 1355–1357, 1998.

Vázquez-Castro, M., F. Pérez-Fontán, and B. Arbesser-Rastburg, Channel modeling for satellite and HAPs system design, *Wireless Communications and Mobile Computing*, 2(3), 285–300, 2002.

Verdin, D. and T.C. Tozer, Generating a fading process for the simulation of land-mobile radio communications, *Electronics Letters*, 29(23), 2011–2012, 1993.

Wang, C.-X., M. Pätzold, and D. Yuan, Accurate and efficient simulation of multiple uncorrelated Rayleigh fading waveforms, *IEEE Transactions on Wireless Communications*, 6(3), 833–839, 2007.

Widiawan, A.K. and R. Tafazolli, High altitude platform station (HAPs): A review of new infrastructure development for future wireless communications, *Wireless Personal Communications*, 42(3), 387–404, 2007.

WIMAX Forum, WIMAX Forum. http://www.wimaxforum.org/, Accessed August 21, 2015.

Xiao, C., Y.R. Zheng, and N.C. Beaulieu, Novel sum-of-sinusoids simulation models for Rayleigh and Rician fading channels, *IEEE Transactions on Wireless Communications*, 5(12), 3667–679, 2006.

Xi Jun, G., C. Zi Li, and H. Yong Jiang, Characteristic analysis on UAV-MIMO channel based on normalized correlation matrix, *The Scientific World Journal*, 2014, 10, 2014.

Xu, H., T.S. Rappaport, R.J. Boyle, and J.H. Schaffner, Measurements and models for 38-GHz point-to-multipoint radiowave propagation, *IEEE Journal on Selected Areas in Communications*, 18(3), 310–321, 2000.

Yang, Y., R. Zong, X. Gao, and J. Cao, Channel modeling for high-altitude platform: A review, in *International Symposium on Intelligent Signal Processing and Communication Systems (lSPACS)*, Cheng Du, China, pp. 1–4, 2010.

Young, D.J. and N.C. Beaulieu, The generation of correlated Rayleigh random variates by inverse discrete Fourier transform, *IEEE Transactions on Communications*, 48(7), 1114–1127, 2000.

Yu, K. and B. Ottersten, Models for MIMO propagation channels: A review, *Wireless Communications and Mobile Computing*, 2(7), 653–666, 2002.

Zajić, A.G. and G.L. Stüber, Three-dimensional modeling, simulation, and capacity analysis of space–time correlated mobile-to-mobile channels, *IEEE Transactions on Vehicular Technology*, 57(4), 2042–2054, 2008.

Zajić, A.G., G.L. Stüber, T.G. Pratt, and S. Nguyen, Envelope level crossing rate and average fade duration in mobile-to-mobile Ricean fading channels, in *IEEE International Conference on Communications (ICC)*, Beijing, China, pp. 4446–4450, 2008.

Zajić, A.G., G.L. Stüber, T.G. Pratt, and S. Nguyen, Wideband MIMO mobile-to-mobile channels: Geometry-based statistical modeling with experimental verification, *IEEE Transactions on Vehicular Technology*, 58(2), 517–534, 2009.

Zang, G., B. Huang, and J. Mu, One scheme of cooperative diversity with two satellites based on the Alamouti code, in *IET 3rd International Conference on Wireless, Mobile and Multimedia Networks (ICWMNN)*, Beijing, China, pp. 151–154, 2010.

Zhan, P., K. Yu, and A.L. Swindlehurst, Wireless relay communications with unmanned aerial vehicles: Performance and optimization, *IEEE Transactions on Aerospace and Electronic Systems*, 47(3), 2068–2085, 2011.

Zvanovec, S., P. Piksa, M. Mazanek, and P. Pechac, A study of gas and rain propagation effects at 48 GHz for HAP scenarios, *EURASIP Journal on Wireless Communications and Networking*, 2008, 1–7, 2008.

Index